FIB AND PHI IN MUSIC

Second in a series of InMusic titles on the topics covered in Fractals in Music: Introductory Mathematics for Musical Analysis

No. 1	Fractals in Music	1999
No. 2	Self-Similarity in Music	
No. 3	Attractors in Music	
No. 4	Fib and Phi in Music	2005
No. 5	Resonance in Music	
No. 6	Randomicity in Music	
No. 7	Dimension in Music	
No. 8	Statistics in Music	
No. 9	Transforms in Music	

FIB AND PHI IN MUSIC

The Golden Proportion in Musical Form

Charles Madden

High Art Press
Salt Lake City
2005

High Art Press
P.O. Box 58661
Salt Lake City UT 84158-0661
Email: HiArtPress@aol.com
Published 2005

Printed in the United States of America.

Printed on acid-free paper.

Publisher's Cataloging in Publication

Madden, Charles (Charles B Jr)
Fib and Phi in music : the golden proportion in musical form / Charles Madden. — 1st ed.
384 p. 5½ x 8½ in. — (InMusic ; no. 4)
Includes bibliographical references and index
LCCN: 99-90421
ISBN-10: 0967172764
ISBN-13: 9780967172767

1. Music—Theory—20th century. 2. Musical analysis. 3. Golden Proportion. I. Title.

MT6.M14F73 2005 781'.1

For Vicki and Krisztina

Who have had to put up

With so much from me

Each in her own way.

And for Paul Larson

Whose interest in phi

Is still lively

After all these years.

Contents

Preface

Fractals in Music: Introductory Mathematics for Musical Analysis was a manifesto for a sequence of books to be produced around the turn of the century proposing a new approach to music theory. *Fib and Phi in Music: The Golden Proportion in Musical Form* is the second release in that sequence (see frontispiece).

Fibonacci numbers (fib) and the golden proportion (phi) are self-similar members of the discipline called fractal geometry, which is a subset of chaos theory, as explained in *Fractals in Music*. They have been used by composers and artists knowingly or intuitively as natural means of organization. Their use to establish temporal structure will be discussed in this book, while their use in tonal structure will be discussed elsewhere.

This work began with a need to address a perceived neglect within the musical community. As *Fractals in Music* showed, numerous claims of composers using Fibonacci numbers and phi to give proportion to musical structure have been put forward in journal articles with little supporting documentation. Nor has there been any notable mention in textbooks. The many dissertations on the topic that came to my attention after writing *Fractals in Music* were a considerable surprise. They will be invaluable to what follows.

The sort of questions that presented themselves to be researched included the following:

What and when did musicians know about fib and phi?

Was phi known and used in ancient times?
How was the knowledge transmitted through time?
Was it strictly mathematical, or did it carry esoteric meaning?
Is there support for the use of phi in science, or is its use confined to cults?
Is phi used carefully and consciously, or strictly by intuition?

Some of the answers involved the stonemasons' guilds, secret societies, the Crusaders, the Freemasons, and the Priory of Sion.

Some of the mathematicians included Leonardo Fibonacci, Luca Pacioli, Jacques Bernoulli and Johannes Kepler.

Some of the painters were Leonardo da Vinci, Piero della Francesca, and Georges Seurat.

Some of the musicians were Guillaume de Machaut, J. S. Bach, Claude Debussy, Béla Bartók, and Karlheinz Stockhausen.

Questions remain, however.

How reliable were claims by Jay Hambidge that the Greeks used phi?
Did Fechner and the nineteenth century scholars influence the musicians?
Is intuitive use of phi simply a question of "a little more than half"?
Can intuitive use of phi be clearly demonstrated?
Is the presence of exact Fibonacci numbers evidence of knowing use?
Do the golden proportions found in the structures of some plants and animals mean that phi has some mystical significance?

All these questions, and more, compelled a long and circuitous journey, through much mystery and intrigue among strongly opposed camps. Some of the hoped-for answers remain shrouded in mystery for reasons that include secrecy, poor documentation, exaggeration, and speculation. Nevertheless, the circumstantial evidence is compelling, and the mere fact that these topics have been so often discussed over the centuries gives reason to suppose that some composers may have applied their own versions of the concepts to their music.

Whether composers have used fib and phi structures knowingly will be explored in this book. The dissertations cited hereafter have tended to focus on statistical averaging to answer this question and have thus found few cases of knowing use. Perhaps reconciling methodological differences among them will result in a clearer picture.

Some of this material was discussed in an introductory way in *Fractals in Music.* Not much can be added to the mathematics presented there, but an attempt is made here to clarify its derivation. Nothing is required beyond high school algebra, geometry, and trigonometry.

Following a development of the history and algebra of phi proportions, the book treats of their use by composers from the medieval period to the postmodern. Although much of the material is mine, other experts have shown the way, as indicated in the appropriate references. The close of the book provides a summary and conclusions.

I have made every effort to remove errors. No doubt some remain, along with unavoidable differences of opinion. I hope that these will lead to discussion and further work.

I make no apologies for not resolving all questions. The remainder may present to some readers opportunities to begin their own research. By all means, they should feel free.

Acknowledgment should be extended to reviewers of *Fractals in Music* Dave Benson, Carola Boehm, Alastair Borthwick, Shawn J. Bulen, François Nicolas, and Bruce Quaglia for helping to focus this work. Thanks also to Myron Patterson and the ever-helpful staff at the University of Utah's excellent Fine Arts Library, and to Arnold Dreyblatt, James Fulkerson, David Hykes, Ben Johnston, Paul Larson, Lou Pine, and my ever-supportive wife, for their help and encouragement.

Chapter 1

Introduction

History of phi

Phi in Egypt

The story of phi begins in Fourth Dynasty Egypt, about 2500 BC.[1] After several experiments that were not altogether satisfactory, Egyptian architects constructed two pyramids that researchers have regarded as extraordinary ever since. These were the Great Pyramid of Khufu (called Cheops by the Greeks) and the pyramid of Khufu's grandson Khafre (Chephren), both on the Giza plateau near Cairo.[2]

Regarded as tombs for the pharaohs by most orthodox Egyptologists, many other purposes have been proposed, from power stations, to shrines for initiation, beacons for travelers or extraterrestrials, alignment markers for surveyors, and even elaborate statements of the earth's dimensions.[3]

One thing that everyone agrees on is the meticulous precision with which they worked. Khufu is oriented to within 3′6″ of arc from true north.[4] Without telescopes, how could they do this? In addition, Khufu contains the phi proportion between its sides and base. With only rudimentary arithmetic and only ropes or sticks for measuring, this seems like an improbable result.

The question of alignment is not difficult to solve, however. All proto-civilizations knew how to find true north. Scientists know this from studying the remains of astronomical observatories all over the world. In every one, the astronomers marked the solstices and equinoxes on the ground; the four compass directions were then easy to find without even sighting on the North Star.[5] This knowledge quickly became a permanent feature of their cultures.

A right angle could be constructed by folding a rope with thirteen equally spaced knots such that three spaces between knots form one side of a triangle, four spaces form another side, and five spaces form the hypotenuse. To correct any errors, the triangle could be turned over and the difference split, thus marking a highly accurate right angle. Several researchers have found murals and papyruses that show knotted ropes and have verified that certain officials were known as "rope stretchers." It is highly likely that this is how the squareness was obtained.[6]

Several pyramids were built on the proportions of the 3–4–5 triangle, possibly as a testament to the rope stretcher's art. The height, half-base, and slant angle of Khafre have been measured as 143.5 and 107.5 meters and 53°10′. The ratio of these—143.5/107.5 = 1.33488 . . .—is very close to 4/3, as verified by the angle.[7]

Khufu has been measured repeatedly since the seventeenth century AD, each time with greater accuracy. Although some of the ideas presented to explain it have been pure speculation and misinformation, several reputable scientists have demonstrated important points to near certainty. Lehner's figures, for example, show that the cross section of a vertical plane through the center of the pyramid contains phi proportions within .06% accuracy.[8] How could this have happened if they had little knowledge of mathematics, as claimed by most historians of mathematics?

Figure 1.1 describes a simple procedure that could have been used by people accustomed to measuring land areas with ropes and pegs, as the rope stretchers had to do every year after the spring floods had wiped out the previous year's boundary markers.

The architect would start with a 1 by 1 square and divide it in half with a vertical line. Placing a peg at A, he[9] would stretch a rope to B and draw an arc on the ground from B to C. Moving his peg to D, he would draw an arc from C to E. This would produce the so-called "pyramid triangle" D–E–F. Line D–E is the slant height of the pyramid; reflected to the right, it outlines the pyramid in profile. This elegant result is but a simple extension of the idea of drawing a circle with a rope and a peg.

Figure 1.1. How an Egyptian architect could have drawn a "pyramid triangle" using a rope and a peg. ϕ is the symbol for phi and $1/\phi$ is its inverse.

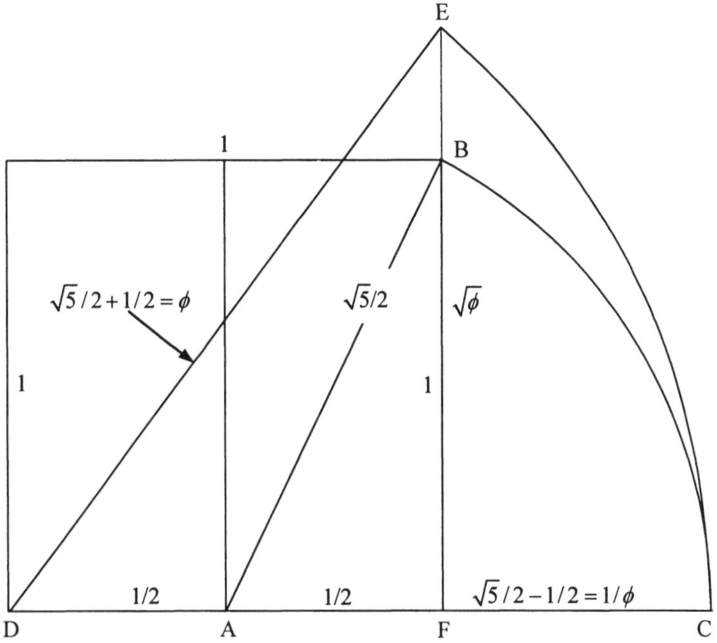

In modern notation, the numbers are as follows: The 1 by 1 box is divided into two boxes of 1 by 1/2. By the Pythagorean theorem, diagonal A–B measures, rounded to three decimal places,

$$\sqrt{\left(\frac{1}{2}\right)^2+1^2}=\sqrt{\frac{1}{4}+1}=\sqrt{\frac{5}{4}}=\frac{\sqrt{5}}{2}=1.118.$$

This value is transferred to the baseline by arc B–C and, if 1/2 is subtracted from it (line A–F), we get

$$\frac{\sqrt{5}}{2}-\frac{1}{2}=\frac{1}{\phi}=0.618 \qquad (1.1)$$

for line C–F, or if 1/2 is added to it (line A–D), we get

$$\frac{\sqrt{5}}{2}+\frac{1}{2}=\phi=1.618\,, \qquad (1.2)$$

where ϕ is phi. This is diagonal D–E and line D–C. Line E–F, the right-hand side of the triangle is, by Pythagoras, again rounding,

$$\sqrt{\phi^2-1^2}=\sqrt{2.618-1}=\sqrt{1.618}=\sqrt{\phi}=1.272\,. \qquad (1.3)$$

The actual measurements of Khufu given in cubits by the best experts are base: 440; height: 280; and slant height: 356.[10] The ratios, then (again, to three decimal places), are 280/220 = 1.273 height to half-base and 356/220 = 1.618, slant height to half-base. This is the pyramid triangle D–E–F shown in figure 1.1. This assessment is verified by the angle of the facing stones as measured by Petrie, 51°50′40″, which is about 1′ of arc from the true value of phi (51°49′39″).[11] A table of trigonometric values supports this conclusion.[12]

Despite the extreme accuracy of the measurements of Khufu, there remain conflicts between investigators because the Egyptians left no written documents to explain it. Of the few surviving documents on mathematics, the *Rhind Papyrus*, dated about 1650 BC, is the most useful and gives formulas for areas and arithmetic, including how to calculate the hypotenuse of a 3–4–5 triangle.[13] Phi is not mentioned, however, and knowledge of how to stretch the ropes is assumed.[14]

The numbers themselves provide overwhelming circumstantial evidence of the presence of phi in Khufu. Although this cannot be taken as proof that the builders knew they were using phi, if there were any other evidence, researchers would be forced to assume they were trying to show that they did know.

Phi in Greece

The story now moves to Greece, with the sixth century BC journey made by Greek philosopher Pythagoras to Egypt and Babylonia to study with the priests and learn their mysteries. It is generally assumed that all knowledge was guarded closely by the priestly classes and not parted with easily. This gives rise to the concept of "ancient mysteries." Those to whom the mysteries were imparted were sworn to secrecy, usually on pain of death for infractions. Secret societies came out of these mystery cults and continue to this day.[15]

After studying in Egypt and Babylonia for twenty years or more, Pythagoras returned to Greece and later to Italy to establish a school of philosophy based on numerology. The most famous tenet was "All is number." Although Pythagoras left no writing of his own, Plato was well versed in the philosophy and described some of it in several books, some of which are now lost. Unfortunately, Plato couched much of his discussion in a kind of doublespeak apparently intended to obscure the details.[16] Addressed to people with inside information, his motto was "Who has ears, let him hear." This was echoed centuries later in the Bible.

Researchers still argue about whether Plato's *Timaeus* discussed phi and we do not know very much about the mathematics that these early Greeks did know. Certainly, the theorem of Pythagoras, in which the square of the hypotenuse of a triangle equals the sum of the squares of the sides, was known long before Pythagoras, and he may have gotten it from the Babylonians.[17]

The usual story is that the study of the pentagon and pentagram (five-pointed star) were of particular interest to the Pythagoreans; that they used the pentagram as an emblem of recognition; and that they knew about the phi proportions it contains. Herz-Fischler, however, wrote that the mathematics were developed by many contributors over several centuries and did not reach their final form until the time of Euclid, some 300 years after Pythagoras.[18]

Figure 1.2 shows the main features of the mathematics of the pentagram, set inside a pentagon. Notice that the legs of the pentagram have length 1, while the inside diagonals have length $1/\phi$.

Across two segments of the diagonal, the length measures

$$1+\frac{1}{\phi}=\phi\,,$$

which is also the length of the edge of the pentagon. The length of a complete diagonal is

$$1+\frac{1}{\phi}+1=\phi+1=\phi^2\,.$$

You may wish to explore how many combinations of 1, $1/\phi$, ϕ, and ϕ^2 there are in this diagram.

Figure 1.2. Phi proportions in the pentagram

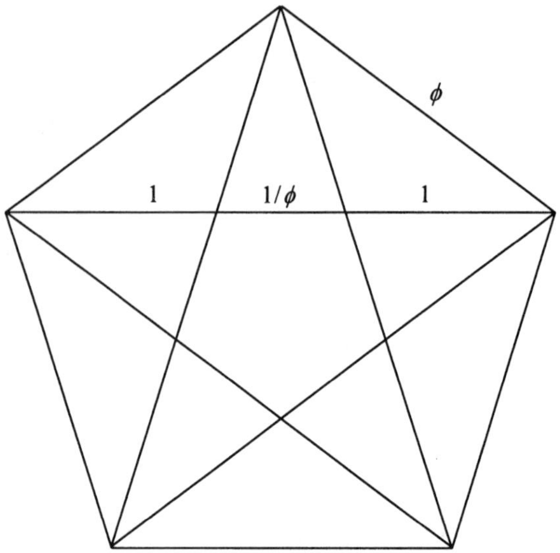

Figure 1.3 shows two smaller pentagrams within the interior pentagon. These are exact replicas of the original, scaled by multiplying all the original values by $1/\phi^2$ and $1/\phi^4$.[19] See if you can determine the sizes of the lines in this figure.

It remains unclear whether Greek sculptors and architects used these proportions in their Golden Age, about 600 to 300 BC. Yale University professor Jay Hambidge asserted that using them was the reason for the superb beauty of Greek works during that period.[20] His system of analysis, however, was so complicated that probably any proportion could be found. Brunés, who was adamant that neither the Greeks nor the Egyptians used the golden proportion, had a solution that was no better, being based on the equally problematic "sacred cut" involving $\sqrt{2}$.[21]

Garth E. Runion's example of a beautiful Greek cup seemed to show phi proportions, but measuring the photograph as marked (at the lip and the base) suggested otherwise.[22] However, redrawing the lines at the edges of the cup and remeasuring the result confirmed phi proportions within the limitations imposed by measuring a photograph with a ruler. This and many similar uncertainties show that photographs and illustrations can be misleading and must be examined carefully.

Figure 1.3. Multiple phi proportions in the pentagram

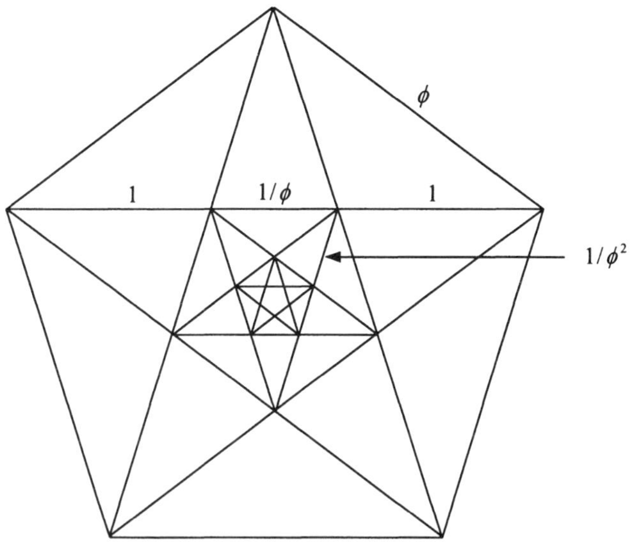

The Parthenon frequently is a focus of disagreement because it is a prominent artifact. Stecchini claimed that "the elevation of the fronts . . . was calculated by the factor phi and that of the flanks . . . by . . . pi," similarly to the Great Pyramid.[23] Hambidge found phi, while Brunés denied it.[24] Colman and Coan found angle 51°30′, which is fairly close to pyramid angle 51°49′39″, while Huntley claimed that the façade fitted neatly into a golden rectangle (measuring 1 by phi).[25]

These solutions are mutually exclusive and none has been proven with certainty. Some of the problems with analyzing the Parthenon have been its poor physical state, the use of drawings that lacked stated dimensions, and the use of photographs taken from distorting angles.

Modern problems of measurement aside, the first unequivocal discussion of phi is found in Euclid's *Elements,* written in the third century BC. This compendium of the known mathematics of the time gave a clear definition of the golden proportion, called by Euclid "extreme and mean ratio."[26] In Book VI, Definition 3, Euclid stated, "A straight line is cut in extreme and mean ratio when the whole line is to the greater segment as the greater to the less."[27] Extreme and mean ratio is called phi (ϕ) in this book and is illustrated in figure 1.4.

Figure 1. 4. Sketch of phi proportion (A + B) : A :: A : B

In this description, phi is defined as (A + B) : A :: A : B.[28] Rewritten in modern algebra, and substituting 1 for A + B and x for A, this is

$$\frac{1}{x} = \frac{x}{1-x} . \tag{1.4}$$

Cross multiplying and collecting terms gives

$$x^2 + x - 1 = 0 . \tag{1.5}$$

Solving for x using the Quadratic formula[29]

$$x = \left| \frac{-b \pm \sqrt{b^2 - 4ac}}{2a} \right| , \tag{1.6}$$

where the coefficients of equation (1.5) are

$$a = 1,\ b = 1,\ c = -1 ,$$

and

$$x = \left| \frac{-1 \pm \sqrt{1+4}}{2} \right| = \left| \frac{-1 \pm \sqrt{5}}{2} \right| = 0.618 \text{ and } 1.618 , \tag{1.7}$$

rounded as before. Phi (ϕ) is assigned to 1.618, while $1/\phi$ is 0.618.

It is important to point out that equations (1.1) to (1.3) and (1.4) to (1.7) are alternative approaches to finding the proportions shown in figures 1.1 to 1.3. Notice that A and B in figure 1.4 correspond to D–F and F–C in figure 1.1 and to 1 and $1/\phi$ in figure 1.2.

Phi is intriguing because it has a mystical quality. It is unique in that it can be expressed in terms of itself. Its inverse, for example, is equal to itself minus 1, that is,

$$\frac{1}{\phi} = \phi - 1, \qquad (1.8)$$

or, in numbers again rounded to three decimal places,

$$\frac{1}{1.618} = 1.618 - 1 = 0.618. \qquad (1.9)$$

The square of ϕ may be expressed as itself plus 1, or

$$\phi^2 = \phi + 1, \qquad (1.10)$$

that is,

$$1.618^2 = 1.618 + 1 = 2.618. \qquad (1.11)$$

Phi is the only number to have these properties and the only proportion to satisfy Euclid's definition.[30]

Phi in Europe

When the Romans took over Greece and Egypt, phi began moving slowly toward the west.

According to most experts, Roman and medieval builders did not use phi, but rather ratios of small integers. These are the numbers ascribed to Pythagoras for musical pitch by Boethius and other early medieval writers, for example, 2:1, 3:2, 4:3, and 5:4 for the octave, fifth, fourth, and third. A few others might also have been tolerable. The famous book by the Roman architect Vitruvius supported this idea and was the bible of most builders until the Renaissance.

About the time of the Holy Land Crusades of the eleventh to thirteenth centuries AD, cathedrals began to appear all over Europe in a distinctive style now known as Gothic. Although this term was originally used pejoratively to mean ugly, the Gothic cathedrals are now thought of as some of the finest work anywhere. No one really knows where this unique style came from. Some authors have alleged that the builders were the Knights Templar, and others have thought that they were the Catholic bishops.[31]

In any event, stonemasons traveled from site to site, keeping their knowledge secret. Did this knowledge include phi?

In studying the floor plans, most researchers have found the ratios of small integers mentioned above. Several have said categorically that phi was not used anywhere. Cowen, however, thought there may be phi in the northern rose window of Chartres.[32]

Most researchers have also denied that phi was used in the façades of these buildings. However, Schneider showed a photograph of the façade of Notre Dame in Paris with a pair of phi calipers superposed that seemed to demonstrate the use of phi, and von Simson mentioned phi in the façade of Chartres.[33] Some authors have claimed that phi was used in medieval music. This will be discussed in chapter 2.

Early in the Renaissance, scholars translated the *Elements* of Euclid into Italian. In addition, painters began a search for realism that led to the mathematical principle of perspective.[34] One of the by-products of these events was the rediscovery of Euclid's extreme and mean ratio.

Luca Pacioli's book about phi, *De Divina Proportione,* possibly illustrated by Leonardo da Vinci, and probably plagiarized from Piero della Francesca, was discussed widely among artists of all kinds, presumably including musicians.[35] An exploration of phi in Renaissance music will appear in chapter 3.

In the eighteenth century, J. S. Bach showed knowledge of phi, and it is possible that Haydn, Mozart, and Beethoven knew of it as well. The Baroque and Classic periods will be explored in chapters 4 and 5.

In the nineteenth century, several researchers began to publicize phi as a special number. Gustav Fechner, Adolph Zeising, and others began asking people what their preferences were in choosing different sizes of paper and so on. They measured windows, tables, paintings, and many other things looking for a pattern. Their unscientific (i.e., uncontrolled and ad hoc) studies found a preponderance for something near phi, but by allowing variations up to 10%, their results are viewed with suspicion by most researchers today. The possible influence on composers of the Romantic period will be discussed in chapter 6.

The books of Jay Hambidge had considerable influence on artists and analysts in the early to middle decades of the twentieth century, including some musicians. It appears that such influence has waned, however, and articles and books are now hard to find. Perhaps this is due to the excessive complexity of the system. It is known that several composers of the twentieth century used phi. This will be discussed in chapter 7.

There have been analyses of works by painter Georges Seurat trying to show that he used the golden proportion. William A. Camfield, for example, showed a photograph of Seurat's painting *Parade*, which—contrary to his own analysis—is not in phi proportion measured as height to width.[36] However, the vertical dimension considered alone *is* divided near phi, as can be seen by measuring vertically to the bottom of the music stand (≈ .610). This is true also of the horizontal dimension measured horizontally to the vertical divider on the trombonist's left (5/8 = .625).[37]

Camfield also claimed that Juan Gris used phi proportion in his painting *Le Lavabo*. That this is true can be seen by measuring Camfield's photograph vertically to the shelf (.613) and along the bottom to a vertical line passing through the two globes (.620).[38]

This seems quite convincing, but Herz-Fischler could not find phi in Seurat, and R. Stanley Johnson felt that for Gris, phi "represented more an ideal than an absolute goal."[39]

Le Corbusier made the final major application of phi to architecture in the middle of the twentieth century. The United Nations Building in New York is alleged to be in phi proportion and many apartment complexes in France and elsewhere consist of "modules" based on phi at various scales. Critic Henry J. Cowan has argued that this system is unsuccessful because the modules fail to fit together without special treatment.[40]

Recent literature on sacred geometry, written by people identified as Pythagoreans, Rosicrucians, and Freemasons, has declared that the ancient mysteries, including phi, were passed down from the Middle East through various secret societies to the present day. These societies are assumed to have been religious organizations, stonemasons, or practitioners of the occult arts of alchemy, astrology, and witchcraft.[41]

Since the members of such organizations were sworn to secrecy, it is difficult to verify these claims. In fact, one of the most forthcoming of the secret organizations is the Freemasons, who, although they use the pentagram in their symbology, do not apparently assign any importance to its proportions.[42]

On the other hand, Paul Foster Case, writing for the Rosicrucians, showed the mathematics of the pentagram clearly, with connections made to the Great Pyramid.[43] The lore assigns great cabalistic value to numbers 5, 8, and 13, and to 4, 7, and 11 (these will soon be shown to be the well-known Fibonacci and Lucas numbers). Also important are the numbers 280 and 220, which just happen to be the height and half-base of the Great Pyramid in cubits, as discussed above.

Whether phi was known in antiquity and passed down by word of mouth, or was rediscovered during the Renaissance, or was a result of the nineteenth-century studies of Fechner and others, it was widely discussed and further developed by mathematicians Pacioli, Bernoulli, and Kepler, and painters della Francesca and Leonardo da Vinci and possibly later art theorists, such as Charles Henry. Whether musicians were involved in such discussions will be studied in ensuing chapters.

Fibonacci Numbers

Here, the story backs up to 1202 AD, when Leonardo Pisano Fibonacci published a book called *Liber Abaci,* which created a sensation in Europe. Having studied with the greatest Arabic mathematicians of the day, he put much of their knowledge into the *Liber Abaci.* This book was an important impetus to establishing the Arabic numerals from one to nine and the Hindu zero in Europe as a replacement for the awkward Greek and Roman numerals.[44]

Today, Fibonacci is remembered primarily for a puzzle he posed about the reproduction of rabbits: How many pairs of rabbits would there be in any given month if there were one immature pair to begin with, each pair required two months to mature, each mature pair bore a single pair each month, and this process continued ad infinitum?

The result, known as the Fibonacci sequence, uses the formula

$$x_n = x_{n-1} + x_{n-2}, \qquad \text{where } x_0 = 0, x_1 = 1; \qquad (1.13)$$

that is, the next number in the sequence is the sum of the previous two numbers. Thus, we have 1 for the first term (given), 1 + 0 = 1 for the second, 1 + 1 = 2 for the third, and so on. The first several terms derived in this manner are 1, 1, 2, 3, 5, 8, 13, 21, 34, 55, 89, ...[45] The related Lucas sequence is derived similarly: 1, 3, 4, 7, 11, 18, 29, 47, 76, ... An infinite number of sequences can be invented in this way, for example: 2, 5, 7, 12, 19, 31, 50, 81, 131, ... Although many of these are useless, they are referred to as generalized Fibonacci numbers, and a few will be seen in later chapters.[46]

Although the *Liber Abaci* contained no mention of phi, the Fibonacci number sequence has direct connections to the golden proportion and is an integral part of this book. In the seventeenth and eighteenth centuries, Johannes Kepler and Robert Simson showed that, in the limit, the ratios of adjacent Fibonacci numbers approach the golden proportion. In the nineteenth century, Eduard Lucas showed that the sequence now named after him does the same.[47] Equation (1.14) and table 1.1 illustrate this.

$$\lim_{n\to\infty}\frac{F_n}{F_{n-1}} \longrightarrow 1.61803\ldots,\ F_{n-1} \neq 0\,. \tag{1.14}$$

Table 1.1. Fibonacci and Lucas sequences and their approximations to the golden mean (rounded to three decimal places)

Fibonacci Sequence	Ratio F_n / F_{n-1}	Lucas Sequence	Ratio L_n / L_{n-1}
1	-	1	-
2	2.000	3	3.000
3	1.500	4	1.333
5	1.667	7	1.750
8	1.600	11	1.571
13	1.625	18	1.636
21	1.615	29	1.611
34	1.619	47	1.620
55	1.617	76	1.617
89	1.618	123	1.618
144	1.618	199	1.618

It is known that Arnold Schoenberg used phi and that Béla Bartók and Joseph Schillinger used Fibonacci numbers to structure their compositions. There is also speculation that Beethoven may have known and used them. Lucas numbers are likely to be found in this context as well. This will be sorted out at the appropriate time.

Organicism

In the early part of the twentieth century, Theodore Andrea Cook and D'Arcy Wentworth Thompson and others showed that the growth of many plants and animals is based on fib and phi.[48] Such plants branch from the main stem in a spiral that spaces the leaves so as to get the most light. The pattern of repetition follows a Fibonacci number of branches along the stem. The Fibonacci counts differ from species to species.

A similar pattern of growth involves the placement of seeds. Pine cone scales and sunflower seeds, for example, spiral around a central axis. Counting the spirals of pine cones shows that there are almost always—

according to Thompson, 92% of cases[49]—the same numbers of spirals in every cone of a particular species. In some species, there are 21 spirals clockwise and 13 counterclockwise, with another subtle spiral lying crosswise to these. Sunflowers typically count out to 89 seeds in one direction and 55 in another, while the giant ones that grow in the American west often measure 144 to 89.

These patterns approximate logarithmic spirals, which can be constructed from arrangements of rectangles related to each other by units of golden rectangles—a rectangle having the value 1 on opposite sides and phi on the adjacent sides—or Fibonacci numbers. Examples follow.

Figure 1.5 shows a golden rectangle constructed with a unit of 1 on the bottom edge of the smallest box and $1/\phi$ on the vertical. Adding a square of side 1 to a golden rectangle makes a larger golden rectangle. Successively adding scaled up squares around the original rectangle makes increasingly larger golden rectangles. If this process is reversed by subtracting squares, it can be continued until the resolution of the equipment is exceeded.[50]

Figure 1.6 shows how to draw a good approximation of a golden (logarithmic) spiral by connecting circular arcs between opposite corners of the squares. This, too, can be continued until the limits of the equipment are reached.

Figure 1.7 shows how to do this using Fibonacci numbers. The smallest two squares have sides measuring 1. The sequence continues by adding squares of size 2, 3, 5, 8, 13, 21, and 34. Looking closely, you will see it is similar, but not identical, to the sequence of golden rectangles. In addition, this spiral is not infinite in the inward direction like the golden spiral, because it terminates at 1.[51]

Figure 1.8 shows the arcs that create the near-logarithmic spiral from Fibonacci blocks.

These spirals are fundamentally different from the famous spiral of Archimedes. Notice that in figures 1.6 and 1.8, the wrapped layers get farther apart as they spiral outward. By contrast, they remain equally spaced in an Archimedean spiral, like rope on a spindle.

Spirals and vortexes are often found in weather patterns and water phenomena. Similar structures occur in the growth of horns in sheep and antelope, and in the shells of mollusks. In idealized configurations (variations averaged out), these shapes are logarithmic. Such structures were used by Debussy, as we will see in chapter 7.

Figure 1.5. Adding squares to golden rectangles makes larger golden rectangles

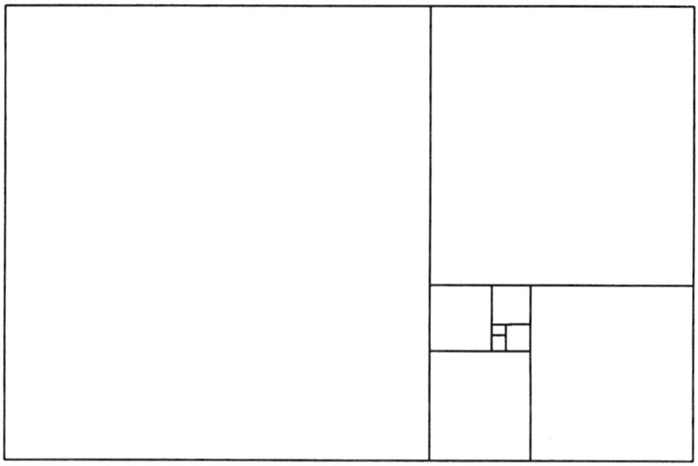

Figure 1.6. Golden spiral formed by connecting arcs in the squares

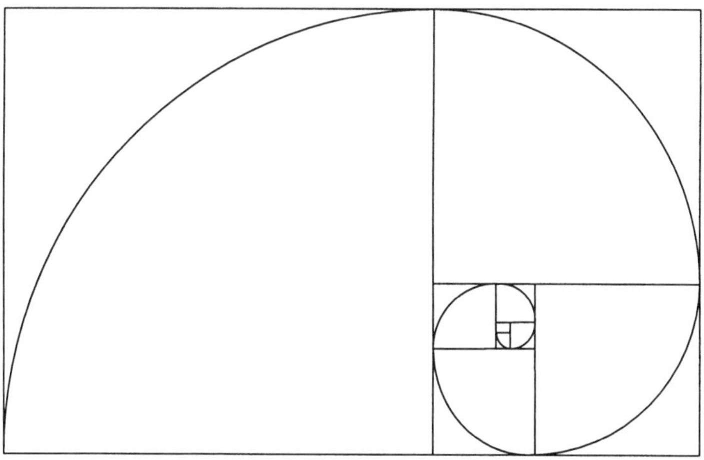

Figure 1.7. Fibonacci blocks forming the outline of a spiral

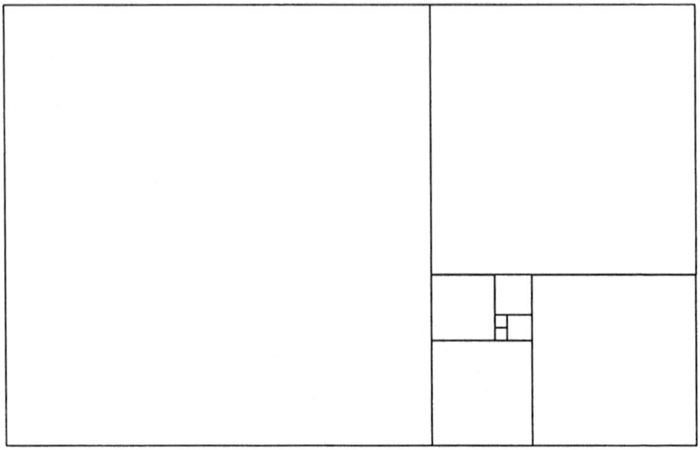

Figure 1.8 Fibonacci spiral

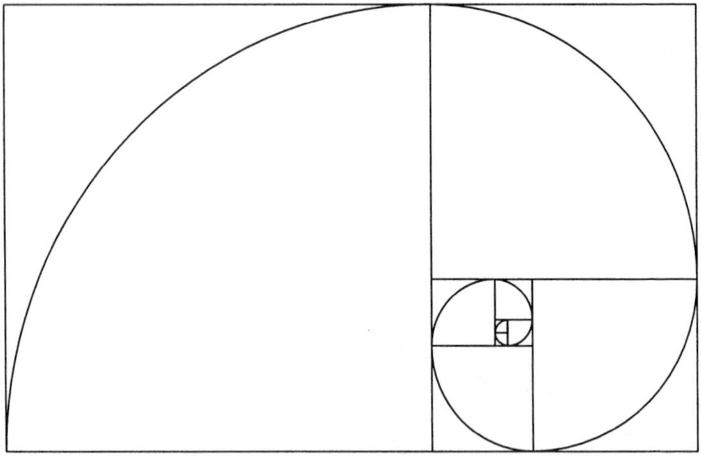

Earlier Studies of Phi in Music

Since the middle of the nineteenth century, many analysts have worked on the topic of fib and phi in music. Although all of their work is valuable, some needs revision and reinterpretation. It will be summarized here and further developed in later chapters.

Allen Arthur Dorfman's 1986 dissertation "A Theory of Form and Proportion in Music" gave an excellent summary of earlier studies, which is further condensed here.[52]

Dorfman reported that Adolph Zeising revealed the earliest known work on phi in music in his 1854 book *Neue Lehre von dem Proportionen des Menschlichen Korpers.* According to a paper entitled "Some Aspects of Beethoven's Instrumental Forms" presented by Gustav Ernest to the Royal Musical Association in 1903, Zeising applied the golden proportion to intervals in the harmonic sequence.[53] A favorite was the ratio 8:5 (a minor sixth in just intonation), which is indeed a Fibonacci number, but it is more than 1% away from 1.618, and only an approximation to phi. Unfortunately, the term is still used inaccurately today. Zeising apparently did not apply phi to musical form.[54]

The little-known Emil Naumann seems to have been the next important researcher. His 1869 book was entitled *Die Tonkunst in der Cultur-Geschichte.* There, the golden proportion was applied to musical form and—according to Gustav Ernest's paper as quoted by Dorfman—"all other manifestations of the human mind," in a "ponderous" and "futile" manner, standing "no chance whatever of being even noticed."[55] Oddly, Ernest noticed.

According to Dorfman, Ernest applied his theory of the golden section to sonata form.[56] He considered it to have the tripartite form *ABA* with *B* probably shorter than *A*. Such groupings as exposition + development/recapitulation, or exposition/development + recapitulation do sometimes come fairly close to phi, as we will see in chapter 5.

Ernest reported examining fifty-five sonata movements of Beethoven and finding that forty-two were significantly close to phi. Although we do not know what tolerances were used, Dorfman's example of Ernest's analysis of Beethoven's *Piano Sonata in C minor, op. 10, no. 1* is structured as exposition (105 m.) + development (61 m.) + recapitulation (117 m.), for a total of 283 measures.[57] The appropriate proportions would be 175 and 108. In this case, the exposition is 105 measures long compared to $1/\phi^2$ at 108 (2.8% error), which might be acceptable under a standard less strict than the one used in this book (one measure). The recapitulation, on the other hand, coming after 166 measures and $1/\phi$ after 175, is nine measures away. This 5.1% difference is far from the standard established in this book.[58] Dorfman reported Ernest as having said that "tolerances" and certain other adjustments—such as taking or not taking repeats—were necessary in this kind of work. This idea is rejected here.

Donald Tovey wrote in his 1935 book *Essays in Musical Analysis* that, although fascinated by the idea "that classical music measures its proportions in 'golden sections,' [I have] become sceptical; there are so many ways of taking your sections that I doubt whether any musical composition can avoid golden ones somewhere."[59] He also felt that theorists were not taking enough care to align their numbers with the actual structures.

Joseph Schillinger's 1941–6 book *The Schillinger System of Musical Composition* included information on the use of Fibonacci numbers by composers to enrich their melodies and rhythms. He also encouraged using phi in larger structures, but did not propose that phi play a role in analysis.[60]

J. H. Douglas Webster was the next author who came forward with a study. His 1950 article "Golden-Mean Form in Music" listed many major works and their proportions.[61] *Fractals in Music* reported that some of this music did indeed have golden proportions to tight tolerances, including works by Beethoven, Mozart, Bach, Mendelssohn, Chopin, Schoenberg, and Bartók.[62] However, since much of the material was not close to phi, the article will be thoroughly reviewed in later chapters of this book.

Ernő Lendvai was mentioned by Edward A. Lowman as having written a book in 1957 entitled *Bartók: sa vie et son ouvre*. The contents appear to be found in English in later books and papers.[63]

Hugo Norden was next, with an article in 1964 entitled "Proportions in Music."[64] His example from J. S. Bach's first fugue from *The Art of the Fugue* is clearly (and apparently knowingly) constructed in phi proportion. More such fugues will be studied in chapter 4.

After this, the frequency of studies accelerated greatly. A 1966 article by Ernő Lendvai "Duality and Synthesis in the Music of Béla Bartók" discussed the use of Fibonacci numbers by Bartók.[65] Although Lendvai's work is frequently disputed, much is of great value. See chapter 7.

Elinor Jane Perry Camp's 1968 dissertation "Temporal Proportion: A Study of Sonata Forms in the Piano Sonatas of Mozart" followed. Dorfman rightly criticized her use of ratios instead of proportions.[66] See the section on Method later in this chapter.

Ernő Lendvai produced a book in English about Bartók's musical forms in 1971 entitled *Béla Bartók: An Analysis of his Music.*[67] It clearly showed the composer's use of the Fibonacci numbers and his interest in the formations of the logarithmic spiral.

Edward A. Lowman's 1971 article "Some Striking Proportions in the Music of Béla Bartók" mentioned Lendvai's 1957 book *Bartók: sa vie et son ouvre* and several pieces discussed in chapter 7 of this book.[68]

Clive Brownley Pascoe's 1973 dissertation "Golden Proportion in Musical Design" analyzed fifty movements from music of several centuries and applied the most complex subdivisional notation to date, going up to the eighth power of $1/\phi$. Dorfman criticized this work because of its lax definition of musical significance as a determinant of structural points.[69]

Robert William Demaree Jr.'s 1973 dissertation "The Structural Proportions of the Haydn Quartets" did not find usage of phi proportions by Haydn.[70]

Jonathan Kramer's 1973 article "The Fibonacci Series in Twentieth-Century Music" showed conclusively that Karlheinz Stockhausen used Fibonacci numbers in his *Klavierstück IX.*[71]

Michael Richard Rogers' 1977 dissertation "The Golden Section in Musical Time: Speculations on Temporal Proportion" was marred by its critical attitude toward other studies.[72] He also proposed omitting certain sections of the music, such as introductions and codas, which is generally not allowed in this book.

James Austin Rothwell's 1977 dissertation "The Phi Factor: Mathematical Proportions in Musical Forms" contained a good development of the subject, but there were many counting and calculation errors.[73]

Roy Howat's 1977 article "Debussy, Ravel and Bartók: Towards Some New Concepts of Form" was the beginning of Howat's revealing analyses of Debussy's work.[74]

George Perle's 1977 article "The String Quartets of Béla Bartók" showed a clear use of phi.[75]

Paul Larson's 1978 article "The Golden Section in the Earliest Notated Western Music" showed that there is a tendency toward the golden proportion in the Kyries of the *Liber Usualis.*[76] This will be discussed in depth in chapter 2.

Derek W. Haylock's 1978 article "The Golden Section and Beethoven's Fifth" was a morass of errors and misunderstandings that misled Trudy Garland to state in 1987 that the first movement of Beethoven's *Fifth*

Symphony was structured in exact Fibonacci numbers.[77] This error has propagated itself throughout the musical community, and even appeared in Michael Hoffman's 1999 dissertation and Dan Brown's best-selling novel *The Da Vinci Code* (p. 96) . It will be discussed in chapter 5.

Newman W. Powell's 1979 article "Fibonacci and the Gold Mean [*sic*]: Rabbits, Rumbas, and Rondeaux" showed persuasively that Fibonacci numbers are found in Machaut's rondeaux.[78] This will be discussed in chapter 2.

Tibor Bachmann and Peter J. Bachmann's 1979 article "An Analysis of Béla Bartók's Music Through Fibonaccian Numbers and The Golden Mean" mentioned Lendvai but not his 1971 book.[79]

Charles Andre Barbera's 1980 dissertation "The Persistence of Pythagorean Mathematics in Ancient Musical Thought" will be more useful in a treatment of pitch organization.[80]

Alexander E. Sidorowicz's 1981 dissertation "The Proportional and Spatial Analysis of the First Movement Sonata-Allegro Form of Mozart, Haydn, and Beethoven" was a massive compilation of data from many works of all three composers that proved to be of fundamental importance to this book.[81]

Margaret V. Sandresky's 1981 article "The Golden Section in Three Byzantine Motets of Dufay" used the notation developed by Powell, and showed the presence of Fibonacci numbers in some of Dufay's music.[82] This will be discussed in chapter 3.

Ernő Lendvai appeared in print again in 1983 with his monumental book *The Workshop of Bartók and Kodály.*[83] Although partly about the Fibonacci numbers in pitch relationships (not a topic of *Fib and Phi in Music*), some of this book was about proportional form. Apparently a compilation of other work, it repeated itself a great deal. Most of the examples need careful verification and will be discussed in chapter 7.

Roy Howat's 1983 book *Debussy in Proportion* and article "Bartók, Lendvai, and the Principles of Proportional Analysis" remain the definitive works on Debussy's use of the Fibonacci numbers.[84]

Scott Anthony Olsen's 1983 dissertation "The Pythagorean Plato and the Golden Section: A Study in Abductive Inference" proposed that the early Greeks may have known about phi and that "the so-called discovery by Leonardo of Pisa regarding the Fibonacci series is at best a rediscovery."[85]

Dorfman gave useful critiques of some of the above writings. He went on to create a theory based on tension and release in relation to phi proportions, which may be useful to future researchers. He also discussed the structure of Chopin's *Preludes, op. 28*.[86]

Karen Anne Bates' 1986 dissertation "The Fifth String Quartet of Béla Bartók: An Analysis Based on the Theories of Ernő Lendvai" was largely about pitch structures and her comments about proportional structure were not persuasive.[87] She did, however, lead us to George Perle's 1977 article mentioned above.

Kenneth Patrick Kirk's 1987 dissertation "The Golden Ratio in Chopin's Preludes, Opus 28" showed that Chopin sometimes used phi, as discussed in chapter 6.[88]

Donald Glenn Adkins' 1987 dissertation "Golden Mean Durational Analysis as a Guide to Orchestral Repeats and Tempos" proposed that golden proportional structure can indicate whether repeats should be taken and how tempos should relate among sections and movements.[89] This work looks useful for analyzing contemporary works.

Ronald Jan Frey's 1994 Master's thesis "Attractiveness of the Golden Section in the Aesthetics of Sound Perception" was mostly about psychoacoustics and tried to show that the minor sixth based on the ratio 8:5 as an approximation to phi is superior to other ratios.[90] This notion is questionable and does not have a direct bearing on the work of this book.

Xiaoman Zhang Wardell's 1996 dissertation "An Examination of Selected Contemporary Works Composed by Means of Numbers" was based partially on phi and included a study of the advanced number theory used by Xenakis.[91]

Dominique Pascale Bardet's 1999 dissertation "Logos as Number and Proportion in Amiens Cathedral and in the Kyrie of the Messe de Nostre Dame by Guillaume de Machaut" suggested that small number ratios were used in the design of Amiens except in the single instance where a phi ratio was the generating figure for a floor plan.[92] How this compares to Machaut's piece will be studied in chapter 2.

Michael Hoffman's 1999 dissertation "Fibonacci Proportions in Composition: An Analysis of an Original Work for Jazz Orchestra" contained Haylock and Garland's wrong result for Beethoven's *Fifth Symphony*.[93] It was well written, however, had good analyses, and showed how a composer thinks.

Charles Madden's 1999 book *Fractals in Music* contained a single chapter devoted to fib and phi, which in light of more recent research proves to need expansion.[94] Presumably *Fib and Phi in Music* will go some way in that direction.

Methods Used in this Book

Definitions

Repeatedly multiplying or dividing a sequence of phi proportions by one of its members is a scaling—or shift—operation, and thus is self-similar. For example, Figure 1.2 showed values

ϕ^2, ϕ, 1, and $1/\phi$.

These could equally have been described as

ϕ, 1, $1/\phi$, and $1/\phi^2$

(division by phi—right shift), or

ϕ^3, ϕ^2, ϕ, and 1

(multiplication by phi—left shift).

The value of a piece of music is regarded as 1. Structural measurements should be portions of 1, that is

1, $1/\phi$, $1/\phi^2$, $1/\phi^3$,

and so on. These are generally called GS (Golden Section), GS^2, GS^3, GS^4, . . . in the analytical literature.

Table 1.2 shows several values for phi and GS. Values greater than 1 will not generally be used in this book, but GS values will often be referred to as "phi proportions."

Phi has many names, among them the golden mean, golden ratio, golden section, golden proportion, and divine proportion. The emphasis here is on the word "proportion."

It is important to understand the difference between a ratio and a proportion. In figure 1.4, A : B and B : A are ratios; that is, they are parts related to each other. A : (A + B) and B : (A + B) are proportions; they relate the parts to the whole. B : A and A : (A + B) are equal and interchangeable if and only if (iff) they are both exactly phi.

Table 1.2. Phi and GS proportions

Power of phi		Power of GS		Numerical Value
$1/\phi^4$	=	GS^4	=	0.146
$1/\phi^3$	=	GS^3	=	0.236
$1/\phi^2$	=	GS^2	=	0.382
$1/\phi$	=	GS	=	0.618
1		1	=	1.000
ϕ			=	1.618
ϕ^2			=	2.618
ϕ^3			=	4.236
ϕ^4			=	6.854

This has been misunderstood by several authors encountered in this study and has resulted in numerous errors. It is important to convert ratios to proportions before calculating any percentages.

For example, A = 3 and B = 2 would be expressed as a ratio in the form

$$A : B = 3 : 2.$$

To be expressed as proportions of a piece of music, however, they would have the form

$$A : (A + B) = 3/(3 + 2) = 3/5 = 0.600$$

and

$$B : (A + B) = 2/(3 + 2) = 2/5 = 0.400.$$

This is vastly different from 3/2 = 1.5 or 2/3 = 0.667. Proportions in this definition must be complementary; that is, paired proportions add to 1. They are computed by dividing each numerator, A and B, by their sum, which is the lowest common denominator.

Thus, this book uses A/(A + B) and B/(A + B) (i.e., the fraction or percentage of the whole) for all proportions. Fibonacci number ratios will be converted to proportions. Thus, ratio 3:5 becomes 3/8 = 0.375. Adjustments have been made to analyses in the dissertations that did not approach proportional analysis this way.[95]

In addition to the GS notation, FN means any Fibonacci number, LN is a Lucas number, and GN is a generalized Fibonacci number. The terms FN, LN, and GN do not necessarily give numerical values, as GS and GS^2 do, but rather identify the sequences to which particular number sets belong. They will sometimes be found in multiples, such as 2FN, where all numbers of the sequence are multiplied by 2.

Table 1.3 shows the additive quality of the phi proportions and may provide a clearer understanding of the phi relationships. Many additive combinations are simpler in form than multiplications, yet remain equivalent. You should memorize these and similar relationships and practice working out their numerical values. They will be needed often in working with the music.

Table 1.3. Additive equivalents of phi proportions

$$1/\phi = \phi - 1$$
$$1/\phi^2 = 1 - 1/\phi = 2 - \phi$$
$$1/\phi^3 = 1/\phi^2\phi = 1/(\phi^2 + \phi) = 1/(2\phi + 1)$$
$$\phi = 1 + 1/\phi$$
$$\phi^2 = \phi + 1$$
$$\phi^3 = \phi^2\phi = \phi^2 + \phi = 2\phi + 1$$

Levels of Accuracy

Since the work of such authors as Fechner and Zeising in the nineteenth century is hard to accept today because of the wide tolerances they used, a stricter standard should be established. Figure 1.9 illustrates how close the FN approximations seem to come to GS.

Each Fibonacci proportion (F_n / F_{n+1}) in the figure approximates GS, but some, such as proportions 1/1 and 1/2, are obviously not close. Proportions 3/5 (= 0.600), 5/8 (= 0.625), and 2/3 (= 0.667) may look close on a small scale like this, but if the whole piece were 1,000 measures—not uncommon in major works—the difference between GS (0.618) and 3/5 (0.600) would be 18 measures. The difference divided by the correct value ($0.018 \div 0.618$) is approximately 3%.

Figure 1.9. Near and far relationships of GS and FN

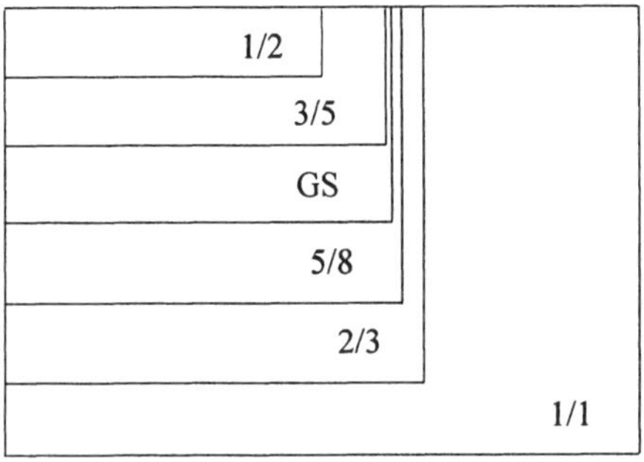

With this degree of looseness, figures should be checked carefully. In the past, an error of 8% (.667 for .618) may have been all right for a loose concept of phi, but today tolerances must be tightened greatly.

Modern computer studies have verified the fib and phi relationships in organic structures with a high degree of precision. Dunlap, Prusinkiewicz and Lindenmayer, and Ridley have demonstrated that models of organic structures require accuracy on the order of .1%.[96] Following their lead, therefore, we improve tolerances from the roughly 10% accepted by Webster and others to .1%: a factor of 100.

Granted, this allows a variance of only 1 point, or 0.617 to 0.619, which may be asking a lot, but some examples herein fit this model. Moreover, an acceptably looser standard of one measure or other unit from GS is not too much to ask. Many examples herein meet this standard. FN, being measured in round numbers, should always be exact.

The many studies of fib and phi in music referred to above have not used a consistent methodology and some analysts have ignored introductions and codas in order to make the theory fit. This, too, needs to be tightened up. Most of the music studied in this book does not benefit from ignoring sections at will. This practice will generally not be allowed.

If analysts work within this tighter framework, they can discover who has used GS or FN rigorously or knowingly, and who has not. They will find that some of the older data is acceptable. Furthermore, they can show that several modern composers used fib and phi knowingly.

Procedures

To find the proportions of a piece, this book uses the following procedure:

Find the number of measures or units
Multiply by GS and GS^2
Round to the nearest whole number of measures or units
Mark the measures or units that correspond to GS and GS^2
Determine whether harmonic or sectional structural points appear there[97]
Subdivide the piece, looking for other phi relations
Be aware that GS and GS^2 measure durations—that is, they mark the ends of sections, not the beginnings.[98]

A calculator will prove handy for determining GS and GS^2. Nowadays they are plentiful and inexpensive. Nevertheless, if necessary, table 1.4 is a reasonable substitute. It comprises the products of digits 0 to 9 and GS. (You may drop leading zeros in decimal fractions.)

Table 1.4. Products of the first ten digits and GS

	Example for use:	
0 × .618 = 0.000		
1 × .618 = 0.618	.618	GS
2 × .618 = 1.236	× 78	Total number of measures or units
3 × .618 = 1.854	4944	(8 × 618)
4 × .618 = 2.472	43260	(70 × 618)
5 × .618 = 3.090	48.204	Add; place decimal point
6 × .618 = 3.708	48.000	Round to nearest whole number
7 × .618 = 4.326		
8 × .618 = 4.944	48 is the GS measure or unit number	
9 × .618 = 5.562	30 is the GS^2 measure or unit number.	

To use the chart, place GS (.618) above and the total number of measures or units below (for example, 78), as shown in the "example for use" on the right. Look up the product of 8 and .618 (4.944) and place it under the line, ignoring the decimal point, as in multiplication. Then look up the product of 7 and .618 (4.326) and place it under the first product, offset one digit to the left. Continue, displacing subsequent digits one more place to the left. Add, place the decimal point, and round to the nearest whole number (do not merely truncate).

This is the measure or unit number corresponding to GS; the complement is GS^2. If there is a structural point within one measure or unit, it is a phi proportion; otherwise, it is not.

It may not always be easy to tell whether you have a structural point. It may look or sound like one because of a pause or a change of texture or dynamics, but not be supported by the harmonic structure. If harmony and GS do not coincide, it may simply mean that the composer was not consciously using GS for structural purposes.

The next step is to look nearby for a structural point and check whether it is an FN, LN, or GN. If there is still nothing, the composer was not thinking in these terms in this piece. To make this clear, the actual proportion in percentage points should be calculated by dividing the duration of a major section by the duration of the whole piece, and a statement of closeness to GS or FN should be given.

If you do find something, try looking for subdividing groupings. Some composers, such as Stockhausen, used FN or GS pervasively. Others, such as Varèse and Schoenberg, used GS only for larger structures. Still others must have relied on intuition.

Chapter Notes

Bracketed abbreviations are references to the bibliography.

[1] *Merriam-Webster's Collegiate Dictionary,* Tenth Edition shows that phi is pronounced fī, like pi, not fee, as some engineers and a famous novel would have it.
[2] [Leh p. 8, 17, 102–104, 108] Judged inadequate were the Bent Pyramid of Khufu's father Sneferu, whose slope began at a high angle of 54°27′44″ that had to be reduced to 43°22′ when it was found to be collapsing, and the Red Pyramid, which was constructed at 43°22′ from the start.
[3] [Bru] [Capt] [Dunn] [Gad99] [Mich69] [Mich77] [Mich01] [Tom]
[4] [Leh p. 108]
[5] In any event, the North Star was far from where it is now. The 26,000-year wobble called the precession of the equinoxes causes the North Pole to draw out a ("small") circle in the northern sky about 46.8° across. In Fourth Dynasty Egypt, the north celestial pole was somewhere on the circle between Vega (where it was 13,000 years ago) and Polaris, where it is now. The staff at Salt Lake City's Hansen (now Clark) Planetarium clarified that 46.8° is the "small circle" referred to, meaning simply that it is not a "great circle."
[6] Wooden triangles could have been used in the same manner.
[7] [Leh p. 17]
[8] [Leh p. 17]
[9] This is not sexism. Only males did this work in Egypt.
[10] [Stecc p. 368] [Tom p. 106, 209, 213] [Leh p. 17]
[11] [Herz00 p. 175, 193] [Leh p. 17] Herz-Fischler and Lehner did not give the measurements in cubits, but their values in meters gave the same result.
[12] Some readers will notice that pi is here as well. Stecchini claimed that the north side was calculated as phi, and the west side as pi. Although this is more detail than we really need, pi and phi do go hand-in-hand here. The ratio of the perimeter to the height is $1760/280 = 44/7 = 6.2857 \ldots \approx 2\pi$. Some theorists think they meant to show that they knew about pi as well as phi. [Stecc p. 368]
[13] [Rhind prob. 58, 59]
[14] Boyer and Kline alleged that some information in the *Rhind Papyrus* came from Imhotep's time, roughly 3500 BC [Boy p. 12] [Kli72 vol. 1 p. 16].
[15] [Ax] [Bai83] [Bai86] [Bai89] [Howard] [Marrs] [Picknett]
[16] [Bru p. 9]
[17] [Boy p. 49]
[18] [Herz98 p. 98]
[19] This is a self-similar fractal comparable to those shown in *Fractals in Music.*

---

[20] [Hamb26] [Hamb32]
[21] [Bru] Brunés's entire book was an effort to refute the use of the golden proportion in ancient times and establish the sacred cut in its place.
[22] [Run p. 87]
[23] [Stecc p. 355, 359]
[24] [Ghy p. 136] [Bru]
[25] [Col p. 207] [Hun p. 63]
[26] Boyer preferred to call the *Elements* "an introductory textbook" [Boy p. 104].
[27] [Kli72 vol. 1 p. 73]
[28] Read "A plus B is to A as A is to B."
[29] Absolute values are used because negative values have no meaning here.
[30] This statement holds only if the values are exact. Approximations will not do. This needs to be stressed, for some authors encountered here have not observed the distinction and have led themselves into error.
[31] [Marrs p. 287] [Star p. 80] [Sim p. 13–41]
[32] [Cowen p. 125]
[33] [Schn p. 135] [Sim p. 155] I do not find phi in these cathedrals using a compass and straight edge in the manner of figure 1.1. However, a Babylonian stele shown in Schneider (p. 133) is clearly a golden rectangle.
[34] [Kli53 p. 126–144]
[35] [Kli72 vol. 1 p. 237] [Herz98 p. 149–50]
[36] [Camf Ill. 1]
[37] Despite the likelihood of errors creeping in at this scale, the general outline seems to support phi in this painting.
[38] [Camf Ill. 2]
[39] [Herz83 p. 111] [JohnsonR p. 23]
[40] [Cowan p. 326]
[41] Although Americans tend to scoff at the notion of secret societies, such secrecy was needed in ancient times, as European survivors of the many Inquisitions well knew. Edward E. Lowinsky presented a very strong case for such need even as late as the sixteenth century [Lowi]. Despite controversy over Lowinsky's main points, this part of his thesis is incontrovertible. See also note 15.
[42] Tons Brunés, who claimed to be writing on behalf of the Freemasons, based his lopsided construction of the pentagram on the square root of two, not phi.
[43] [Case p. 38–9, 60–1, 117–19, 173]
[44] [Kli72 vol. 1 p. 209]
[45] To see that this is correct, notice that the immature beginning pair is still immature in the second month, but mature in the third month and bears a pair, making one mature and one immature pair. In the fourth month, the original pair bears a pair, making one mature and two immature pairs. In the fifth month, the original pair bears a pair, as does the now mature second pair, making two mature pairs

---

and three immature pairs. You can easily continue this to see that in the sixth month there are three mature pairs and five immature pairs, and so on.

[46] Powell claimed that musicians of the medieval period could not have dug the additive series out of the *Liber Abaci*, but might have known about it from Nicomachus, who wrote in the second century AD. However, Nichomachus's argument was complicated as well, and not particularly obvious. [Pow p. 239]

[47] [Dunlap p. 43, 52] [Liv p. 97, 101]

[48] [Cook p. ix, 107, 120–1, 417–8] [Thom p. 922]

[49] [Thom p. 923] It is thus a tendency, not a law [Liv 115].

[50] [Boy p. 111–12]

[51] [New p. 216, 218]

[52] It is surprising that at least part of this dissertation has not been published. It contains much valuable information.

[53] As shown in *Fractals in Music*, the so called harmonic series is mathematically a sequence [Mad p. 29]. Thus, it is referred to as such from now on.

[54] [Dor p. 83]

[55] [Dor p. 84]

[56] [Dor p. 99–104]

[57] [Dor p. 102]

[58] See following section on methods.

[59] [Tov35 p. 19] [Tov45 p. 112]

[60] [SchilS]

[61] [Webster]

[62] [Mad]

[63] [Lowm p. 527]

[64] [Nord64]

[65] [Len66]

[66] [Camp] [Dor p. 114]

[67] [Len71]

[68] [Lowm]

[69] [Pas] [Dor p. 120, 124] See also Rogers.

[70] [Dem]

[71] [Kra73]

[72] [Rog]

[73] [Rothwell]

[74] [Howat77]

[75] [Per]

[76] [Lar]

[77] [Hayl] [Gar87 p. 38] [Gar95 p. 116]

[78] [Pow]

[79] [Bachmann]

[80] [Barb]
[81] [Sid]
[82] [San]
[83] [Len83]
[84] [Howat83] I have not been able to acquire the article.
[85] [Olsen p. 145]
[86] [Dor]
[87] [Bat]
[88] [Kirk]
[89] [Adkins]
[90] [Frey]
[91] [War]
[92] [Bard]
[93] [Hof]
[94] [Mad]
[95] The point is intentionally over explained so as to make it patently clear. Please think about it, as even the best analysts have erred here.
[96] [Dunlap p. 133] [Pru p. 101]
[97] The term "structural point" is defined as harmonic and sectional structure, as taught in standard analysis courses. Dynamics and texture may support this, but are not definitive of the form for our purposes.
[98] A common error committed by the analysts studied here has been to mark the beginnings, thus making the durations off by one unit.

Chapter 2

Medieval

Liber Usualis

The medieval period left a massive record of Gregorian chant called the *Liber Usualis*. This music is still used by the Roman Catholic Church. All 1,900 or so pages are readily available in English and other languages at libraries, bookstores, and the internet.

An important grouping within the book consists of thirty chants known as Kyries. These always contain nine lines arranged as three renditions of "Kyrie eleison," followed by three renditions of "Christe eleison," and three more renditions of "Kyrie eleison."

If the three lines of a Kyrie or Christe section are identical in pitch and phraseology, they are simply marked "×3" at the end of the section, meaning "to be repeated three times." If all three sections are in this condition, a three-line set results. In some cases, the third section contains two identical Kyries marked "×2" and a single line that is varied with inclusions or extensions. This results in a four-line set.

A third presentation results in a nine-line set, in which inclusions and extensions are present throughout. All told, then, there are $30 \times 9 = 270$ lines arranged in sets of three, four, or nine lines. This simple-seeming *ABA* form can result in some rather complex structures, as we will see in the analysis of Kyrie III.

The Kyries

Paul Larson briefly reported a study of all thirty Kyries of the collection.[1] He found a tendency toward structures based on phi, justifying a similar study of other music from the medieval period. Apparently, this suggestion has not been followed up.

Larson's discussion of Kyrie IV as reported in *Fractals in Music* was not encouraging.[2] The conclusion was that the examples were too far from .382 or .618 to be acceptable. That assessment may have been hasty, for the division actually is at one note from GS, which is close enough although the percentages are poor.

But Larson did not show that a structural point in the fourth line falls directly on GS^2. This success suggests a need for further study.

A transcription of all thirty Kyries follows.[3] They are marked with the Golden Section (GS) notation introduced in chapter 1.[4] Because they are very short, structural divisions are expected to take place at the end of a note, rather than the end of a measure. Although rounding makes tolerances very loose, they cannot be otherwise at this scale. The proportions of the structural divisions are shown in parentheses at the bar lines.

Following the transcriptions, table 2.1 summarizes the occurrence of GS and GS^2 at and near structural points for each line. The percentage of lines having structural points at GS or GS^2 is 63/270 = 23%, and of those near (one note away) GS or GS^2 is 125/270 = 46%. (Twice as many are *near* as are *at* GS because there are two sides on which to fall.) This is encouraging and seems to suggest a tendency toward phi structures.

However, the range of proportions at which GS falls is .58 (Kyrie IX) to .65 (XV) and of those at which GS^2 falls is .30 (X) to .43 (XVI). This is a rather wide range and will not be acceptable for longer pieces. However, because of the shortness of these pieces, it is necessary to accept this error with a statement of the percentages.

It is interesting that GS^2 falls at major structural points about as often as GS does. Thus, the lines are about equally divided between long–short and short–long. The greatest concentration of lines defined by GS or GS^2 is found in Kyrie XVIII, containing eight such lines. Several have none.

Table 2.2 shows how the three parts of each Kyrie are divided. It shows the number of notes in each section and their proportions. The GS^2 column shows GS^2 rounded to the nearest note. Only Kyrie XV is divided at GS and three others are at 3:2, while Kyrie XIV is one note away from GS. This small quantity is not encouraging with regard to phi.

(Lux et origo)

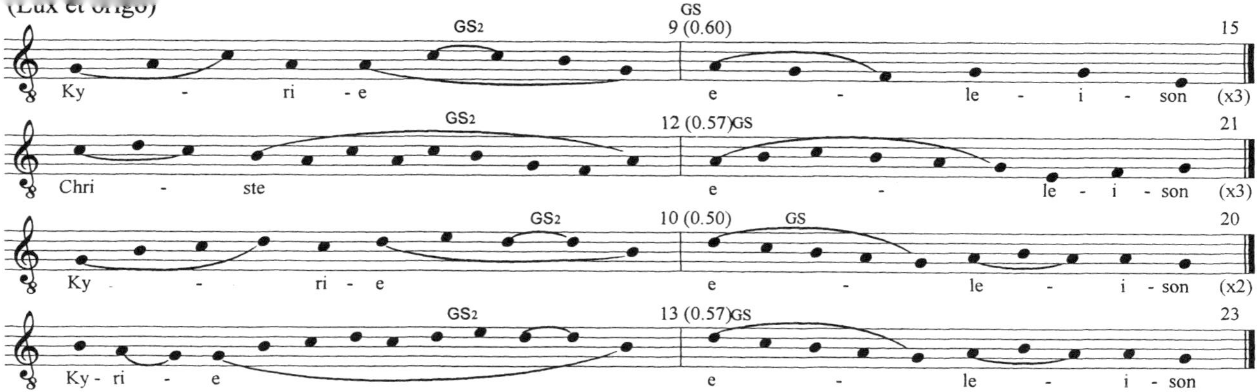

Kyrie II: For Solemn Feasts. 1.
(Kyrie fons bonitatis)

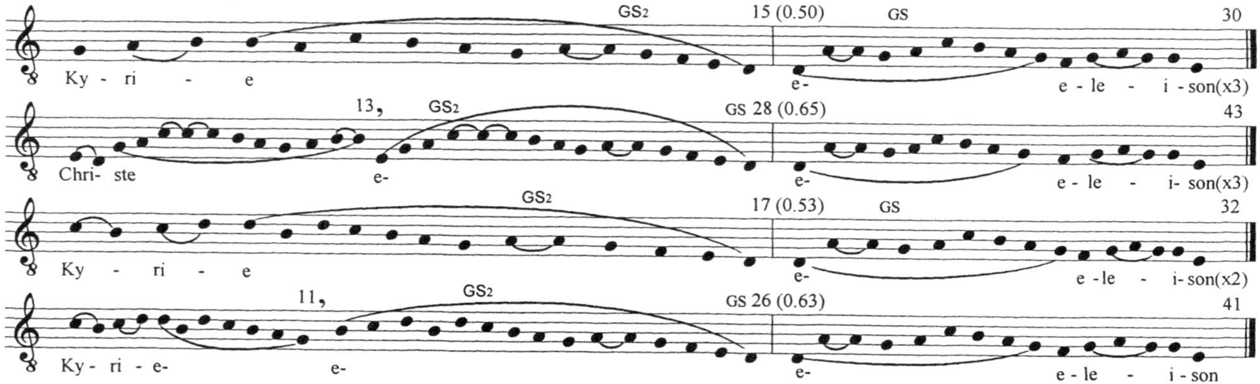

Kyrie III: For Solemn Feasts. 2.
(Kyrie Deus sempiterne)

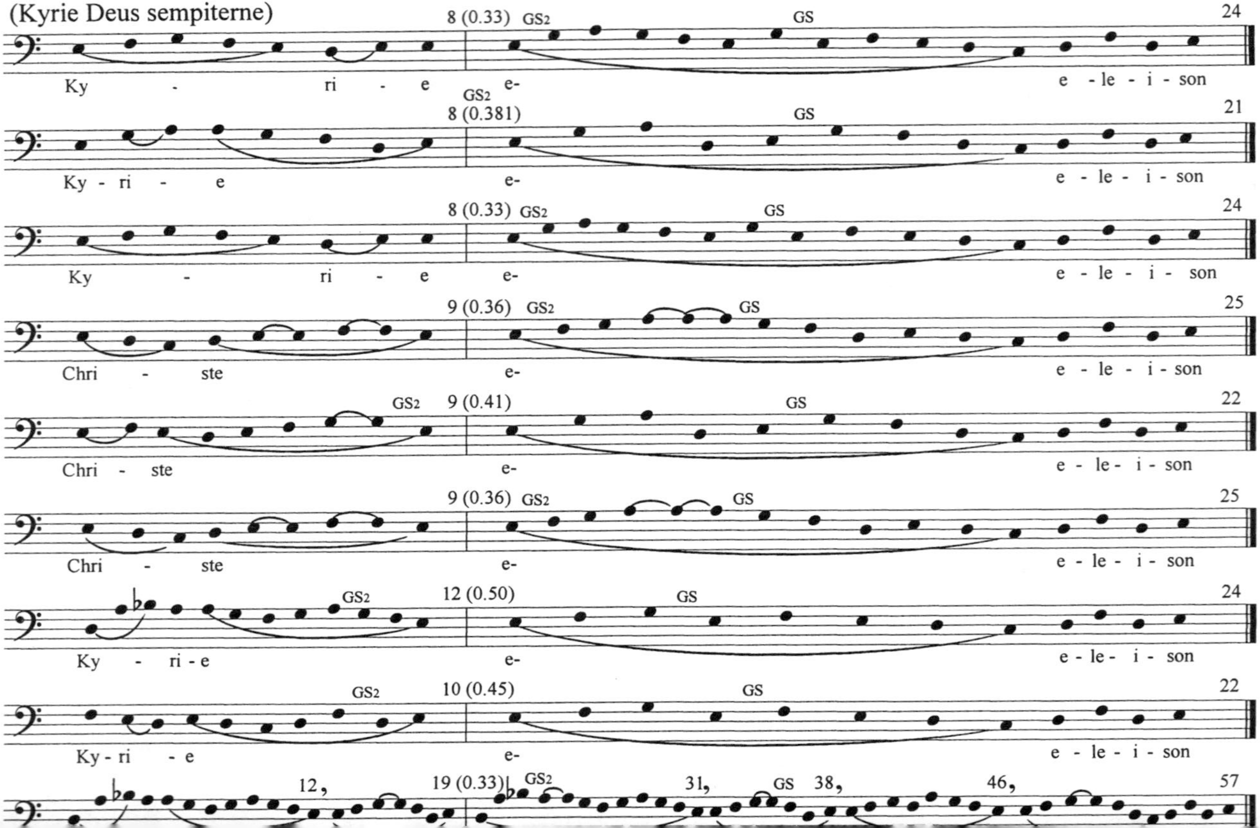

Kyrie IV: For Doubles. 1.
(Cunctipotens Genitor Deus)

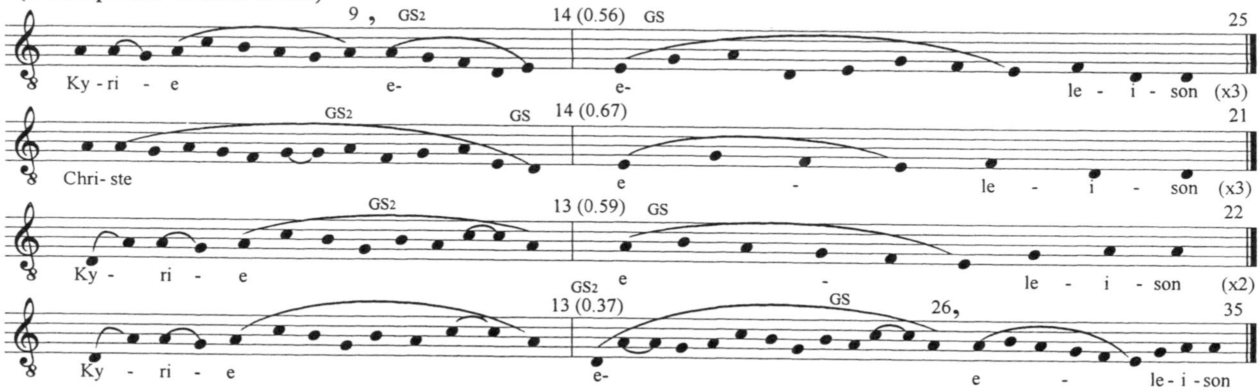

Kyrie V: For Doubles. 2.
(Kyrie magnae Deus potentiae)

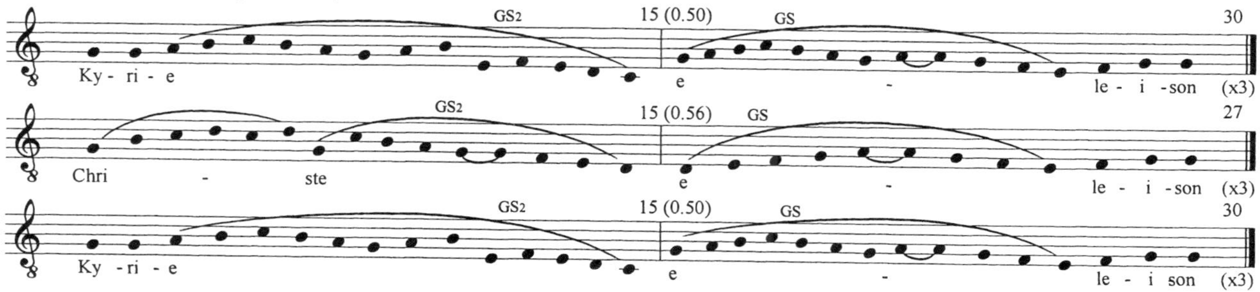

Kyrie VI: For Doubles. 3.
(Kyrie Rex Genitor)

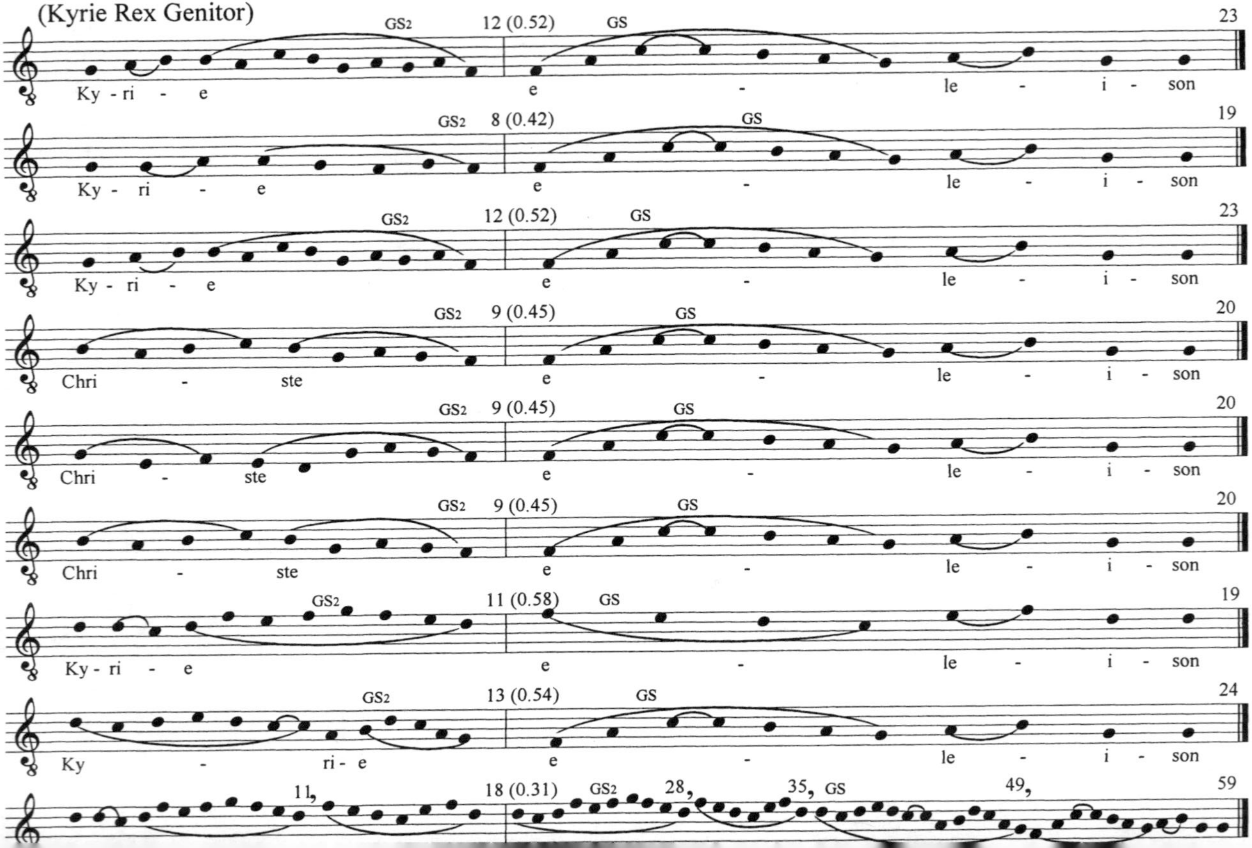

Kyrie VII: For Doubles. 4.
(Kyrie Rex splendens)

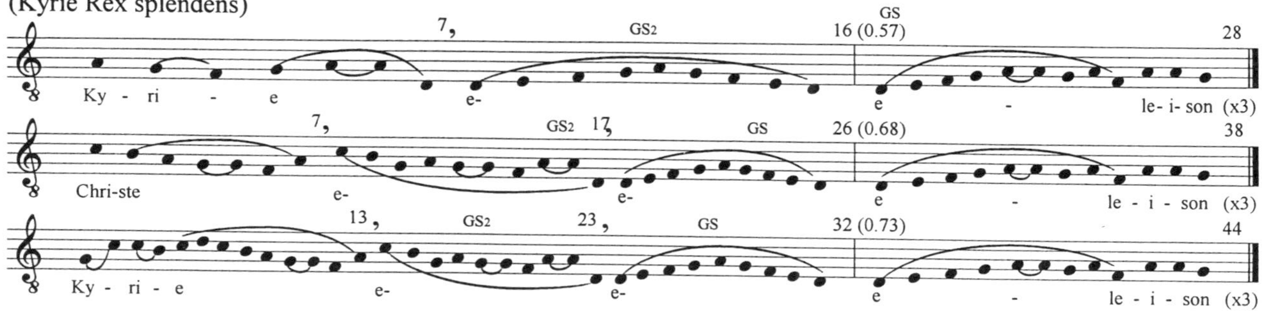

Kyrie VIII: For Doubles. 5.
(De Angelis)

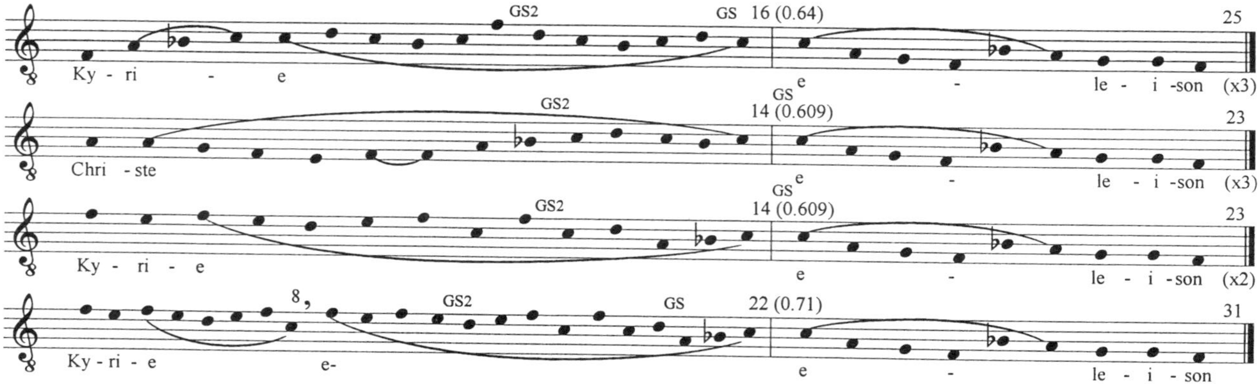

Kyrie IX: For Feasts of the Blessed Virgin. 1.
(Cum jubilo)

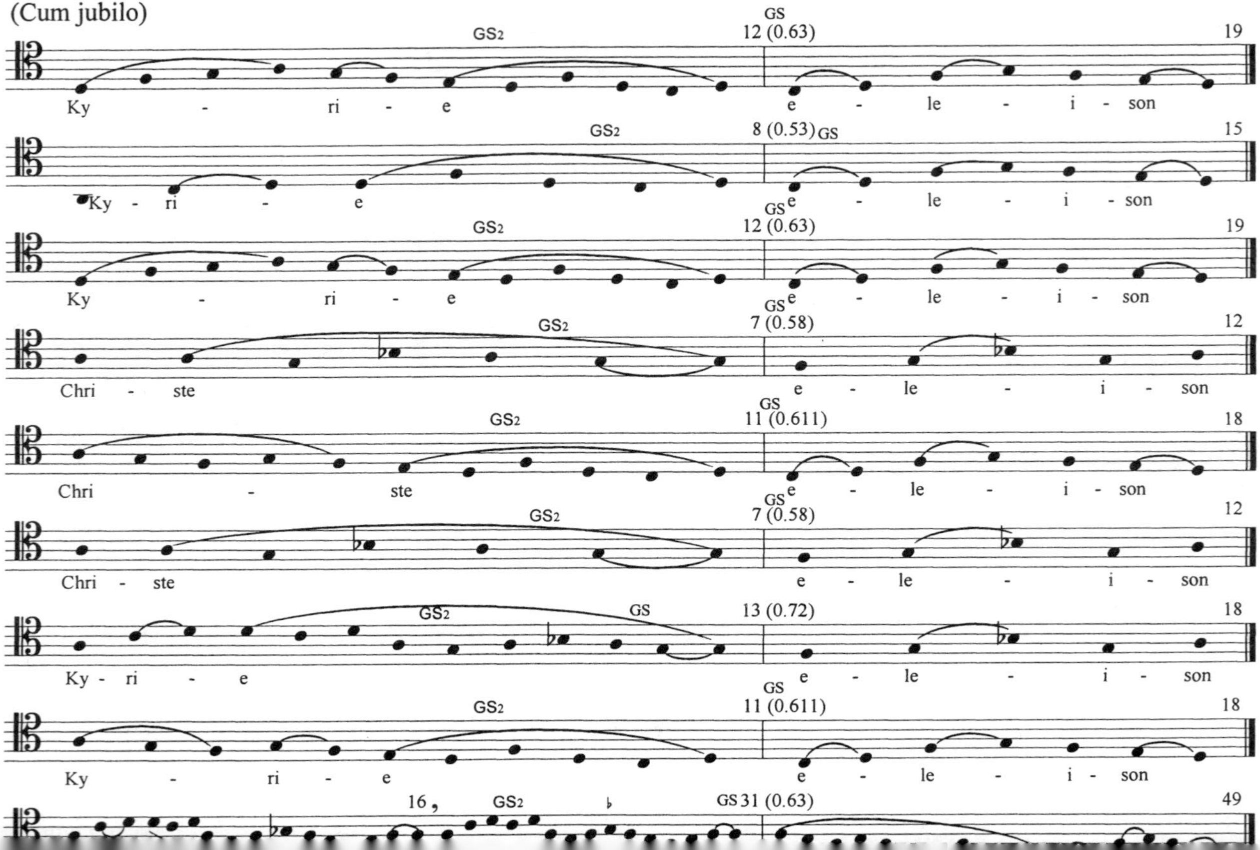

(Anne Pater)

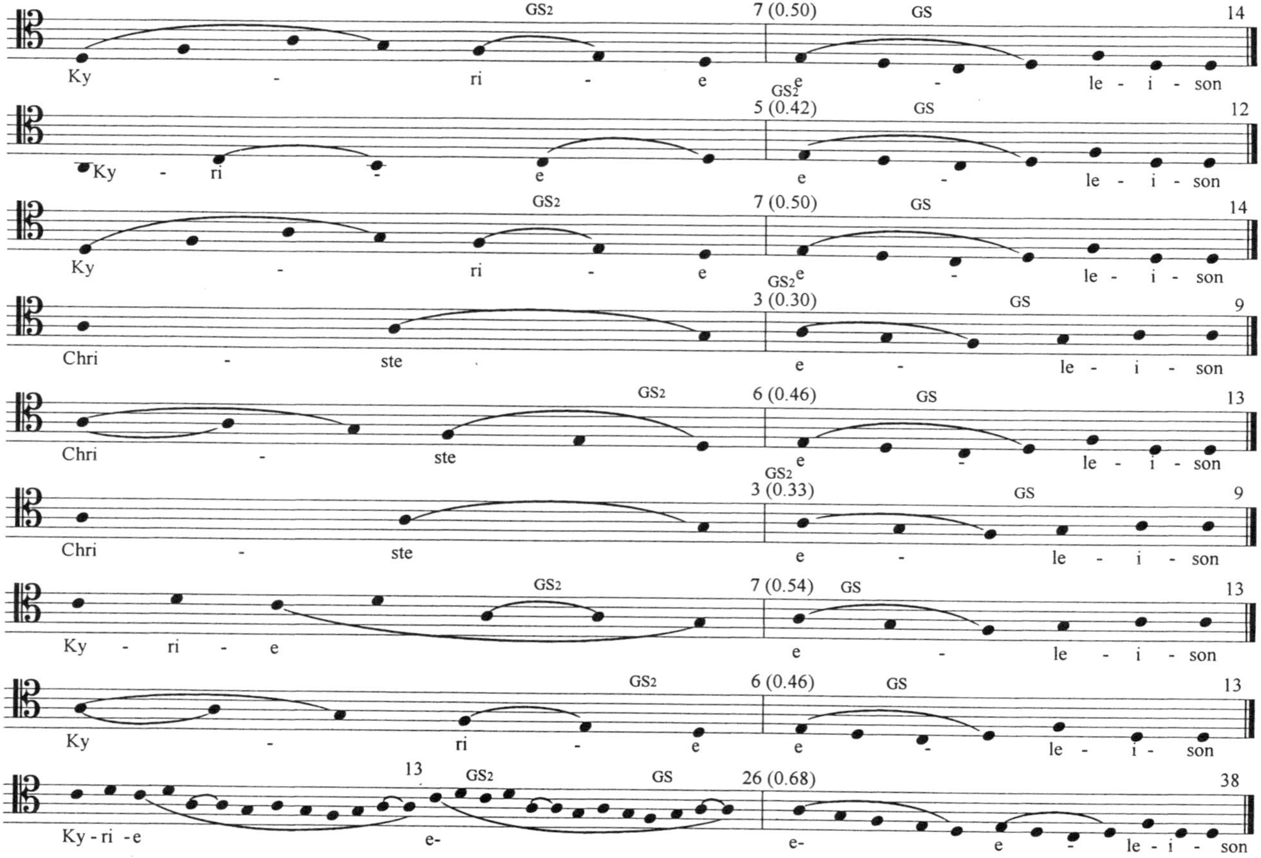
GS2 7 (0.50) GS 14
Ky - ri - e e - le - i - son
GS2 5 (0.42) GS 12
Ky - ri - e e - le - i - son
GS2 7 (0.50) GS 14
Ky - ri - e e - le - i - son
GS2 3 (0.30) GS 9
Chri - ste e - le - i - son
GS2 6 (0.46) GS 13
Chri - ste e - le - i - son
GS2 3 (0.33) GS 9
Chri - ste e - le - i - son
GS2 7 (0.54) GS 13
Ky - ri - e e - le - i - son
GS2 6 (0.46) GS 13
Ky - ri - e e - le - i - son
13 GS2 GS 26 (0.68) 38
Ky - ri - e e- e- e - le - i - son

Kyrie XI: For Sundays throughout the Year.
(Orbis factor)

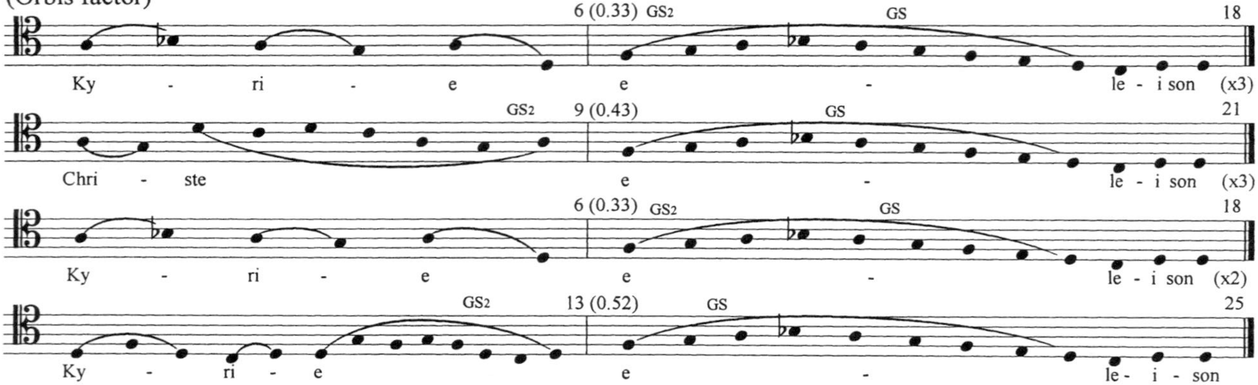

Kyrie XII: For Semidoubles. 1.
(Pater cuncta)

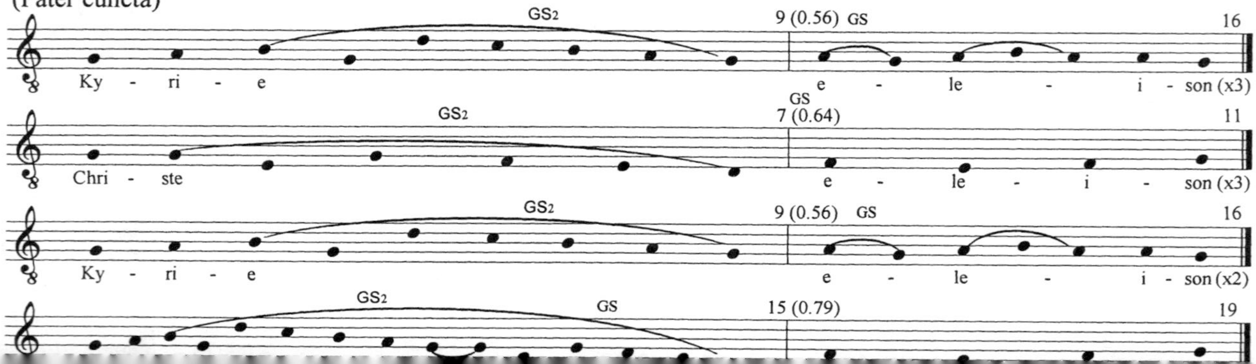

(Stelliferi Conditor orbis)

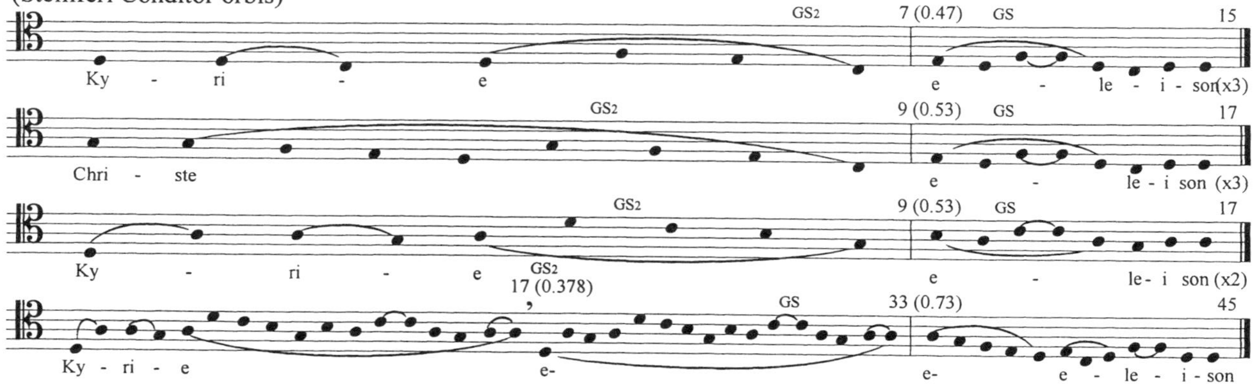

Kyrie XIV: During Octaves except those of the Blessed Virgin.
(Jesu Redemptor)

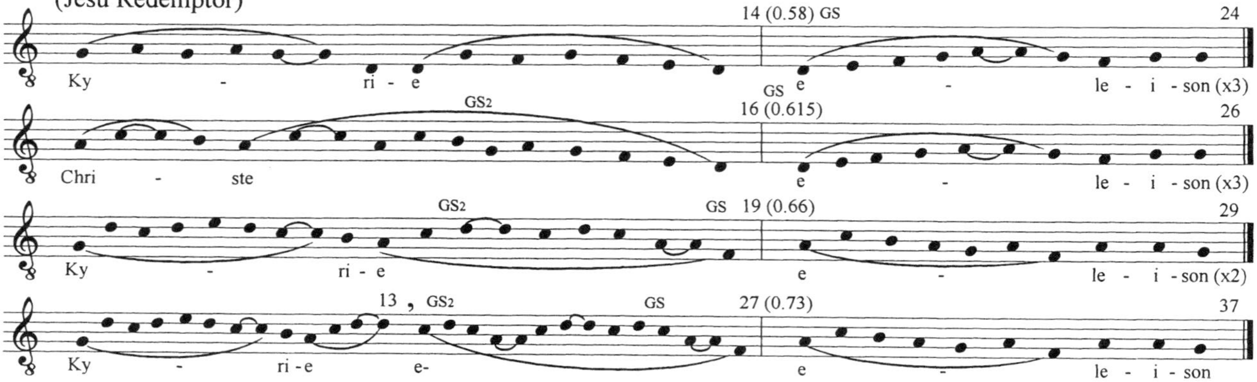

Kyrie XV: For Simple Feasts.
(Dominator Deus)

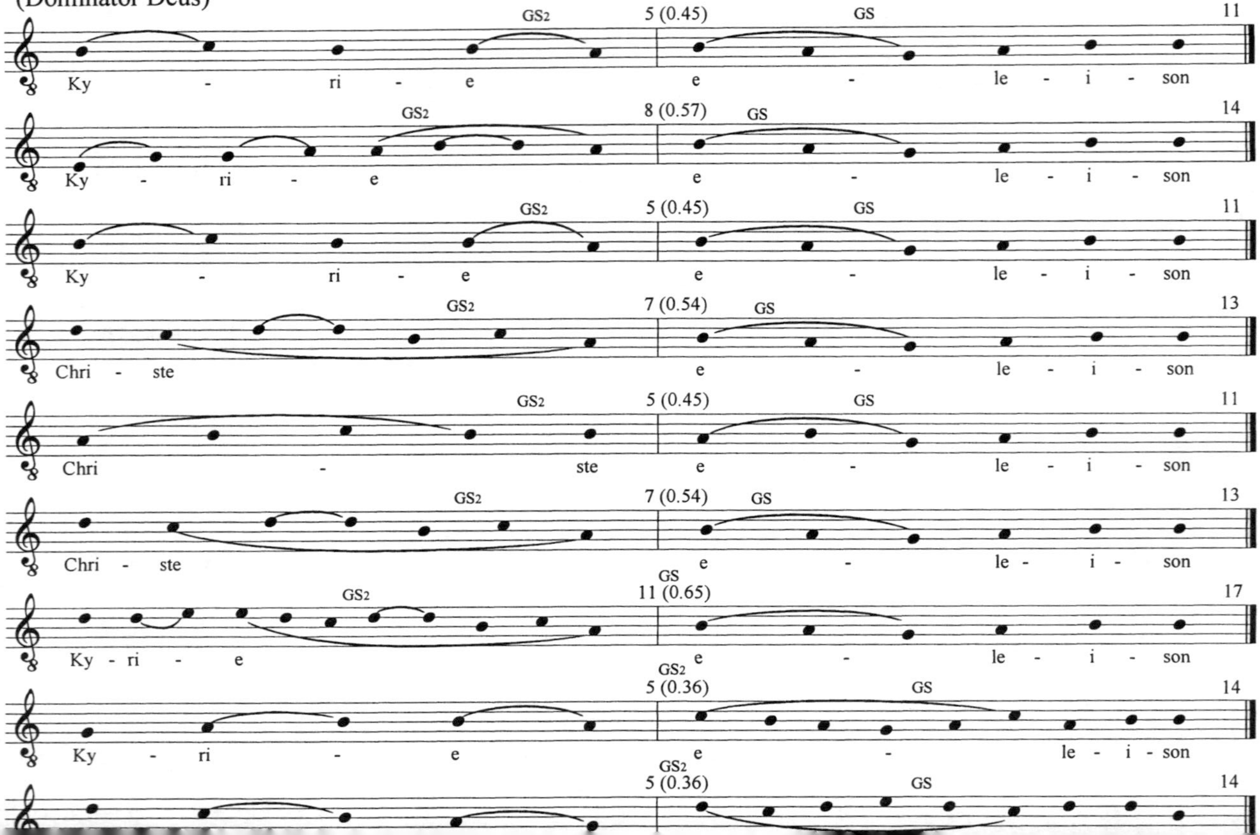

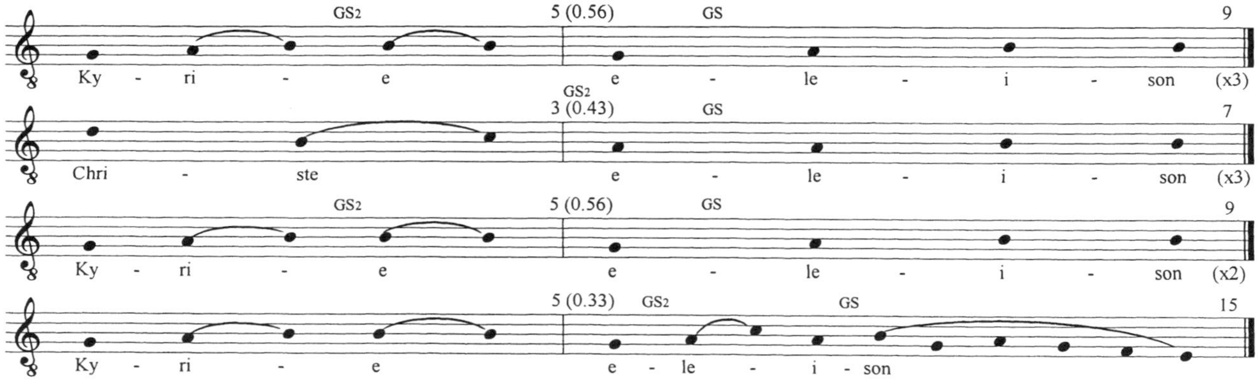

Kyrie XVII: For the Sundays of Advent and Lent.

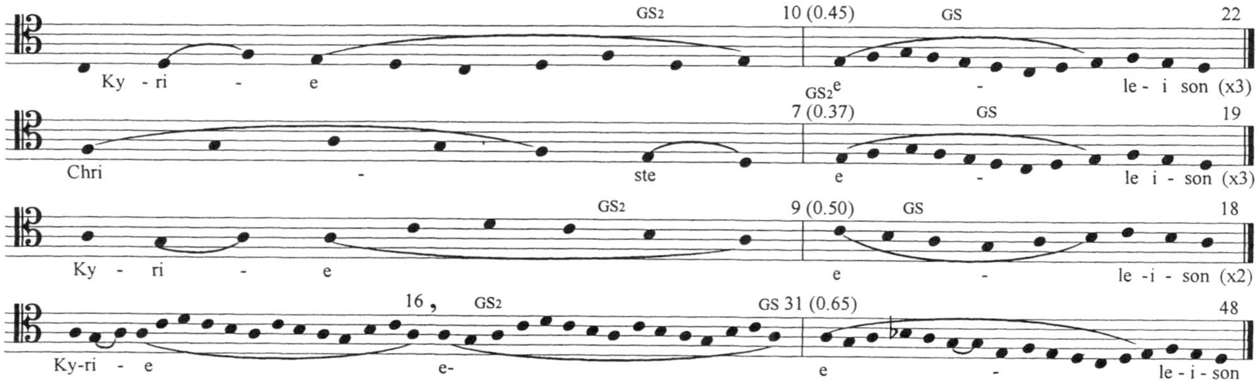

Kyrie XVII-alt: For the Sundays of Advent and Lent.
(p. 61 of 1997 English edition)

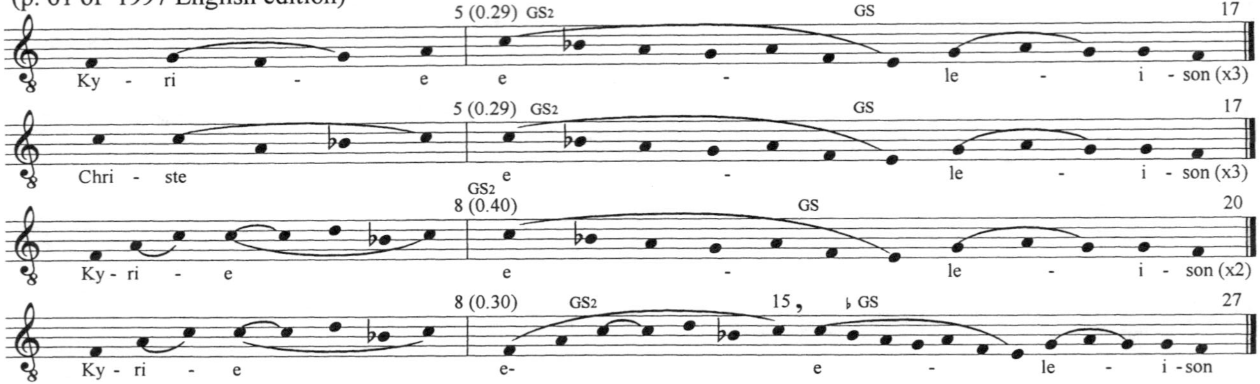

Kyrie XVIII: For the Ferias of Advent and Lent, as well as for Vigils, Ember Days, and Rogation Days.
(Deus genitor alme)

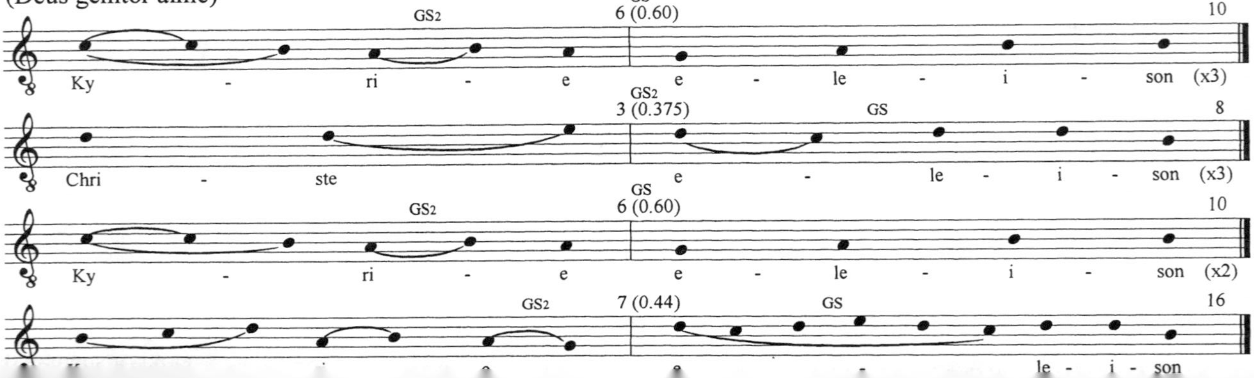

(Clemens Rector)

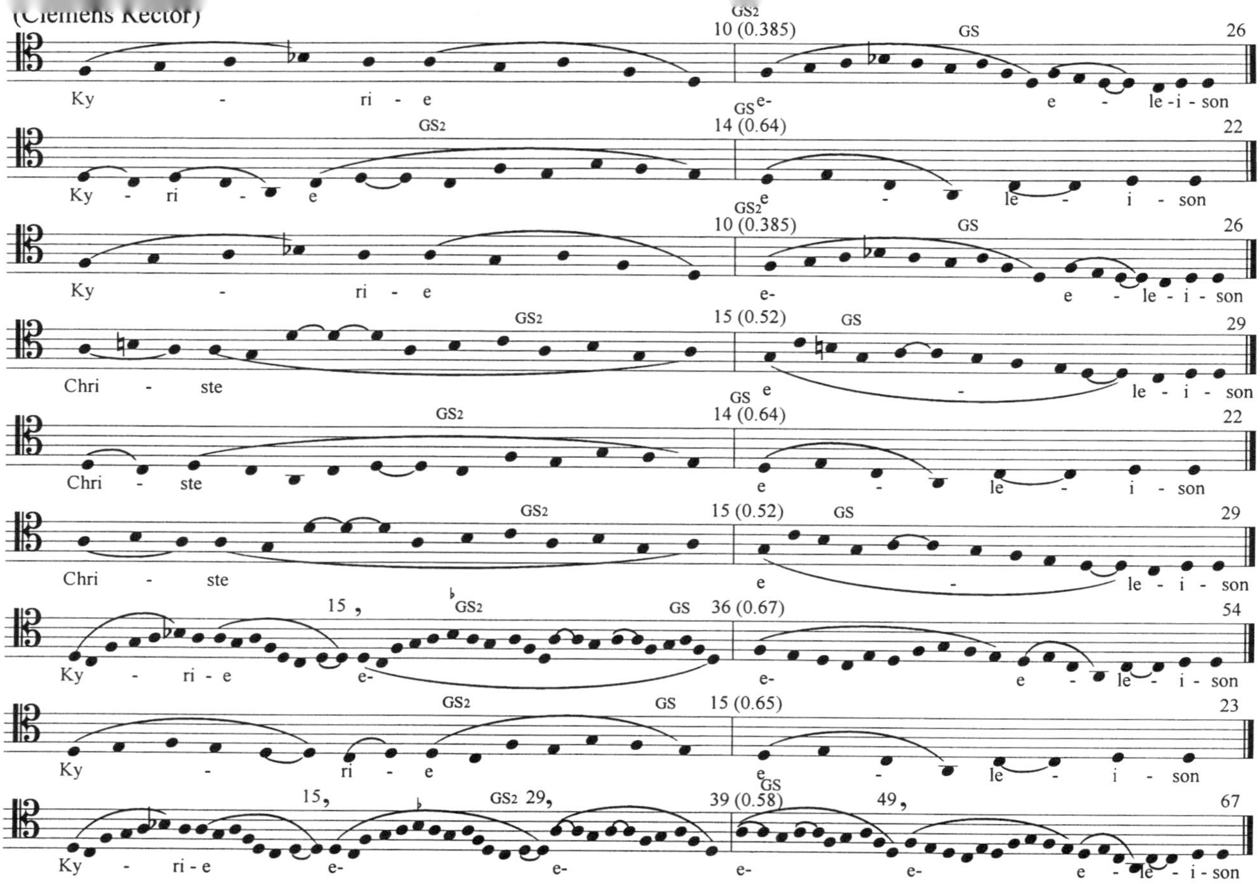

Chants "Ad Libitum." II.
(Summe Deus)

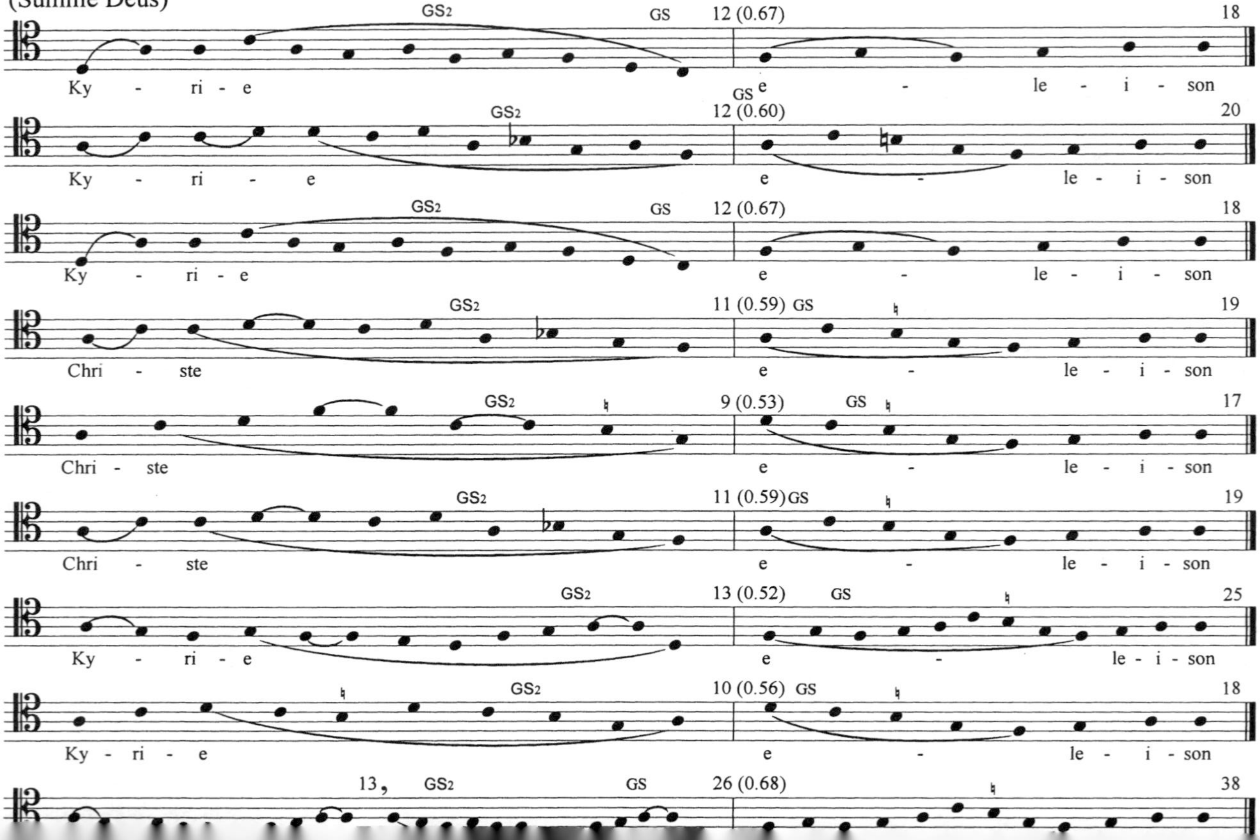

(Rector cosmi pie)

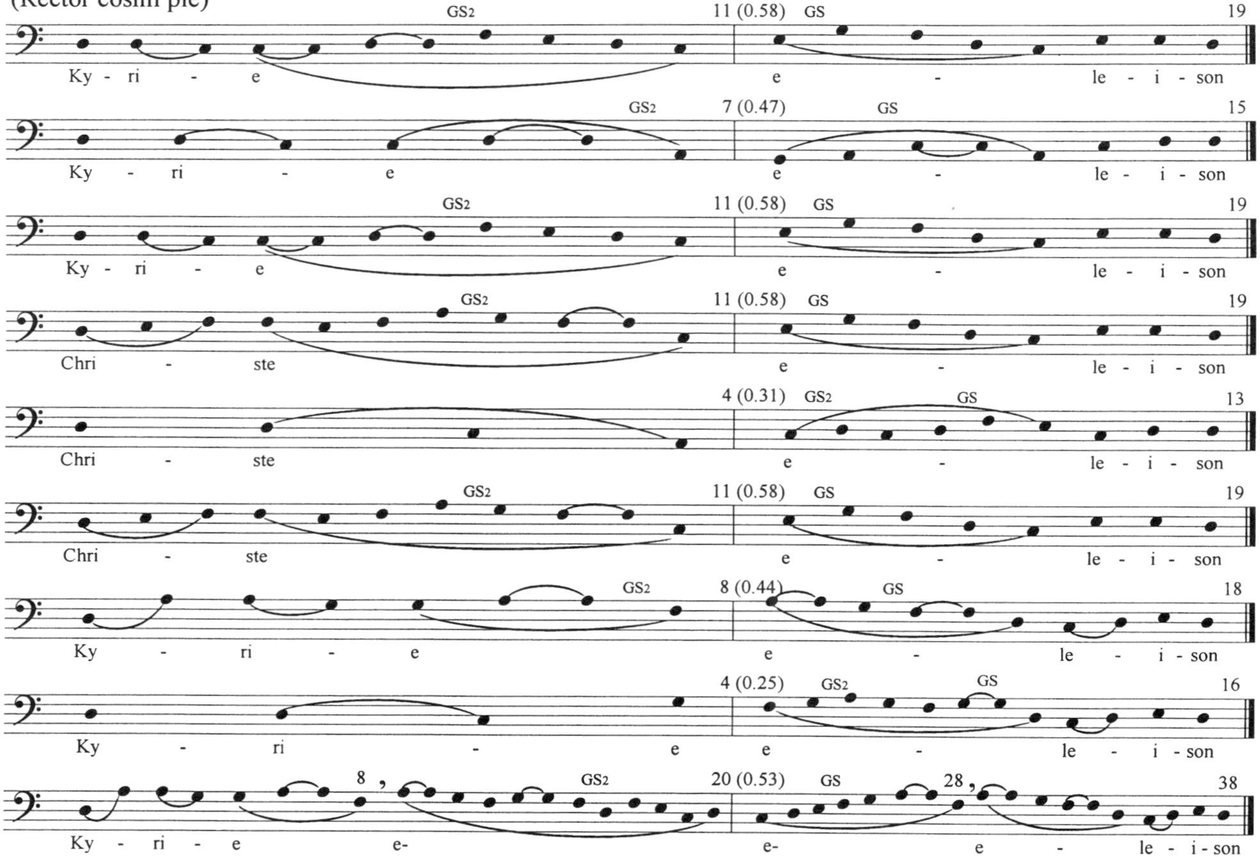

Chants "Ad Libitum." IV.
(Kyrie altissime)

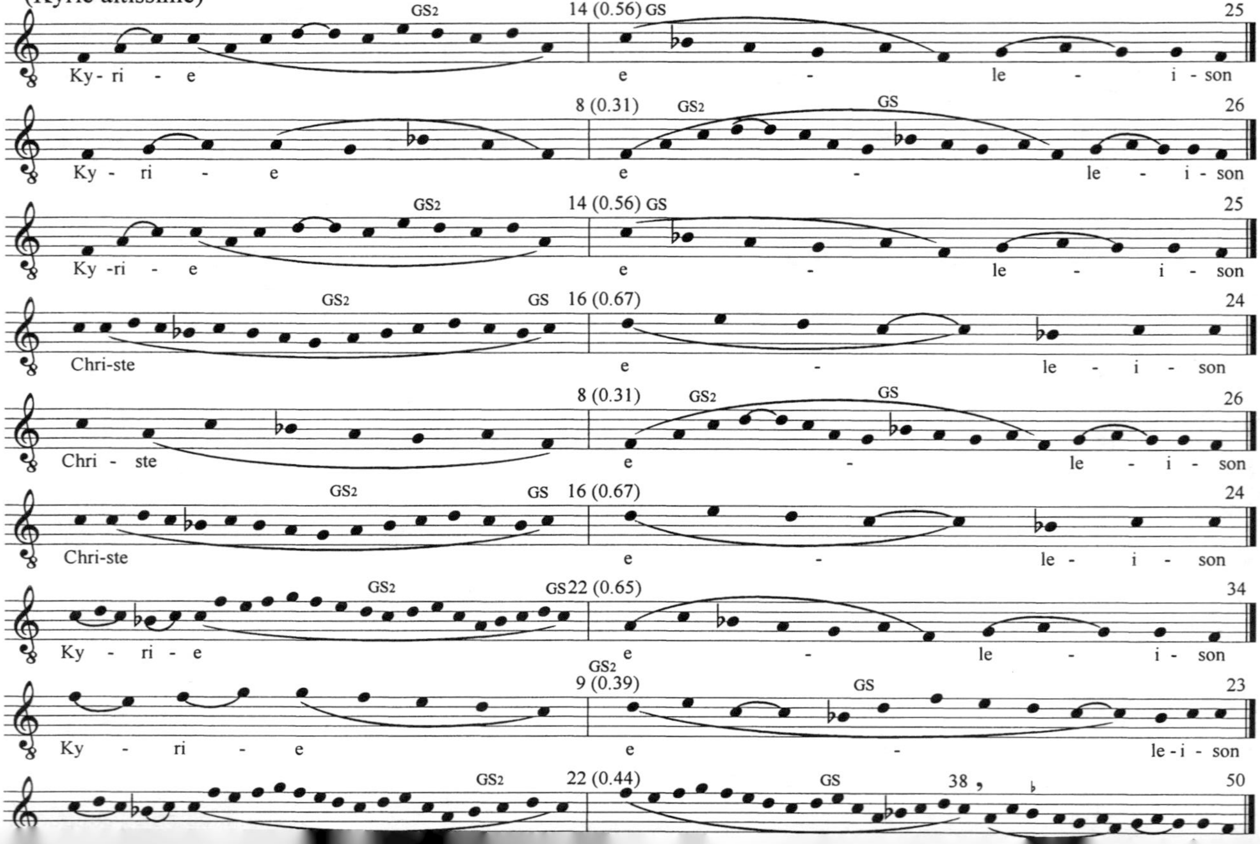

(Conditor Kyrie omnium)

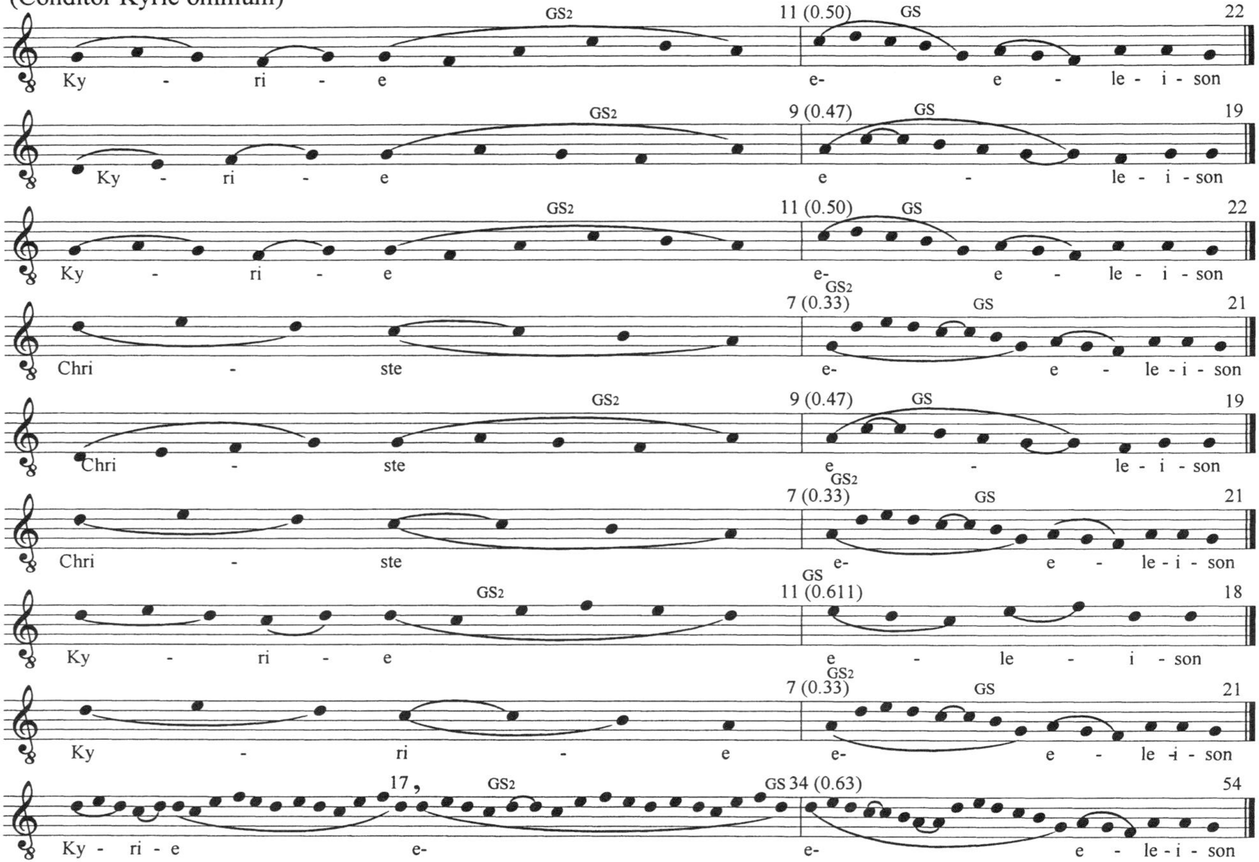

Chants "Ad Libitum." VI.
(Te Christe Rex supplices)

(Splendor aeterne)

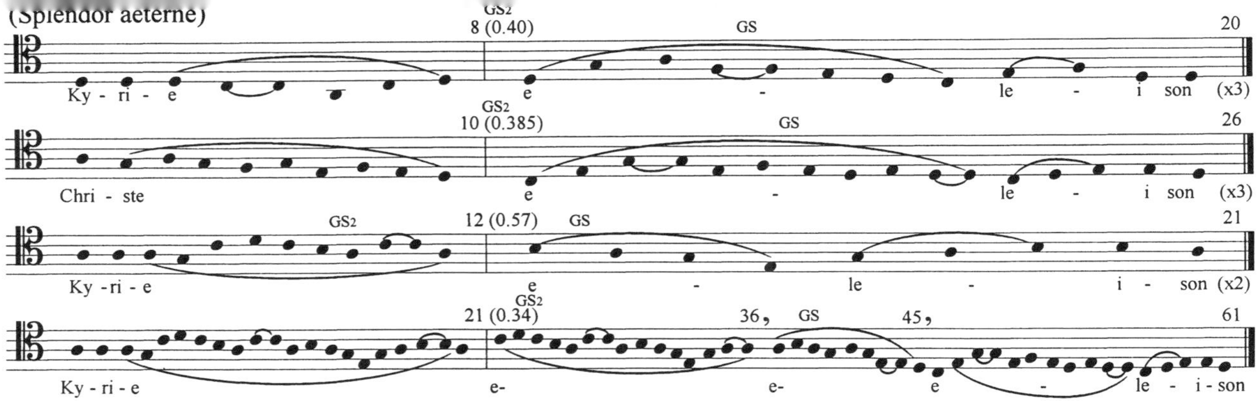

Chants "Ad Libitum." VIII.
(Firmator sancte)

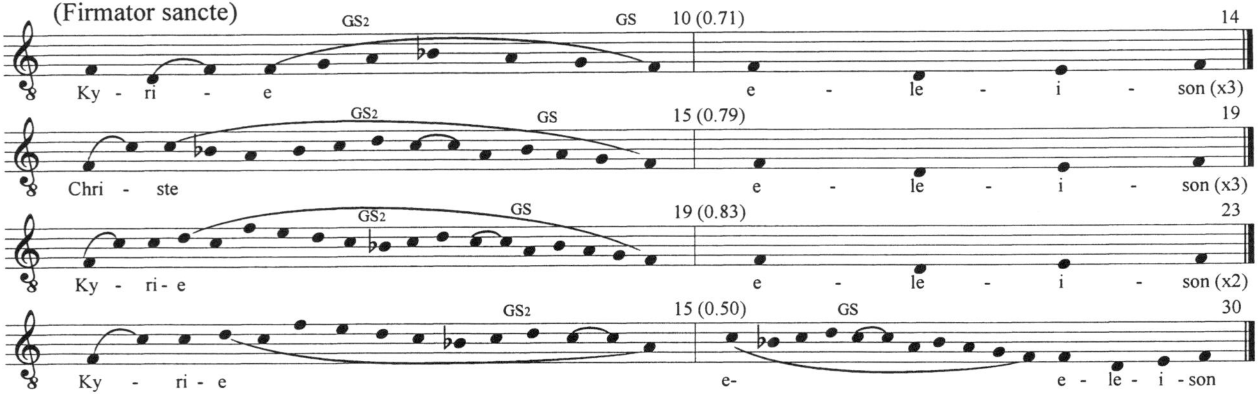

Chants "Ad Libitum." IX.
(O Pater excelse)

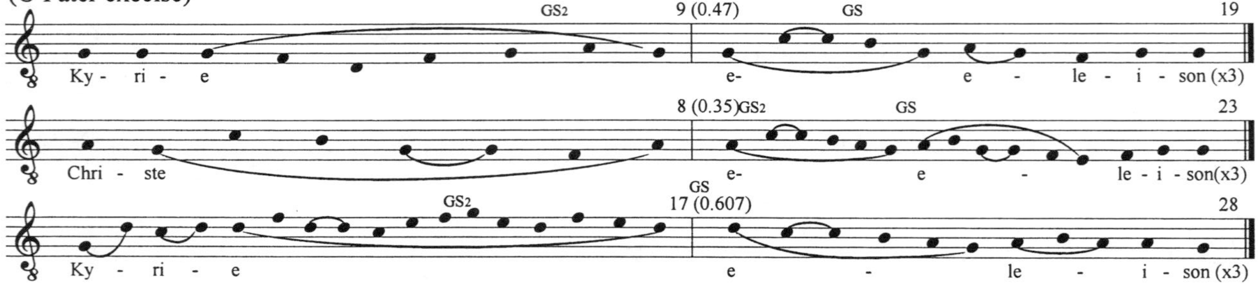

Chants "Ad Libitum." X. (For Sundays throughout the year.)
(Orbis factor)

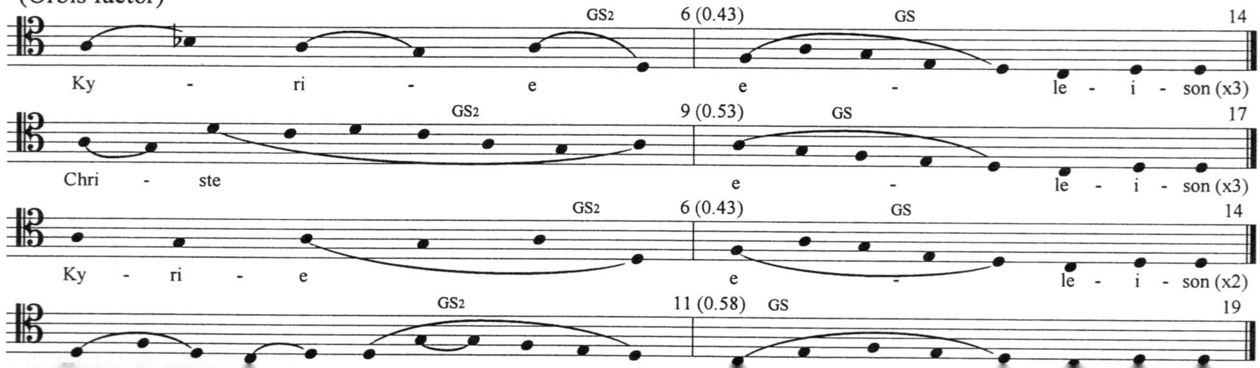

Chants "Ad Libitum." XI. (For the Sundays of Advent and Lent.)
(Kyrie Salve)

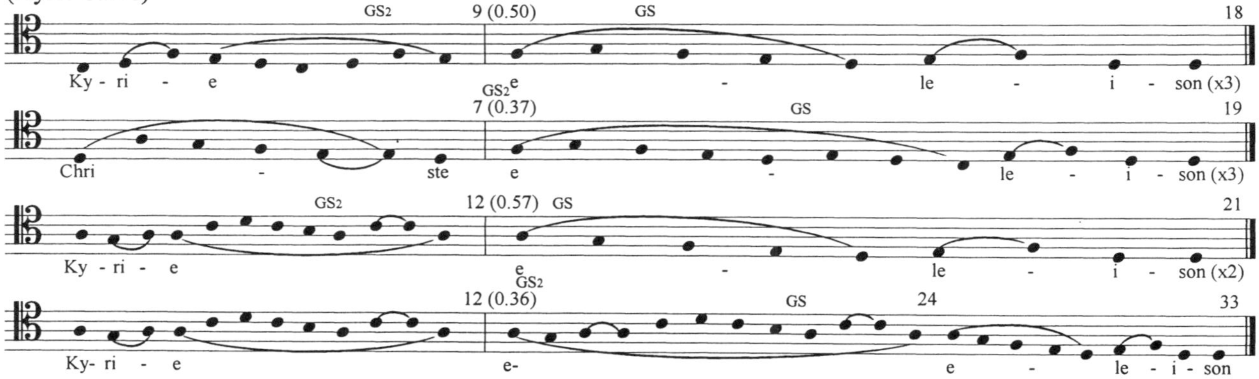

Table 2.1. Kyries summary. Lines with structural points at or near GS or bisected

Piece	Number at GS/ GS^2	Number near GS/ GS^2	Number bi-sected
Kyrie I	3 / 0	4 / 0	2
Kyrie II		4 / 0	3
Kyrie III	0 / 1	0 / 5	1
Kyrie IV	0 / 1	8 / 0	
Kyrie V			6
Kyrie VI		1 / 4	
Kyrie VII		3 / 0	
Kyrie VIII	5 / 0	3 / 0	
Kyrie IX	6 / 0	2 / 0	
Kyrie X	0 / 3	1 / 2	2
Kyrie XI		0 / 8	
Kyrie XII	3 / 0	5 / 0	
Kyrie XIII		0 / 3	
Kyrie XIV	3 / 0	5 / 0	
Kyrie XV	1 / 2	3 / 3	
Kyrie XVI	0 / 3	8 / 1	
Kyrie XVII	0 / 3	1 / 0	2
Kyrie XVII-alt	0 / 2	0 / 6	
Kyrie XVIII	5 / 3	0 / 1	
Kyrie Ad-Lib I	2 / 2	1 / 0	
Kyrie Ad-Lib II	1 / 0	5 / 0	
Kyrie Ad-Lib III		4 / 3	
Kyrie Ad-Lib IV	0 / 1	5 / 0	
Kyrie Ad-Lib V	1 / 0	1 / 3	2
Kyrie Ad-Lib VI		5 / 0	1
Kyrie Ad-Lib VII	0 / 6	2 / 0	
Kyrie Ad-Lib VIII		3 / 0	1
Kyrie Ad-Lib IX	3 / 0	0 / 3	
Kyrie Ad-Lib X		1 / 5	
Kyrie Ad-Lib XI	0 / 3	2 / 1	3
Total	33 / 30 = 63	77 / 48 = 125	23

The five Kyries analyzed below show how their simple three-part forms were developed by using insertions, deletions, extensions, and variations.

Table 2.2. Summary of the structure of each Kyrie as a whole (notes)

Kyrie	Kyrie I	Christe	Kyrie II	GS^2	Result
I	45 (.263)	63 (.368)	63 (.369)	65	Two
II	90 (.278)	129 (.398)	105 (.324)	124	
III	69 (.283)	72 (.295)	103 (.422)	93	
IV	75 (.346)	63 (.290)	79 (.364)	83	
V	90 (.345)	81 (.310)	90 (.345)	100	
VI	65 (.286)	60 (.264)	102 (.450)	87	
VII	84 (.255)	114 (.345)	132 (.400)	126	3:2
VIII	75 (.339)	69 (.312)	77 (.349)	84	
IX	53 (.295)	42 (.233)	85 (.472)	69	
X	40 (.296)	31 (.230)	64 (.474)	52	
XI	54 (.303)	63 (.354)	61 (.343)	68	
XII	48 (.397)	22 (.182)	51 (.421)	46	Two
XIII	45 (.257)	51 (.292)	79 (.451)	67	
XIV	72 (.294)	78 (.318)	95 (.388)	94	One
XV	36 (.305)	37 (.314)	45 (.381)	45	GS
XVI	27 (.333)	21 (.259)	33 (.408)	31	Two
XVII	66 (.319)	57 (.275)	84 (.406)	79	
XVII-alt	51 (.302)	51 (.302)	67 (.396)	65	Two
XVIII	30 (.333)	24 (.267)	36 (.400)	34	3:2
Ad-Lib I	74 (.248)	80 (.269)	144 (.483)	114	
Ad-Lib II	56 (.292)	55 (.286)	81 (.422)	73	
Ad-Lib III	53 (.301)	51 (.290)	72 (.409)	67	
Ad-Lib IV	76 (.296)	74 (.288)	107 (.416)	98	
Ad-Lib V	63 (.290)	61 (.281)	93 (.429)	83	
Ad-Lib VI	87 (.364)	63 (.264)	89 (.372)	91	Two
Ad-Lib VII	60 (.249)	78 (.324)	103 (.427)	92	
Ad-Lib VIII	42 (.240)	57 (.326)	76 (.434)	67	
Ad-Lib IX	57 (.271)	69 (.329)	84 (.400)	80	3:2
Ad-Lib X	42 (.300)	51 (.364)	47 (.336)	53	
Ad-Lib XI	54 (.290)	57 (.307)	75 (.403)	71	

Kyrie V

In Kyrie V, the structural points are at or near .50. Although this condition occurs in 8.5% (23) of the lines in this collection, it is not a strong enough tendency to suggest a conscious technique.

Its structure is a simple *ABA*. In the Kyrie, the question—or first phrase—is obviously *a*. We call its answer *b*, which is somewhat similar to *a* for the first half-phrase and quite different for the second. The first phrase of the Christe is different from the Kyrie, making it part *c*, but the second phrase is similar enough to *b* to enable calling it *b'*. The similar closing, much like a refrain, is called rounding and occurs in several pieces in this collection. The entire Kyrie is repeated verbatim in the second set of Kyries. Thus, the form is *Aab Bcb' Aab.*

Part *b'* was constructed by transposing the first four notes of *b* down a fourth and excising the next three (b_1–a_1–g_1). The excision causes the proportion of the Christe to change from .50 to .56 because it now has a smaller divisor. Thus, the proportion moves closer to GS, although it is still almost 10% away. However, the proportion of Kyrie V as a whole still is not phi, as shown in table 2.2.

Kyrie XVI

The divisions of Kyrie XVI are far from phi. In the Christe, GS^2 and GS are only one note apart. Such is to be expected in a melody that contains only seven notes, but there is a feeling of unease with a system that is so loose, since .43 is more than 10% away from .382. The proportion changes drastically (from .56 to .33) at the bar line in the last line, due to the long extension in *c* that provides a much larger divisor (15 versus 9). The proportion of the piece overall is far from phi, as shown in table 2.2.

Despite being the shortest piece of the collection (containing only ninety notes), the subdivisions of the ABA form are a little more complicated than those of Kyrie V.

Part *a* consists of five notes intoned on the word "Kyrie." This is followed by part *a'* four notes long intoned on "eleison." The whole line is repeated three times, with *a* changed to *a'* by excising one note (b_1). This seems quite evident.

This is followed by part *b* three notes long intoned on "Christe" followed by part *a''* intoned on "eleison," repeated three times. Here, *a'* is changed to *a''* by changing g_1 to a_1.

The second set of Kyries is identical to the first (*aa'*) but repeated only twice, followed by a single rendition of part *a* followed by interpolation (a_1 and c_2) and extension (g_1–a_1–g_1–f_1–e_1) that turns *a'* into *c*. Thus, the form is *Aaa' Bba'' A'aa'ac.*

Spare use of material results in simplicity in this piece. Repetition is literal in all renditions of the word "Kyrie" and for renditions of the first and third sets of "eleison," with only a small change in the second set. Only the last "eleison" is different from the others.[5]

Kyrie XVIII

The second-shortest piece in the collection, Kyrie XVIII contains the greatest concentration of structural points at GS or GS^2 (8). Specifically, the first and third lines divide at GS (.60, i.e., FN 3:2). This is 3% away from .618. The second line divides at GS^2 (.375, FN 3:5). This is less than 2% away from .382. These values are close enough. The fourth line divides at .44 because its construction is completely different from the Kyrie lines. It uses melodies borrowed from the Christe, into which it interpolates several additional notes. Overall, the structure is divided at 3:2. Thus, Fibonacci numbers figure prominently in this piece.

The answer to the question posed by the first Kyrie is significantly different from *a*, and should be called *b*. Both phrases of the Christe are also different from *a* and *b* and should be called *c* and *d*. This is followed by a literal repetition of the first line in the second set of Kyries.

In the fourth line, the Kyrie is thought of as a variant of the first line (thus *a'*), with d_2 and g_1 as neighbor notes and b_1 out of place. The last phrase is called *d'* because of the interpolation.

Thus, the structure of the piece is *Aab Bcd Aaba'd'*.

Kyrie VII

In Kyrie VII, no structural points fall on GS or GS^2, but the Kyrie as a whole is divided at 3:2.

In this piece, the structure becomes more complicated. There are more variations, the parts are lengthened by insertions and extensions, and rounding becomes more apparent.

Elements of part *a* are found in every line, with small modifications such as added or dropped notes. These modifications make it difficult to be sure the structure is correctly analyzed.

Part *a,* the first phrase, is seven notes in length. It is followed by two phrases, *b* and *b'* totaling twenty-one notes, that are identical in all lines. Parts *b* and *b'* are entirely different from part *a*.

The Christe begins with what at first looks like *c*, because of an apparently different shape, but the similarities of content and shape to *a* (a_1–g_1–f_1–g_1–a_1–d_1) make it look more like a variant of *a*. Hence, calling the first phrase of the Christe *a',* and the second phrase *a'',* we see it repeated in the second phrase of the Kyrie in the third line, and note the similarity in the first phrase of that line, with several notes added at the beginning. It then seems that the first phrase of the third line should be called *a'''*.

This brings about a tripartite form of *Aabb' A'a'a"bb' A"a"'a"bb',* which is simpler-looking than the above explanation seems. The extensions, interpolations, and excisions are beautifully balanced by the rounding phrases *b* and *b'.* This technique of adding parts of phrases gets quite complicated in many of the Kyries.

Kyrie III

Kyrie III has one occurrence of GS^2 (.381) in the second line, which is only one point from GS^2. This is also FN 8:13. All other lines divide far from phi, as does the piece as a whole.

This Kyrie has the most complicated structure of any that will be examined here. It is rather like a rondo, with numerous recurring refrains, or perhaps like a set of variations. Many questions remain.

It is a masterpiece of subtlety, with at least five elements and six variations that weave in and out of three tripartite forms. Every line has a refrain of four notes (d–f–d–e) that rounds out the phrase.

Phrase *a* is followed by distinctive phrase *b.* This is followed in the second line by variants *b'* and *b".* The third line is a literal repeat of the first line. Thus, the form of the first section is *Aab Bb'b" Aab.*

The first Christe begins with a distinctive phrase *c.* It is followed by *b"'.* The second Christe begins with another distinctive phrase, *d,* and is followed by *b",* from the second line. The third Christe is a repeat of the first. Thus, this structure is *Ccb"' Ddb" Ccb"'.*

In the return to the Kyries, line seven, there is a distinctive phrase *e,* followed by *b"".* This is followed by *b""'* and a repeat of *b"".* Finally, the ninth line interpolates many variants of previous lines in a literal repetition of *e,* variant *a',* a repeat of this, *e',* and *b""".* Thus, the third tripartite section would be *Eeb"" B'b""'b"" E'ea'ea'e'b""".*

The overall result of these partial forms is a rather messy-looking *AabBb'b"Aab Ccb"'Ddb"Ccb"' Eeb""B'b""'b""E'ea'ea'e'b""",* which can be summarized as *ABA CDC EB'E'.* This piece is satisfying because its unity and variety are very well distributed.

Conclusions concerning the *Liber Usualis*

This fascinating study suggests that there may be some support for the notion that the golden proportion is an innate urge in the production of music, but only if we admit loose tolerances of up to 10%.

The argument of chapter 4 of *Fractals in Music* and chapter 1 of *Fib and Phi in Music* is that such a tolerance is too loose. Where Fechner and Zeising in the nineteenth century and Hambidge in the twentieth could get away with such tolerances, today's computer-graphics experts cannot allow loose approximations. The same must be said for our studies of phi in music.

No one is prohibited to claim that the golden proportion is prominent in Gregorian chant. After all, 23% of lines *at,* and 46% *near* GS or GS^2, at first seems like a rather strong tendency, but it should be remembered that 31% of structural points are *not* at or near GS or GS^2. Also, only one of the structural points identified as GS or GS^2 lies near enough to the golden proportion to be considered strong. Thus, it is a weak tendency at best. This is supported by the data shown in table 2.2 for the whole pieces. It seems likely that the composers were not thinking in terms of fib or phi, which made their way into the Kyries purely by accident.

Machaut

Messe de Nostre Dame

Some very strong indicators of phi proportion in medieval music will be presented in this section.

Dominique Pascale Bardet wrote a dissertation in 1999 entitled "Logos as Number and Proportion in Amiens Cathedral and in the Kyrie of the Messe de Nostre Dame by Guillaume de Machaut."[6] Although the bulk of the musical portion of the study was about isorhythm, the Kyrie was inadvertently found to be divided within one measure of GS^2.

Bardet's logos revolved around the musical ratios passed down from Pythagoras as musical consonances: small-number ratios 1:1, 1:2, 2:3, and the like. This notion has been confirmed by many other authors and need not be disputed. However, Stephen Murray's book *Notre Dame, Cathedral of Amiens* was cited as finding that the generating square at the center of the transept crossing was used to create a golden rectangle, from which the rest of the plan was developed.[7]

Murray explained that this rectangle was developed from a 50 foot square that marked the centers of the main piers by adding other piers placed 30 feet to the sides of them, thus creating what he took to be golden rectangles with ratio 5:3. (He said that doing this all the way around created a larger square measuring 30 + 50 + 30 = 110 feet on a side, which was used to derive other parts of the floor plan.) Explaining further that 5:3 = 5/3 = 1.667 was the medieval equivalent of phi, he implied that the builders were using a loose version of phi.[8]

Since 50 ft/1.667 = 30 ft (5:3) and 50 ft/1.618 = 30.9 ft (phi), the 50 to 30 ratio is .9 ft smaller than the phi ratio. If intended to be phi, this would have been an error of 27.4 cm—almost 11 inches. But, the piers have been measured carefully and found on average to be within a centimeter of their correct locations, with no error greater than 9.5 cm. If the builders were attempting to approximate phi, the measurements would have been much closer to, and on both sides of, 30.9 ft.[9]

Thus, we conclude that the builders used Fibonacci sequence 3:5:3—not phi—in the construction. This is a balanced construction similar to one used by Claude Debussy in his music six hundred years later. Whether they knew it was a Fibonacci sequence or were simply using a small number ratio that musicians would see as a major sixth is an interesting question that remains unanswered.

The Kyrie from Machaut's *Messe de Nostre Dame* is not structured in Fibonacci numbers. Rather, Bardet clearly showed—without mentioning phi—that it is divided within one measure of GS^2, as shown in figure 2.1.[10] Here, the Christe begins after measure 81, while Kyrie II begins after 147, as shown at top. Section lengths are given below the line, with actual proportions in parentheses, while GS and GS^2 are given at bottom.

Figure 2.1. Machaut's *Messe de Nostre Dame* Kyrie (Bardet) (Measures)

Kyrie			Christe			Kyrie II		Kyrie III
27	54	81	103	125	147	164	181	210
27	27	27	22	22	22	17	17	29
	81 (.386)				129 (.614)			
	GS^2 = 80				GS = 130			
One from GS^2								

It is easy to see that the *Messe* is not constructed in either ratio 5:3 (5/8 = .625) or 3:5 (3/8 = .375). These would cause divisions after 131 and 79 measures respectively. Thus, they would be on the wrong side of GS. Consequently, the conclusion must be that the structures of the cathedral and the *Messe* are only related in the most general way. How Machaut came to structure it this way is anyone's guess at present.

Although Bardet's purpose was merely to show that Greek thinking and number systems were found in two relics of medieval Europe, she actually provided a result more profound than intended.[11]

Newman W. Powell analyzed Machaut's *Motet no. 6* and found it suffused with Fibonacci numbers.[12] However, this is without foundation. Not only was it miscounted as consisting of 99 measures, but it is a three-part polyphonic piece with elisions at the end of every phrase and shows no clear divisions. On the basis of the repeating structure of the cantus firmus and the punctuation of the two melodic voices (semicolons and periods), it is clear that phi is not present, as shown in figure 2.2.[13]

Figure 2.2. Machaut's *Motet no. 6* (Madden) (Measures)

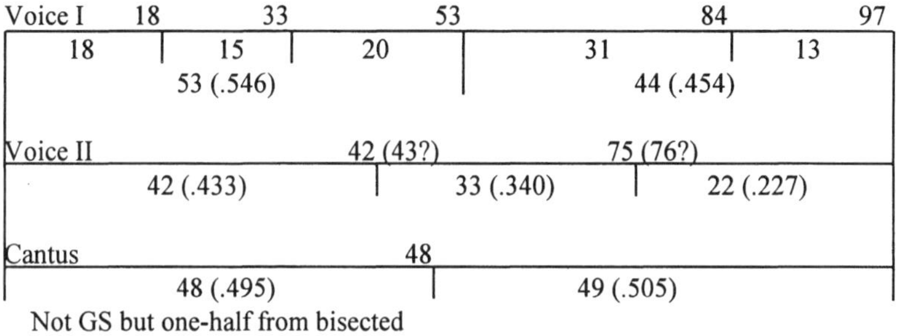

The *Rondeaux*

Powell correctly suggested the presence of Fibonacci numbers in the rondeaux of Machaut. His analysis will appear as figure 2.11.[14] All of the rondeaux will be analyzed here, except number 16 because it is text only.

These rondeaux have a standard form, taking two melodies in the order ABA AABAB, which often gives a ratio of 5:3.[15] When parts A and B are the same length, as they are in numbers 14, 17, and 19, the 3:5 ratio is automatic (as are 5:3 and 4:4). Even when A and B do not have the same length, the tendency remains for the pieces to divide near GS^2.

Figure 2.3. Machaut's *Rondeau no. 1* "Doulz Viaire Gracieus"

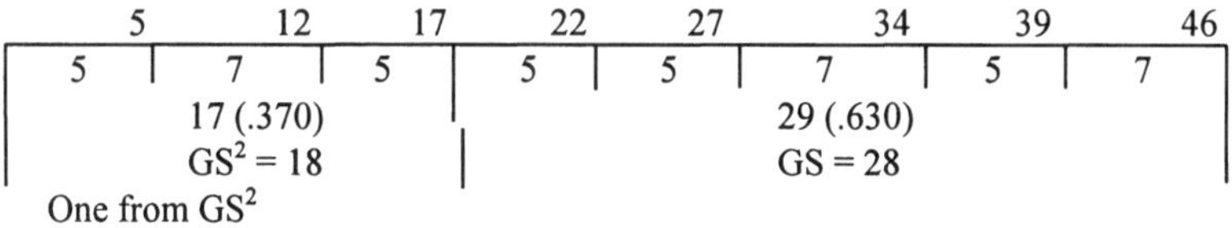

Figure 2.4. Machaut's *Rondeau no. 2* "Helas! pour quoy"

18	35	53	71	89	106	124	141
18	17	18	18	18	17	18	17
	53 (.376)				88 (.624)		
	$GS^2 = 54$				$GS = 87$		

One from GS^2 and .001 from 3:5.

Figure 2.5. Machaut's *Rondeau no. 3* "Merci vous pri"

27	46	73	100	127	146	173	192
27	19	27	27	27	19	27	19
	73 (.380)				119 (.620)		

GS^2:GS

GS^2

Figure 2.6. Machaut's *Rondeau no. 4* "Sans cuer, dolens"

24	43	67	91	115	134	158	177
24	19	24	24	24	19	24	19
	67 (.379)				110 (.621)		
	$GS^2 = 68$				$GS = 109$		

One from GS^2

Rondeau no. 5 has each verse repeated three times.

Figure 2.7. Machaut's *Rondeau no. 5* "Quant j'ay l'espart"

69	135	204	273	342	408	477	543
69	66	69	69	69	66	69	66
	204 (.376)				339 (.624)		
	$GS^2 = 207$				$GS = 336$		

Not GS but .001 from 3:5

Figure 2.8. Machaut's *Rondeau no. 6* "Cinc, un, treze"

18	31	49	67	85	98	116	129
18	13	18	18	18	13	18	13
	49 (.380)				80 (.620)		

GS^2:GS

GS^2

Figure 2.9. Machaut's *Rondeau no. 7* "Se vous n'estes"

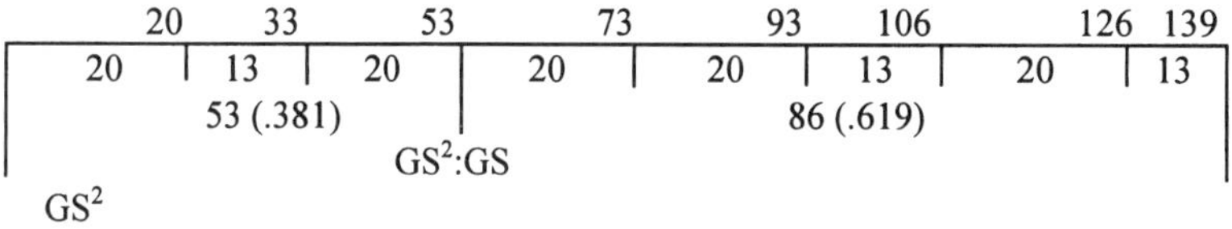

Figure 2.10. Machaut's *Rondeau no. 8* "Vos doulz resgars"

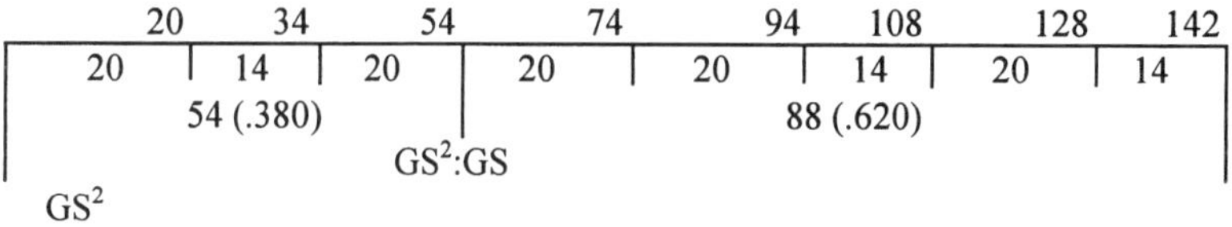

Figure 2.11. Machaut's *Rondeau no. 9* "Tant doucement" (Powell)

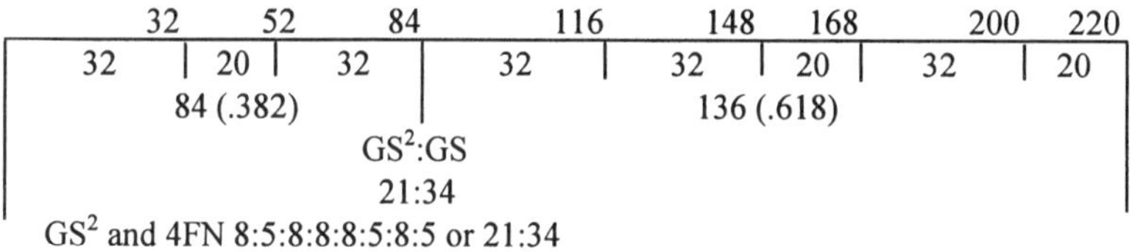

Numbers 10, 11, and 13 repeat the A verses but not the Bs.

Figure 2.12. Machaut's *Rondeau no. 10* "Rose, liz, printemps"

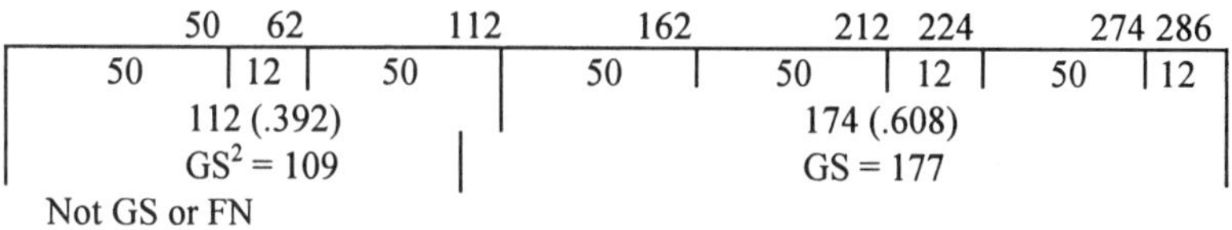

Figure 2.13. Machaut's *Rondeau no. 11* "Comment puet"

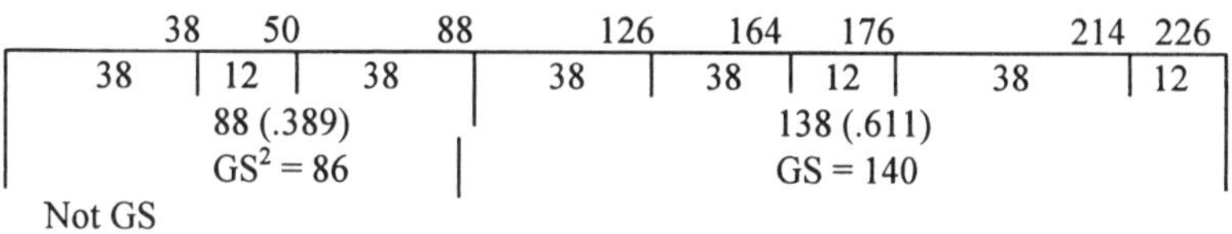

Figure 2.14. Machaut's *Rondeau no. 12* "Ce qui soustient moy"

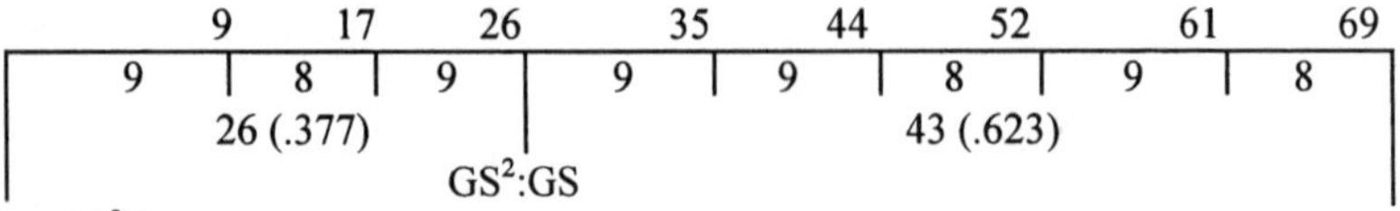

GS² but percentages are poor

Figure 2.15. Machaut's *Rondeau no. 13* "Dame, se vous n'avez"

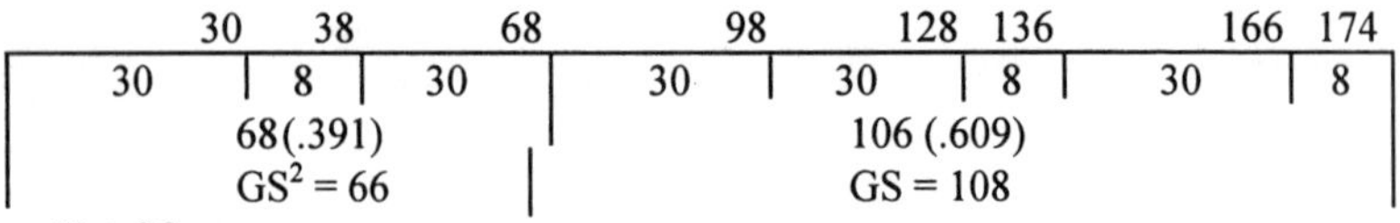

Not GS

Figure 2.16. Machaut's *Rondeau no. 14* "Ma fin est mon commencement"

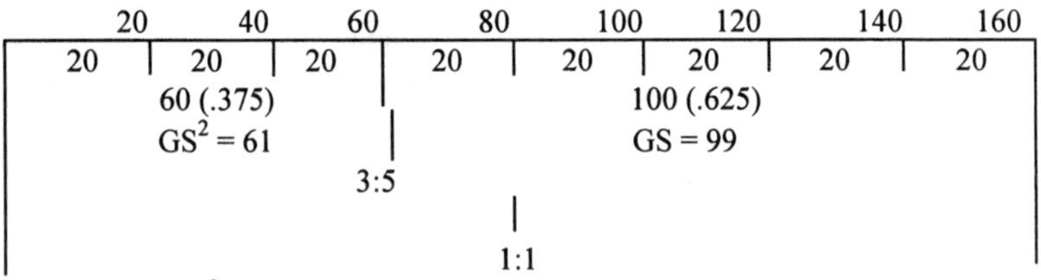

One from GS², bisected, and 20 FN 3:5

Figure 2.17. Machaut's *Rondeau no. 15* "Certes, mon oueil"

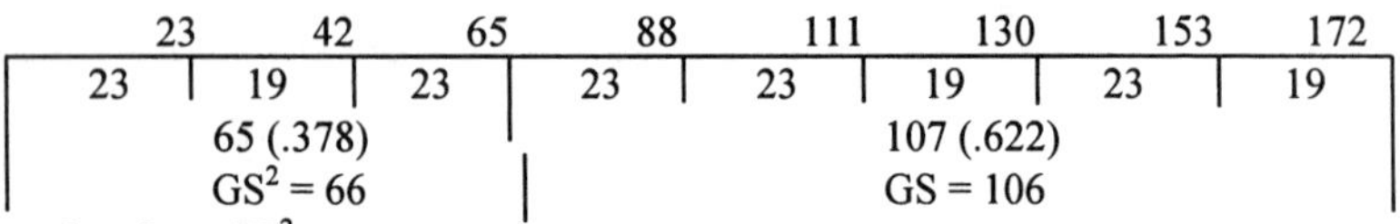

One from GS²

Figure 2.18. Machaut's *Rondeau no. 17* "Dix et sept, cinq" (Counted in beats)

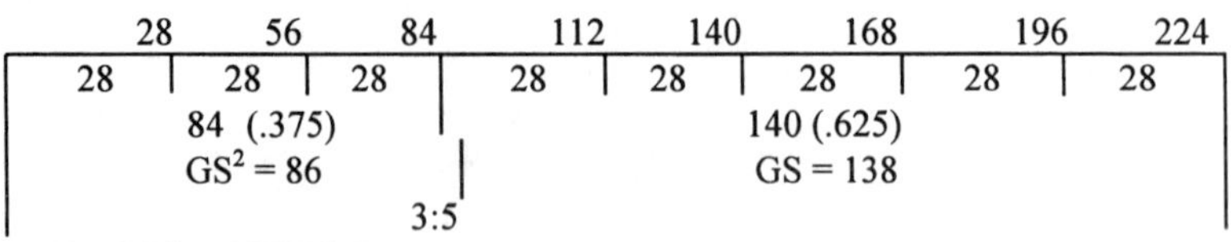

Not GS but 28FN 3:5

Figure 2.19. Machaut's *Rondeau no. 18* "Puis qu'en oubli"

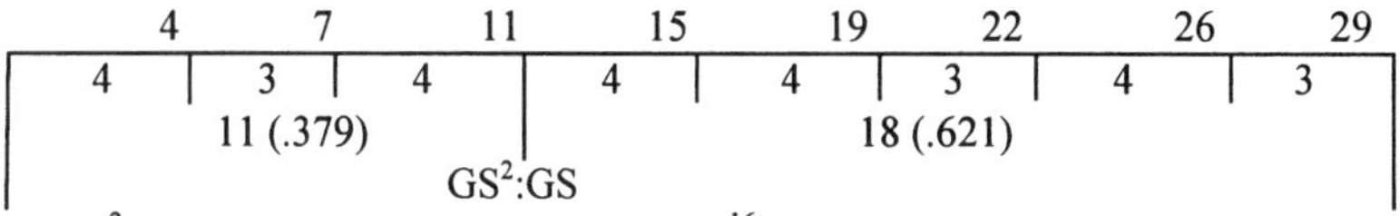

GS^2 and LN 4:3:4:4:4:3:4:3 or LN11:18[16]

Figure 2.20. Machaut's *Rondeau no. 19* "Quant ma dame"

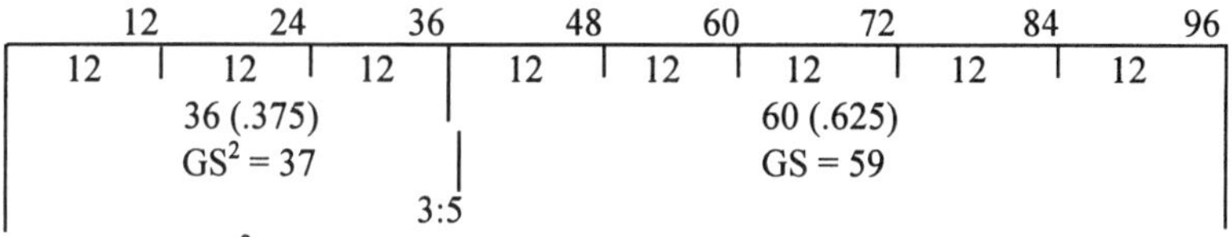

One from GS^2 and 12FN 3:5

Figure 2.21. Machaut's *Rondeau no. 20* "Douce dame"

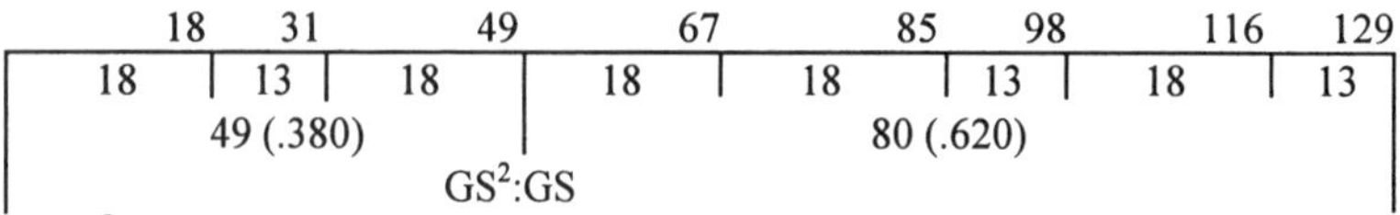

GS^2

Figure 2.22. Machaut's *Rondeau no. 21* "Quant je ne voy"

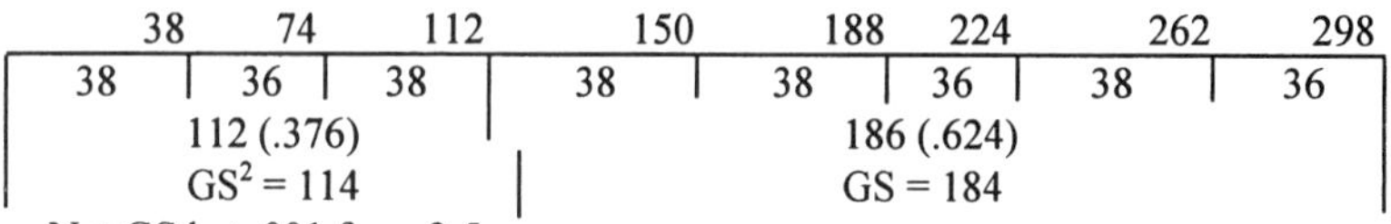

Not GS but .001 from 3:5

Figure 2.23. Machaut's *Rondeau no. 22* "Dame, mon cuer"

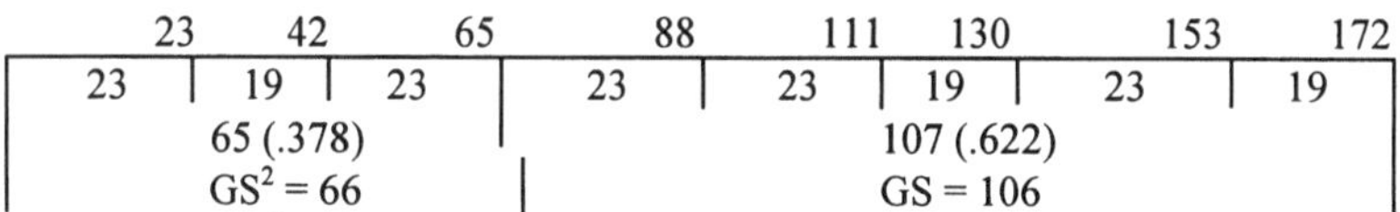

One from GS^2

A summary of the rondeaux appears in table 2.3.

Table 2.3. Summary of Machaut's rondeaux

Number	At GS^2	One meas.	Two meas.	Three meas.	FN/LN	Other
1		.370				
2		.376				.001 from 3:5
3	.380					
4		.379				
5				.376		.001 from 3:5
6	.380					
7	.381					
8	.380					
9	.382				4FN	21:34
10				.392		
11			.389			
12	.377					Poor
13			.391			
14		.375			20FN	3:5, 1:1
15		.378				
16	-	-	-	-	-	-
17			.375		28FN	3:5, 1:1
18	.379				LN	11:18
19		.375			12FN	3:5, 1:1
20	.380					
21			.376			.001 from 3:5
22		.378				

The *Virelais*

Following a suggestion that Machaut's virelais may be structured similarly, all of them have been analyzed, except 13 and 14 which are not in the correct form. Several of these pieces are thrown into question by having half-measure pickups. Redistribution does not usually improve them.

Figure 2.24. Machaut's *Virelai no. 1*

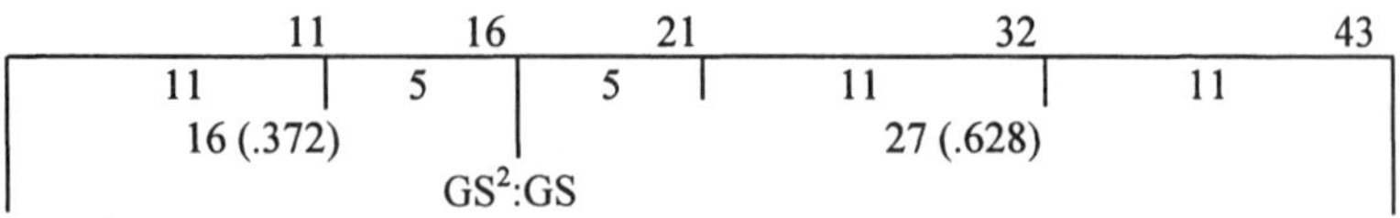

GS^2 but percentages are poor

Figure 2.25. Machaut's *Virelai no. 2*

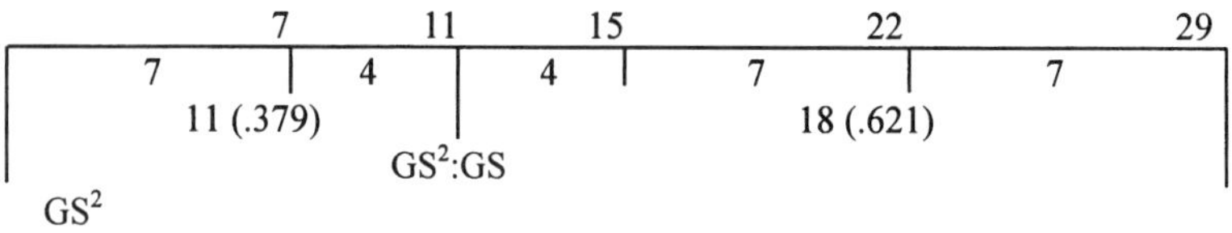

Figure 2.26. Machaut's *Virelai no. 3*

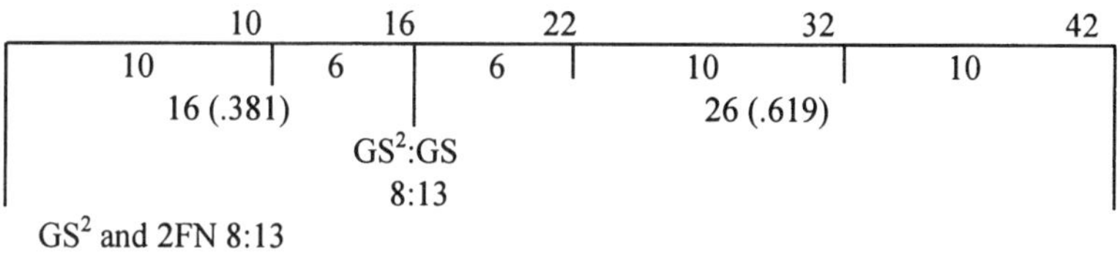

Figure 2.27. Machaut's *Virelai no. 4*

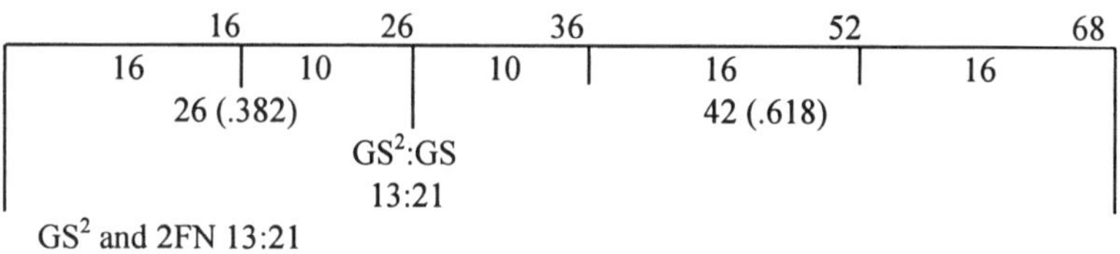

Figure 2.28. Machaut's *Virelai no. 5*

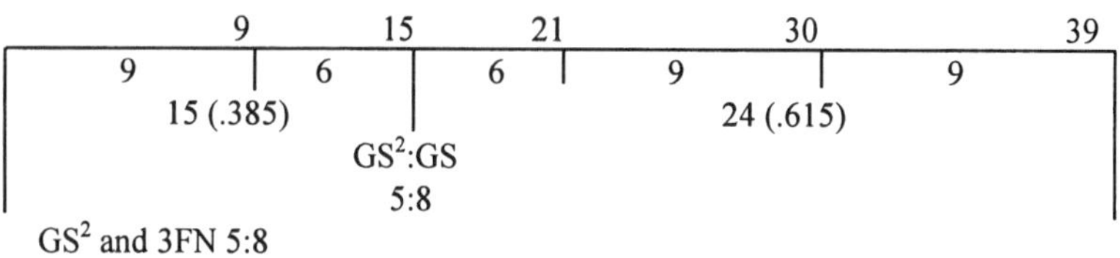

Figure 2.29. Machaut's *Virelai no. 6*

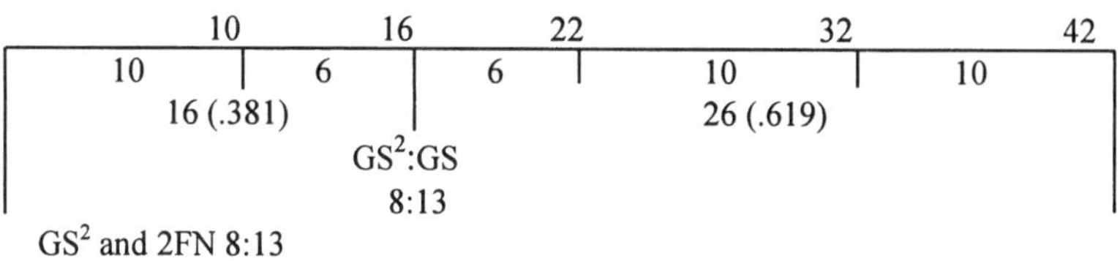

Figure 2.30. Machaut's *Virelai no. 7*

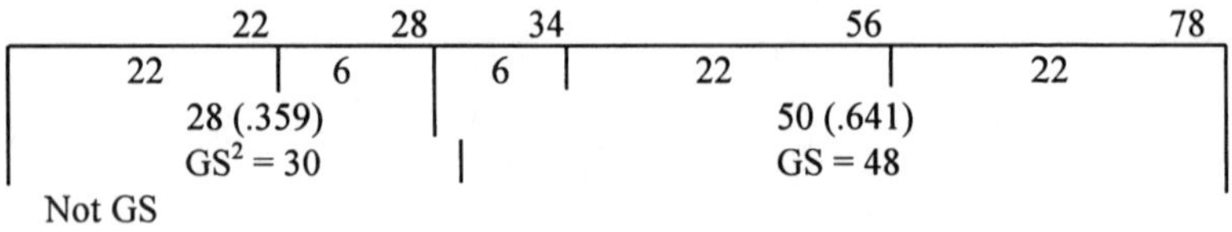

Figure 2.31. Machaut's *Virelai no. 8*

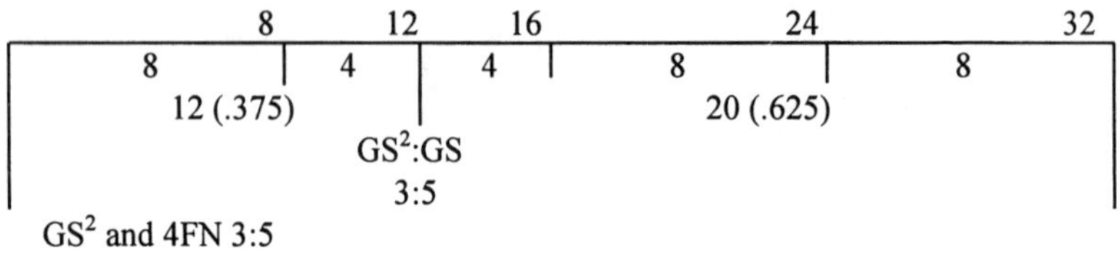

Figure 2.32. Machaut's *Virelai no. 9*

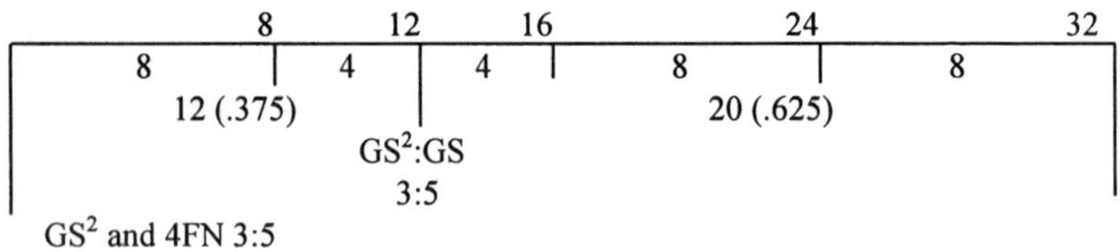

Figure 2.33. Machaut's *Virelai no. 10*

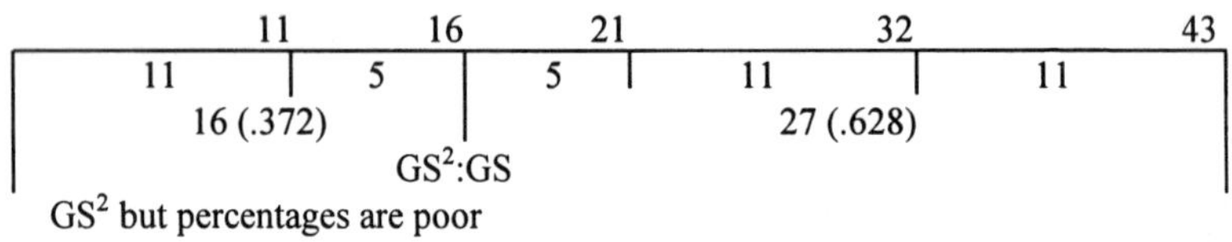

Figure 2.34. Machaut's *Virelai No. 11*

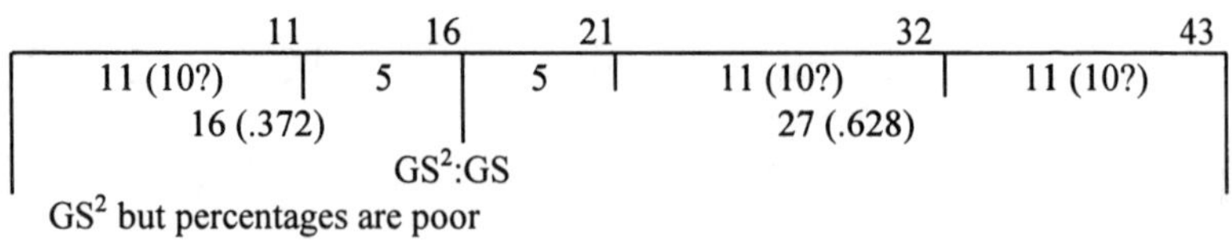

Figure 2.35. Machaut's *Virelai no. 12*

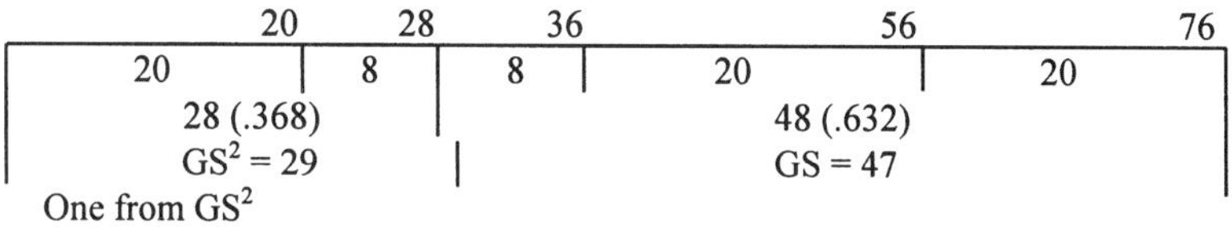

Figure 2.36. Machaut's *Virelai no. 15*

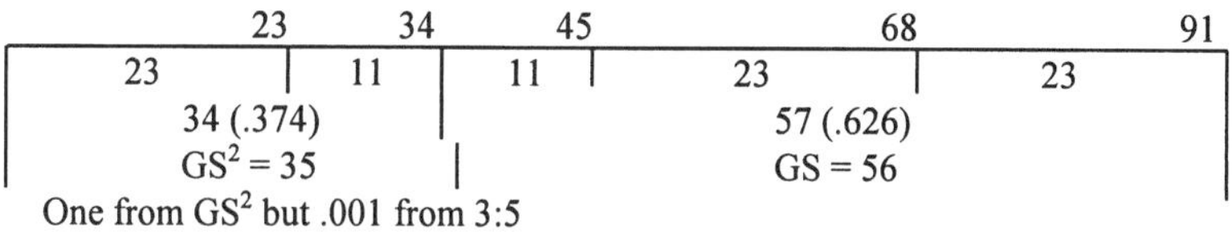

Figure 2.37. Machaut's *Virelai no. 16*

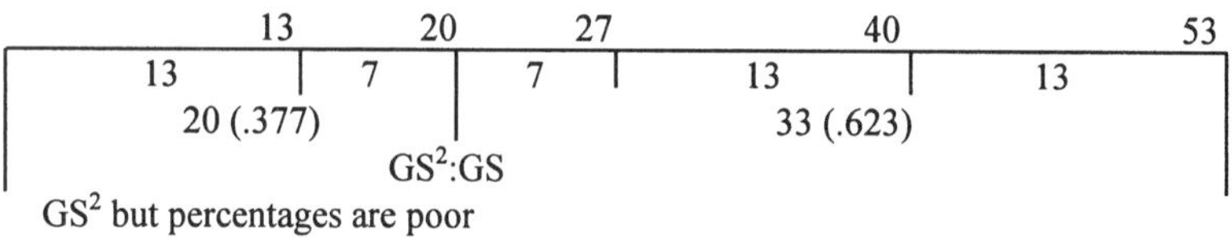

Figure 2.38. Machaut's *Virelai no. 17*

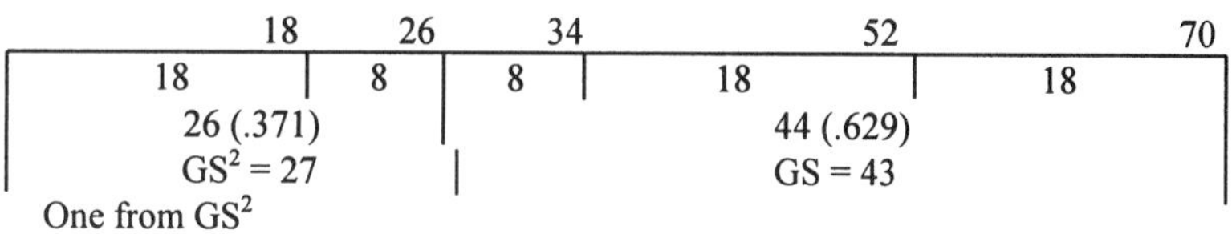

Figure 2.39. Machaut's *Virelai no. 18*

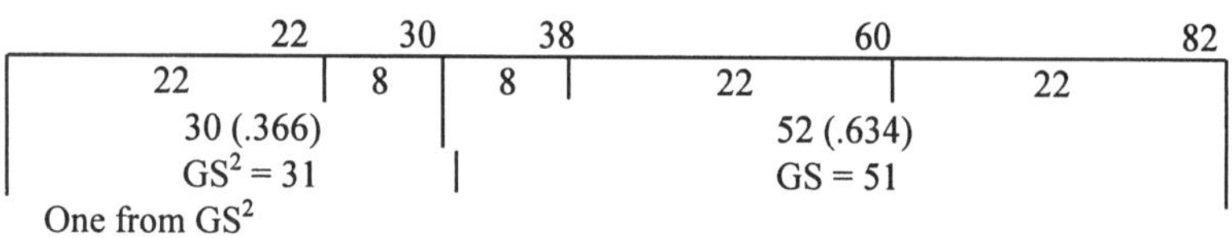

Figure 2.40. Machaut's *Virelai no. 19*

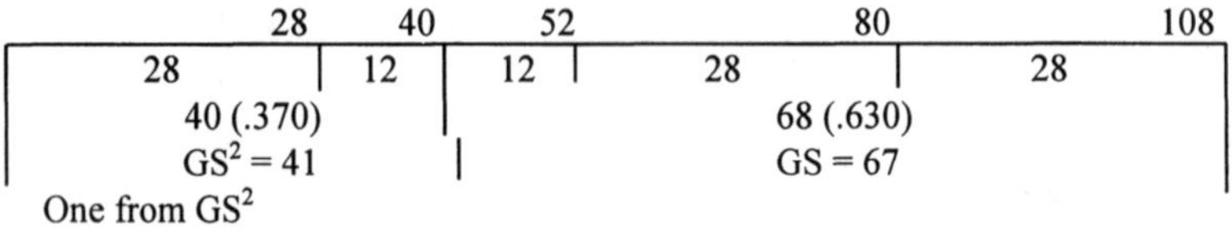

Figure 2.41. Machaut's *Virelai no. 20*

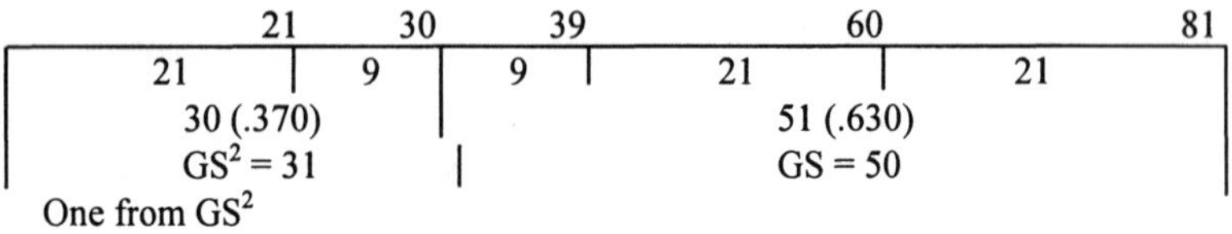

Figure 2.42. Machaut's *Virelai no. 21*

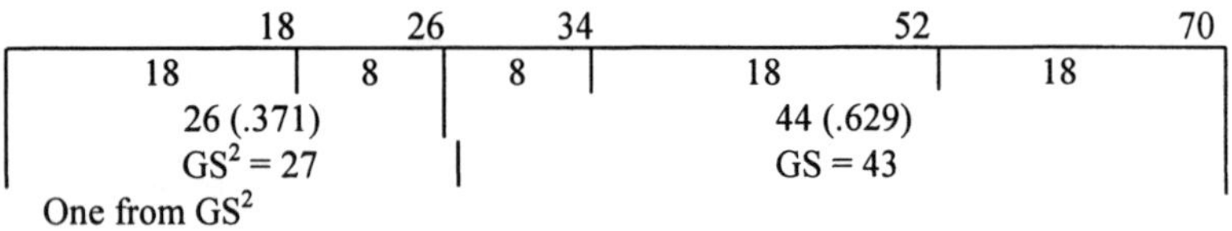

Figure 2.43. Machaut's *Virelai no. 22*

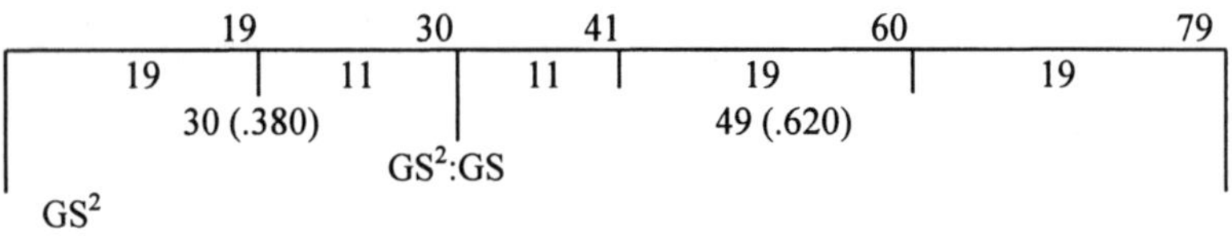

Figure 2.44. Machaut's *Virelai no. 23*

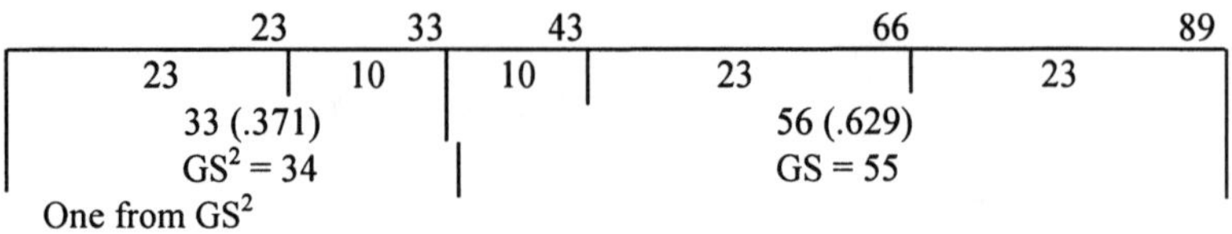

Figure 2.45. Machaut's *Virelai no. 24*

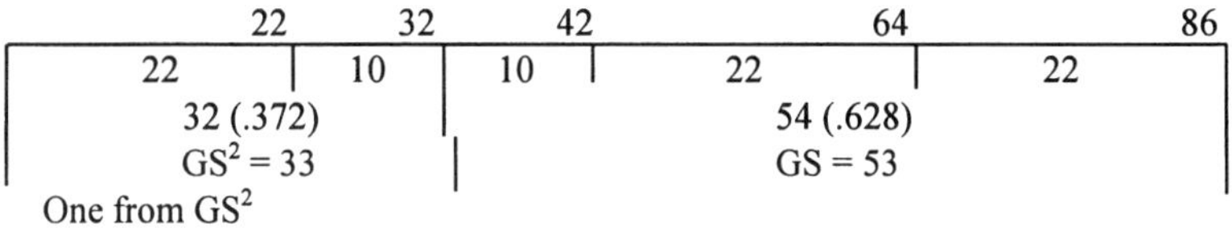

Figure 2.46. Machaut's *Virelai no. 25*

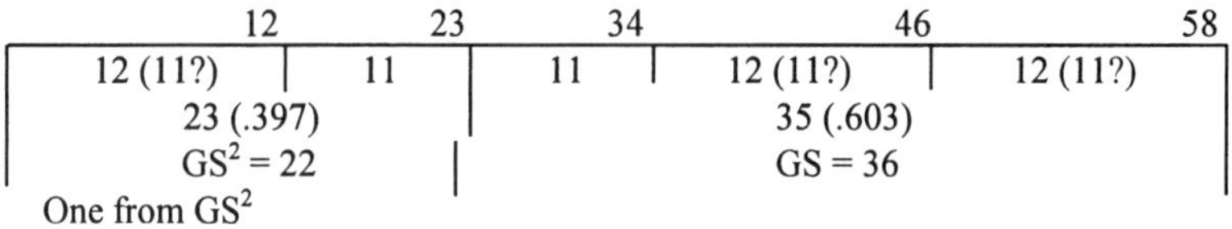

Figure 2.47. Machaut's *Virelai no. 26*

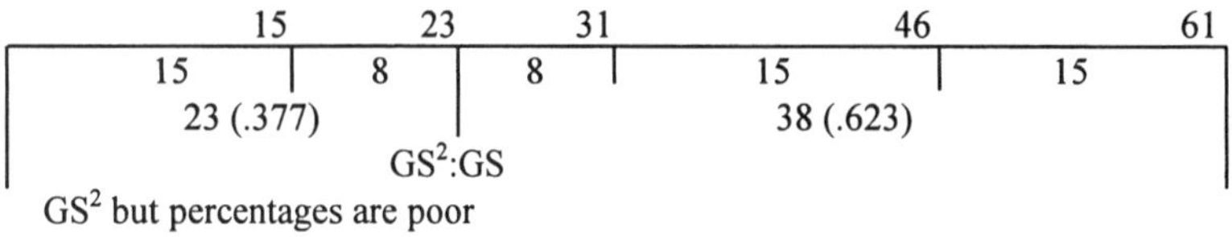

Figure 2.48. Machaut's *Virelai no. 27*

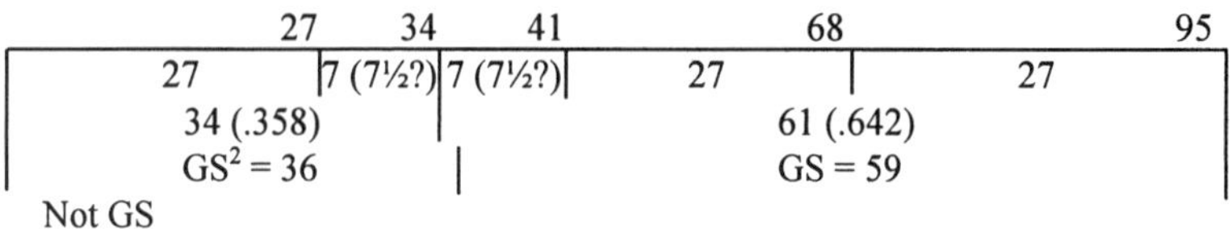

Figure 2.49. Machaut's *Virelai no. 28*

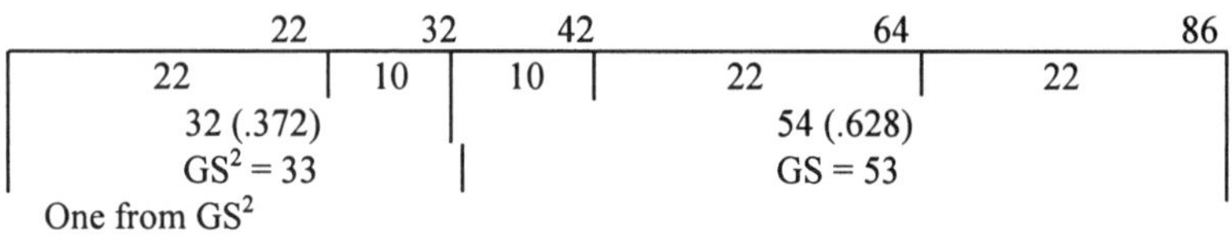

Figure 2.50. Machaut's *Virelai no. 29*

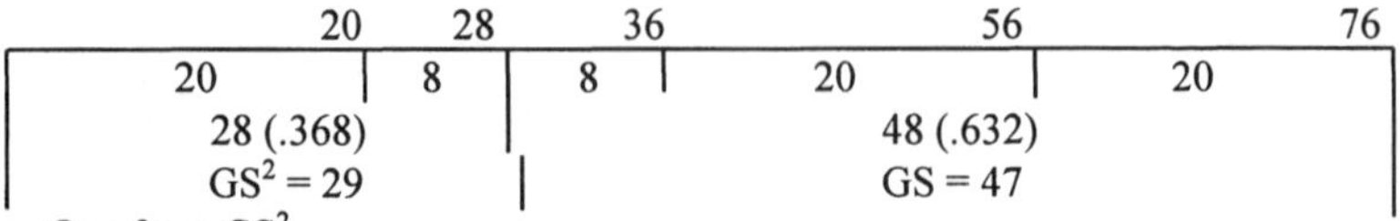

One from GS²

Figure 2.51. Machaut's *Virelai no. 30*

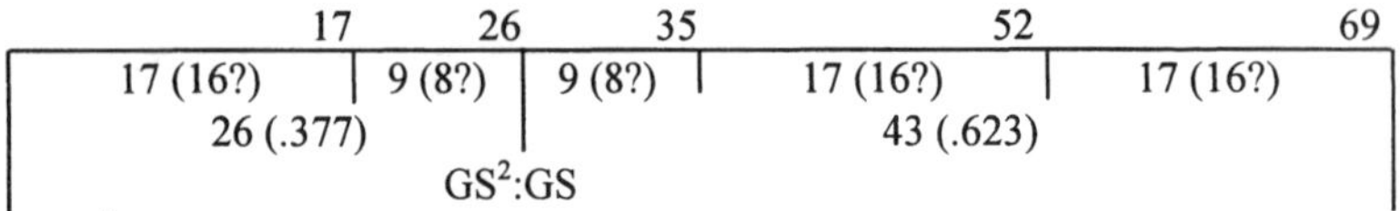

GS² but percentages are poor

Figure 2.52. Machaut's *Virelai no. 31*

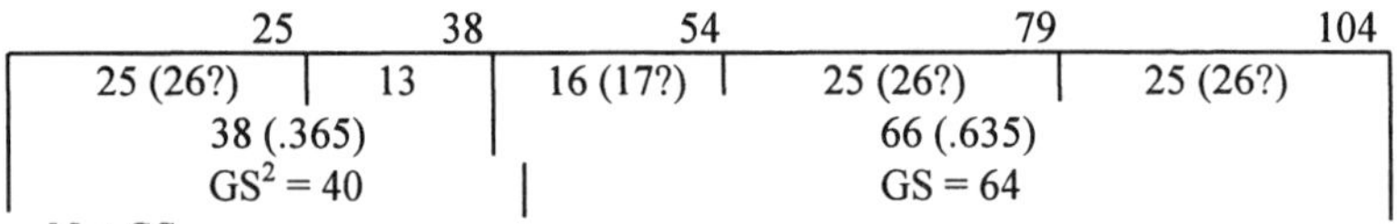

Not GS

Figure 2.53. Machaut's *Virelai no. 32*

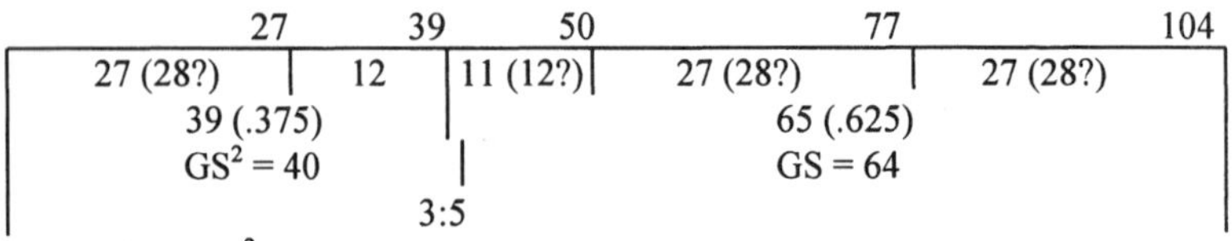

One from GS² and 13FN 3:5

Figure 2.54. Machaut's *Virelai no. 33*

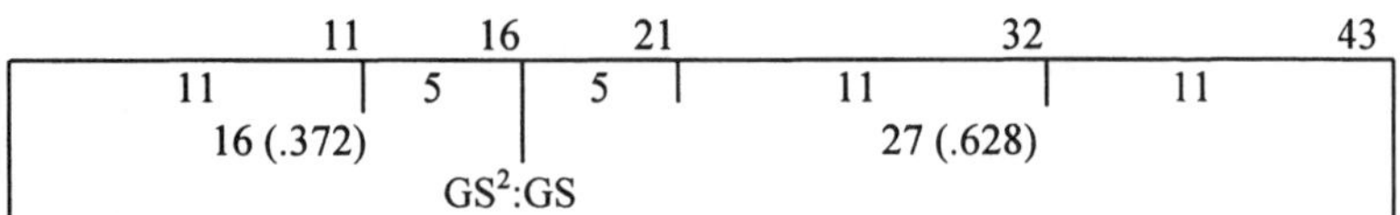

GS² but percentages are poor

Table 2.4 summarizes the virelais.

Table 2.4. Summary of Machaut's virelais

Number	At GS[2]	1 meas.	2 meas.	FN	Other
1	.372				Poor
2	.379				
3	.381			8:13	
4	.382			13:21	
5	.385			5:8	
6	.381			8:13	
7			.359		
8	.375			3:5	
9	.375			3:5	
10	.372				Poor
11	.372				Poor
12		.368			
13	-	-	-	-	-
14	-	-	-	-	-
15		.374			.001 from 3:5
16	.377				Poor
17		.371			
18		.366			
19		.370			
20		.370			
21		.371			
22	.380				
23		.371			
24		.372			
25		.397			
26	.377				Poor
27			.358		
28		.372			
29		.368			
30	.377				Poor
31			.365		
32		.375		3:5	
33	.372				Poor

Conclusions concerning Machaut

As can be seen in tables 2.3 and 2.4, the rondeaux and virelais are heavily represented by GS and FN, all but two being within two measures of GS. Several are exactly divided by Fibonacci numbers and others are very close to 3:5. On the other hand, only the Kyrie of the *Messe de Nostre Dame* was divided in phi proportions. Whether other late medieval composers used fib and phi is an open question at present.

Chapter notes

[1] [Lar p. 513]
[2] [Mad p. 74]
[3] Being unable to find any transcriptions of the Kyries, I ventured to transcribe them myself. Having no particular expertise in medieval notation, I can only hope that any errors are not fatal.
[4] Due to limitations of the software, GS^2 has had to be rendered as GS2 in the illustrations.
[5] This difference gives rise to the question, Since the mode of this piece has been Mixolydian to this point, why does it suddenly veer off to Phrygian?
[6] [Bard]
[7] [Bard p. 128]
[8] [Mur p. 40, 187]
[9] [Mur p. 160] This is not unnecessary nitpicking, given the great amount of arguing that has taken place over these structures.
[10] [Bard p. 154]
[11] Unfortunately, although I expected to find further approximations of fib or phi in the *Messe,* I was unable to.
[12] [Pow p. 242]
[13] Punctuation does not seem to be a reliable indicator of form, which is never obvious in the motets. Opportunities for argument are abundant, so I leave the dubious pleasure of further analyses to others. Also, Powell's elaborate and hard to read diagrams cannot be recommended as a desirable format.
[14] [Pow p. 242–253]
[15] [Wilk]
[16] Machaut could not have known about Lucas numbers unless the principle of a generalized additive sequence was known from Nicomachus or elsewhere, as Powell suggested. [Pow p. 236]

Chapter 3
Renaissance

Introduction

Although chapter 2 showed little evidence that early medieval composers had used fib or phi, it was evident that late in the period Machaut, at least, used fib and phi regularly.

Two articles and several dissertations have attempted to show that Guillaume Dufay and other composers of the Renaissance used fib and phi to very high accuracy.

The two articles to be discussed at the beginning of this chapter have prompted a study of all the motets of Dufay from the *Opera omnia* Volume I, Part 2. We found that Dufay used fib and phi, but not with any consistency.

The dissertations covered some of the same ground, and suggested studies of other composers. As it turned out, many composers of the Renaissance used fib and phi regularly. But why, since none of the theorists of the period are known to have discussed them? Was this a remnant of older forms, the small numbers carried down from Pythagoras, or the influence of Pacioli's *De Divina Proportione*?

The reason is by no means obvious.

Guillaume Dufay

Two important researchers have reported studies of some of the motets of Guillaume Dufay. Newman W. Powell's article "Fibonacci and the Gold Mean [*sic*]: Rabbits, Rumbas, and Rondeaux" showed a spectacular analysis of Dufay's *Motet number 13* "Ecclesie militantis."

Margaret V. Sandresky's article "The Golden Section in Three Byzantine Motets of Dufay" discussed the works *Motet no. 1* "Vasilissa, ergo gaude;" *Motet no. 2* "Balsamus et munda;" and a *Chanson motet* "Lamentatio Sanctae Matris Ecclesiae Constantinopolitanae."

These studies made important points, but their conclusions must be disputed on several grounds. Both authors tended to miscount measures and to make unwarranted assumptions. Powell assumed that division into GS can be supported by a weak structural point while neglecting a much stronger structural point that supports a nearly bisectional structure. Sandresky may have used another edition of the motets that gave a different measure count.

Brief analyses will be given in order of the thirteen motets of the *Opera omnia* of Dufay, accompanied by the relevant issues.

Motets

Sandresky showed *Motet no. 1* "Vasilissa, ergo gaude" as measuring 8:13:5:8 = 34 longs.[1] Although beautiful, it is miscounted: the *Opera omnia* shows nine longs instead of eight in the last section. This changes the analysis, for 8:13:5:9 is not Fibonaccian, although 21:14 = 3:2 is.[2]

Sandresky's study is shown in figure 3.1a and Madden's in 3.1b. The revised version shows that GS and GS^2 are within one long and that the whole structure divides at 7FN 3:2. Thus, Sandresky's incorrect count is accepted—advisedly—for its beauty, but such things should be handled carefully.

Figure 3.1a. Dufay's *Motet no. 1* "Vasilissa, ergo gaude" (Sandresky) (Longs)

8	21	26	34
8 (.235) GS^3	13 (.383) GS^2	5 (.147) GS^4	8 (.235) GS^3
21 (.618)		13 (.382)	
	GS:GS^2		

GS and FN 8:13:5:8, but the last section is miscounted.

Figure 3.1b. Dufay's *Motet no. 1* "Vasilissa, ergo gaude" (Madden) (Longs)

8		21	26	35
8 (.229) GS^3	13 (.371) GS^2		5 (.143) GS^4	9 (.257) $GS^3 = 8$
21 (.600) GS = 22			14 (.400) $GS^2 = 13$	
		3:2		

One from GS and 7FN 3:2

Sandresky's study of Dufay's *Motet no. 2* "Balsamus et munda" is shown in figure 3.2a, Madden's in figure 3.2b.[3] She must have used a different edition, for her change of meters at B1 is not supported by the *Opera omnia*, despite the acceleration of the taleas. The count also differs; she omitted the last six counts shown in the score.

Figure 3.2a. Dufay's *Motet no. 2* "Balsamus et munda" (Sandresky) (Semi-breves)

A1		68	78 A2			146	156 B1	195 B2	234
26 (.382)	16 (.235)	26 (.383)	10	26 (.382)	16 (.235)	26 (.383)	10	39 (.167)	39 (.167)
42 (.617)				42 (.617)				78 (.334)	
156(.666)									

GS in A1, A2 by discarding ten semi-breves. B2 miscounted.

Figure 3.2b. Dufay's *Motet no. 2* "Balsamus et munda" (Madden) (Semi-breves)

A1			78 A2				156 B1	195 B2	234	240
26	16	26	10	26	16	26	10	39	39	6
78 (.325)				78 (.325)				84 (.350) $GS^2 = 92$		
156 (.650) GS = 148										

Not GS but contains 2FN 13:8:13:5 in A1 and A2.

Sections A1 and A2 are divided at FN 13:8:13:5 in both accounts, although Sandresky does not mention it. Her sections A1 and A2 are divided into GS (GS^3 plus GS^2) and GS^2 after deleting ten semi-breves in each section. The reason for these deletions is not clear and is not approved for use in this book. Without the deletions, GS is not present.

Figure 3.3, illustrating Dufay's *Motet no. 3* "Apostolo glorioso," is close to GS in an unusual formulation, although the percentages are poor. The first section is one measure from GS^4 followed by two A sections one from GS in length and two compressed B sections equal to GS^3. Thus, the structure is one measure from GS, flanked by other components of GS^2.

Figure 3.3. Dufay's *Motet no. 3* "Apostolo glorioso" (Madden)

	13 A1	43 A2	73 B1	83 B2 95
13 (.137)	30 (.316)	30 (.316)	10 (.105)	12 (.126)
	60 (.632)		22 (.231)	
$GS^4 = 14$	GS = 59		GS^3	

Unusual formulation close to GS

Figure 3.4, *Motet no. 4* "O sancte Sebastiane," shows an interesting structure involving five nearly equal lengths: Intro, A, A, A, B. Each section and subsection is far from GS.

Figure 3.4. Dufay's *Motet no. 4* "O sancte Sebastiane" (Madden)

Intro	28 A1	55 A2	82 A3	109 B1	118 B2	127 B3 137
28 (.204)	27 (.197)	27 (.197)	27 (.197)	9 (.066)	9 (.066)	10(12?) (.073)
		81 (.591)			28 (.205)	

Nicely balanced but not GS

Figure 3.5, Dufay's *Motet no. 5* "Supremum est mortalibus bonum," is counted in measures of unequal length complicated by numerous tempo changes. In addition, grand pauses occur at measures 10, 56, 106, and in each measure from 107 to 116. If measures 107 to 116 were doubled due to the fermatas, measure 85 would be GS, but it is unsupported by a strong structural point.

Figure 3.5. Dufay's *Motet no. 5* "Supremum est mortalibus bonum" (Madden)

10	25	40	56	70	85	100	106	116	127
10	15 (16?)	15 (14?)	16	14 (15?)	15	15	6	10	11
(.079)		46 (.362)			50 (.394)			(.079)	(.086)

Not GS

Figure 3.6, *Motet no. 6* "Rite maiorem Iacobum canamus" is nearly bisected, and the first section bisected again. It is far from fib and phi.

Figure 3.6. Dufay's *Motet no. 6* "Rite maiorem Iacobum canamus" (Madden)

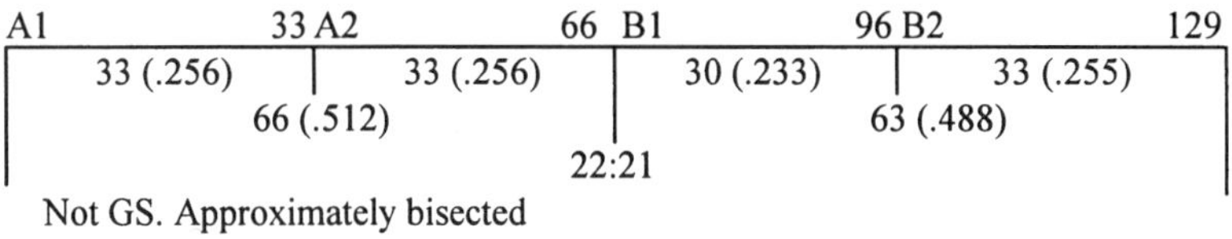

Figure 3.7. Dufay's *Motet no. 7* "O gemma, lux et speculum" (Madden)

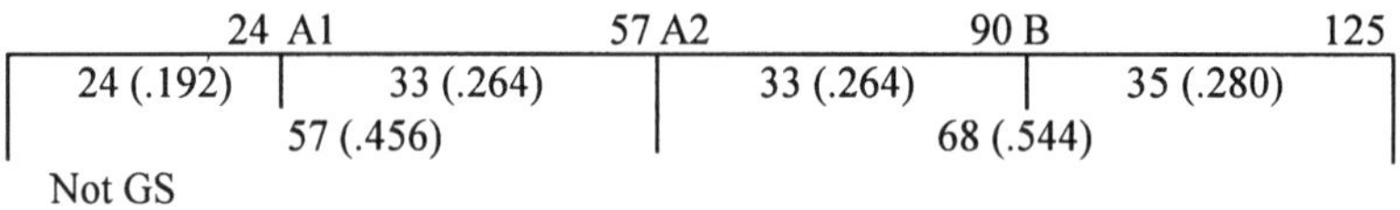

Figure 3.8, *Motet no. 8* "Moribus et genere" is divided into sections A, B, and C, where A and B are each about 2/5 of the piece, while C is about 1/5. This is fairly close to FN 2:2:1, but not persuasively so.

Figure 3.8. Dufay's *Motet no. 8* "Moribus et genere" (Madden)

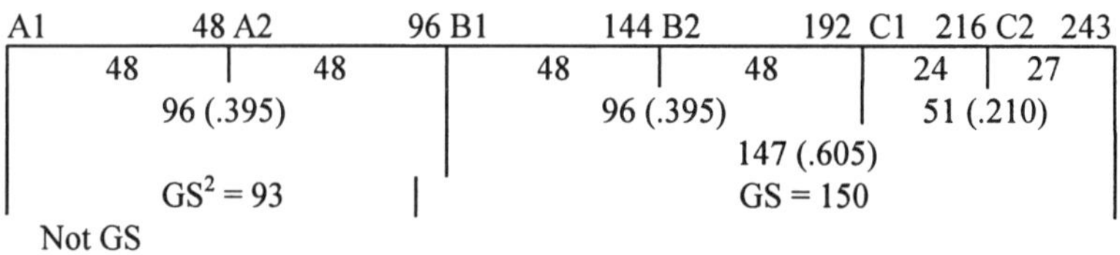

Figure 3.9, *Motet no. 9* "Fulgens iubar ecclesie dei" divides at one measure from a weak structural point.

Figure 3.9. Dufay's *Motet no. 9* "Fulgens iubar ecclesie dei" (Madden)

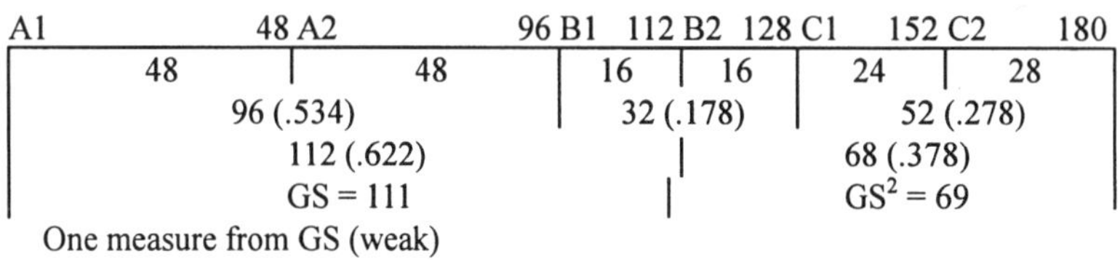

Figure 3.10 *Motet no. 10* "Salve flos tusce gentis" is divided so that section B and sections D and E are compressed into half the space of sections A and C. This gives ratios of 2:1:2:1. It is also bisected. It would be interesting to know if this tight structuring was conscious.

Figure 3.10. Dufay's *Motet no. 10* "Salve flos tusce gentis" (Madden)

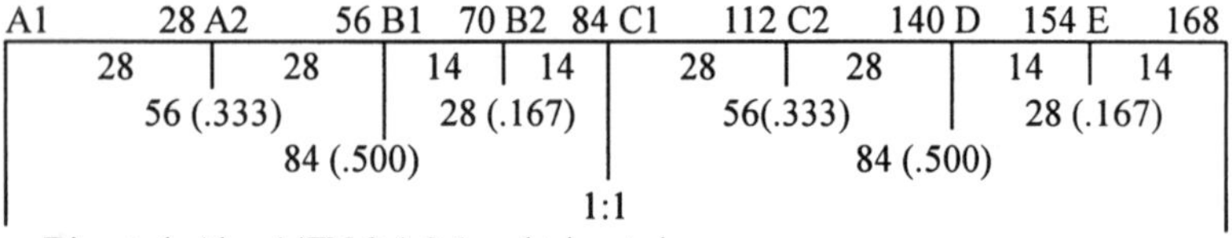

Bisected. Also 14FN 2:1:2:1 and trisected

Figure 3.11, *Motet no. 11* "Nuper rosarum flores" divides into a short section for approximately one-third of the piece and a long section for approximately the other two-thirds. It is not close to fib or phi, but is less than one measure from trisected.

Figure 3.11. Dufay's *Motet no. 11* "Nuper rosarum flores" (Madden)

Not GS

In figure 3.12, *Motet no. 12* "Magnam me gentes laudes paciare," the last two sections of 17 plus 20 measures happen to be GS^3, but nothing else in the piece suggests either fib or phi.

Figure 3.12. Dufay's *Motet no. 12* "Magnam me gentes laudes paciare" (Madden)

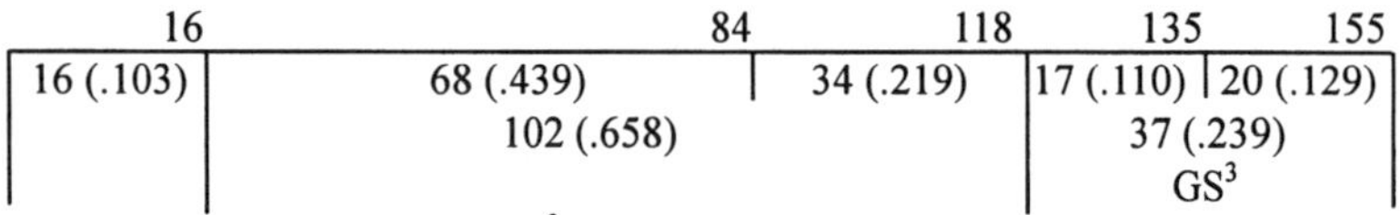

Not GS but last section is GS^3

The first problem with Powell's analysis of *Motet no. 13* "Ecclesie militantis," shown in figure 3.13a, is a small but important discrepancy of 369 beats instead of the 367.5 beats shown in the *Opera omnia.*[4] Is this a miscount, a difference in editions, or a fudge to agree with Lucas sequence 3LN (3 × 1, 3, 4, 7, 11, 18, 29, 47, 76, 123 = 369)?

Figure 3.13a. Dufay's *Motet no. 13* "Ecclesie militantis" (Powell) (Beats)

228 (weak) 369
228 (.618) 141 (.382)
3LN 76:47
GS:GS²
GS and 3LN but structurally weak and disputed

Even if the count were correct, and despite the impressive Lucas sequence and the prominent GS division after 228, the analysis would have to be disputed, for it is based on a weak structural point where most of the voices continue unabated. Beat 180, which nearly bisects the piece, is structurally much stronger with changes of meter and four of the five voices beginning new phrases. This is shown in figure 3.13b.

Figure 3.13b. Dufay's *Motet no. 13* "Ecclesie militantis" (Madden) (Beats)

72 180 (strong) 252 369
72 108 72 117
180 (.488) 189 (.512)
2:3 20:21 8:13
About one measure from bisected and FN in the sections

But if the count is 367.5, as we believe, then FN in the left-hand side is incidental because it is not found in the right. See figure 1.13c.

Figure 3.13c. Dufay's Motet no. 13 *Ecclesie militantis* (Madden) (Beats)

72 180 (strong) 252 367.5
72 108 72 115.5
180 (.490) 187.5 (.510)
About one measure from bisected

Thus, this analysis does not show phi proportions to any important extent. The conclusion to be drawn must be that GS is present only if we are prepared to accept weak structural points as support, or approximately correct counts. Other music to follow will show much stronger support for GS.

Powell presented an interesting analogy of a rumba to explain the Fibonacci sequences in the taleas of the lower voices. See figure 3.13d. These are units of 108:72:108 beats (3:2:3), further divided into 2:1 subunits.[5] This may be a worthwhile representation, and notice that it emphasizes the structural point at 180, but it does leave part of the piece out.

Figure 3.13d. Dufay's *Motet no. 13* "Ecclesie militantis" showing the rumba analogy (Powell) (Beats)

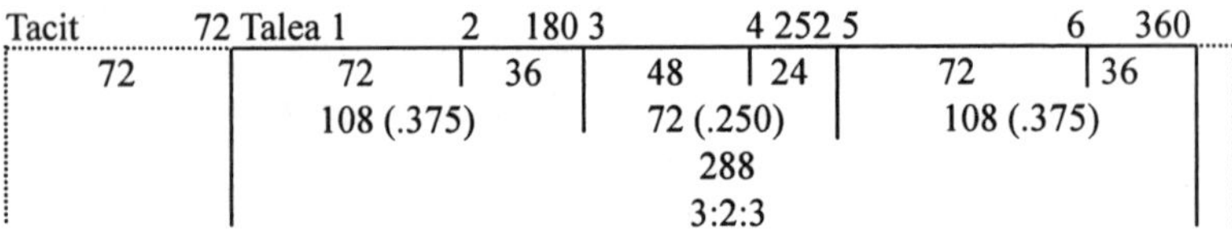

Part of it is FN

Powell's illustrations were spectacular due to the elaborate geometric figures that were perhaps meant to show how Dufay derived his musical structure.[6] Though apparently correct, they are hard to read, and only an experienced geometer would understand them. In their complexity, they raise the question whether Dufay would have made such constructions.

Other Works of Dufay

Of the works of Dufay in the *Harvard Anthology of Music*, only the Kyrie of the *Missa L'Homme armé* is close enough to the golden proportion, as shown in figure 3.14, counted in beats.[7]

Figure 3.14. Kyrie from Dufay's *Missa L'Homme armé* (Beats)

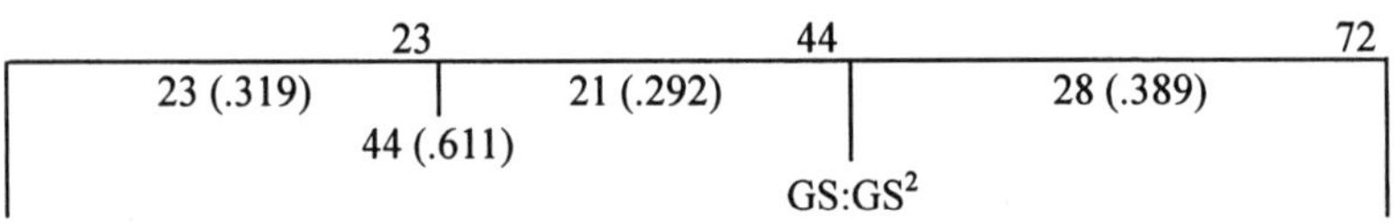

GS, although the percentage is poor

Table 3.1 summarizes the Dufay study.

Table 3.1. Summary of the Dufay study

Motet No.	At GS or GS^2	One unit	Too far	Other
1	.618			Sandresky; miscounted
		.600		Madden; 3:2
2			.666	Sandresky; disputed
			.650	Madden; FN 13:8:13:5
3		.632		Approximately GS^4:GS:GS^3
4				5 approximately equal sections
5			.441	Reminiscences of GS
6			.512	Approximately bisected
7			.456	
8			.395	Approximately 2:2:1
9		.622		Measure 112 weakly structural
10			.500	Bisected, trisected, 2:1:2:1
11			.329	Approximately 1:2
12			.761	Final sections equal GS^3
13	.618			Powell; GS, 3LN, disputed
			.488	Madden; 2:3 and 8:13
			.490	Madden; not fib or phi
				Powell; rumba analogy
Missa	.611			Poor

Conclusions regarding Dufay

Motet no. 1 is miscounted but is close to GS and uses FN. *No. 2* contains FN. *No. 9* is weakly structural at one measure from GS. *No. 13* divides disputably at GS at a weak structural point. *Missa L'Homme armé* is the only Dufay work to strongly show GS, but even there, the percentage is poor.

Other works of Dufay published in *HAM* do not show evidence of phi. Thus, Dufay seldom wrote music structured in the golden proportion. It would be surprising if someone found that he used it pervasively.

John Dunstable

Michael Richard Rogers' dissertation "The Golden Section in Musical Time" borrowed an example from an unpublished paper by Rothwell entitled "A Little Past Half" showing Dunstable's *Polyphonic* (plainsong) *setting no. 35* "Ave Maris Stella" clearly divided into GS and FN.[8]

Figure 3.15a. Dunstable's *Polyphonic setting no. 35* "Ave Maris Stella" (Rogers/ Rothwell) (Quarter notes)

13	26	34	42	55
13 (.236) GS^3	13 (.236) GS^3	8 (.146) GS^4	8 (.146) GS^4	13 (.236) GS^3
	34 (.618)		21 (.382)	
		GS:GS^2		

GS and FN 13:13:8:8:13, but disputed

However, the score begs for another interpretation.[9] The piece is miscounted and the strong cadence is not at beat 34, but rather 27. Thus, the structure is close to bisected rather than phi. See figure 3.15b.

Figure 3.15b. Dunstable's *Polyphonic setting no. 35* "Ave Maris Stella" (Madden) (Quarter notes)

15	27	42	55
15 (.273)	12 (.218)	15 (.273)	13 (.236)
27 (.491)		28 (.509)	

One-half from bisected

Since there is a dispute, all the polyphonic settings of Dunstable have to be analyzed. *Number 36* "Magnificat" is a series of phrases marked by quarter rests and double bars. It is not fib or phi. See figure 3.16.

Figure 3.16. Dunstable's *Polyphonic setting no. 36* "Magnificat" (Measures)

6	22	43	59	78	103	120	145	161	179	193	212
6	16	21	16	19	25	17	25	16	18	14	19
78 (. 368)					134 (.632)						
GS^2 = 81					GS = 131						

Not GS

Figure 3.17. Dunstable's *Polyphonic setting no. 37* "Ave Regina Celorum"

43	80	108
43 (.398)	37 (.343)	28 (.259)
	65 (.602)	
GS^2 = 41	GS = 67	

Not GS

Number 38 "Regina Celi Letare" is shown in figure 3.18. It is two measures from GS. The last section happens to be GS^4.

Figure 3.18. Dunstable's *Polyphonic setting no. 38* "Regina Celi Letare"

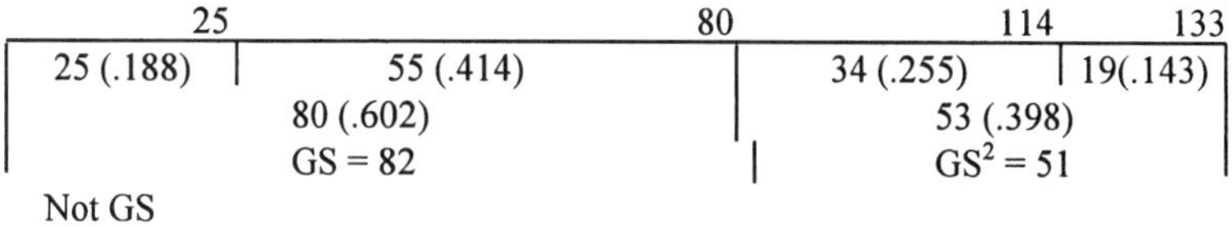

Number 39 "Crux Fidelis" is one from bisected. See figure 3.19.

Figure 3.19. Dunstable's *Polyphonic setting no. 39* "Crux Fidelis"

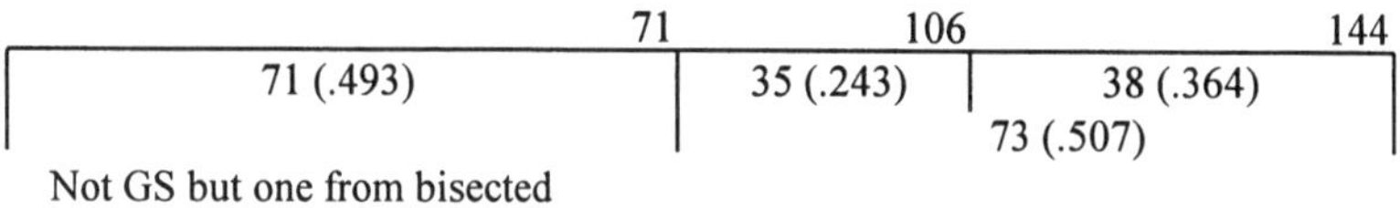

Thus, none of the polyphonic settings has GS proportions, but GS can be found in the motets, as we will see.

Motets

Motet no. 40 "Alma Redemptoris Mater" is illustrated in figure 3.20. It is far from GS, although the middle section is one from GS^3.

Figure 3.20. Dunstable's *Motet no. 40* "Alma Redemptoris Mater"

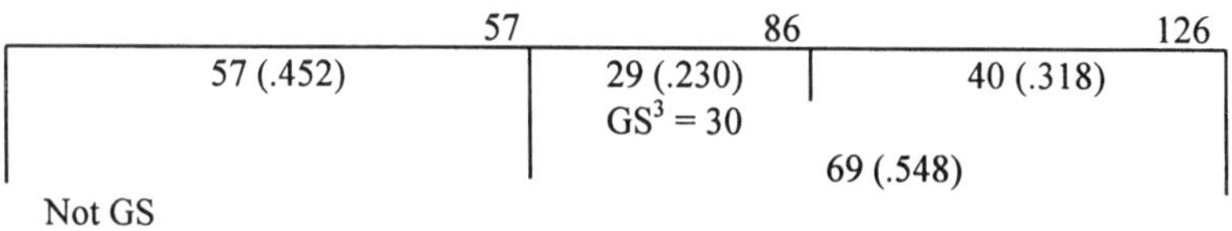

Motet no. 41 "Beata Dei Genitrix" is presented in figure 3.21. It is clearly structured in phi proportion. LN 47:76:123 derives from the additive sequence 1, 3, 4, 7, ... This is curious because Dunstable could not have known of the Lucas sequence by that name. It seems to be an isolated instance in his work.

Figure 3.21. Dunstable's *Motet no. 41* "Beata Dei Genitrix"

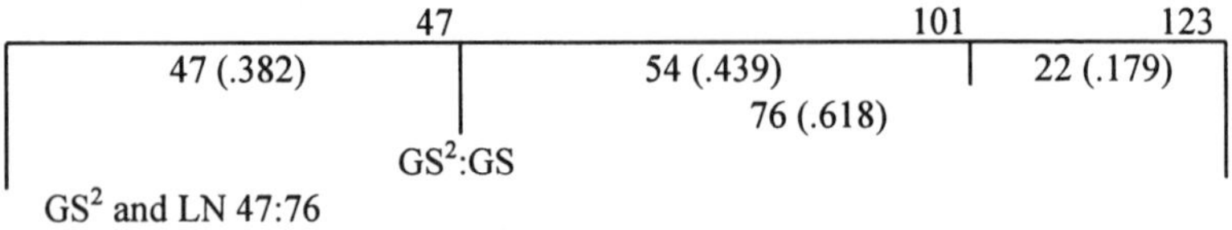

Motet no. 42 "Beata Mater" is fairly close to FN 3:2:2. See figure 3.22.

Figure 3.22. Dunstable's *Motet no. 42* "Beata Mater"

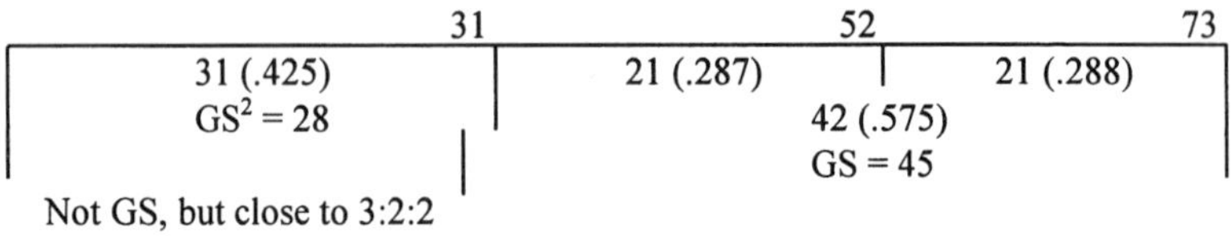

Figure 3.23. Dunstable's *Motet no. 43* "Gloria Sanctorum"

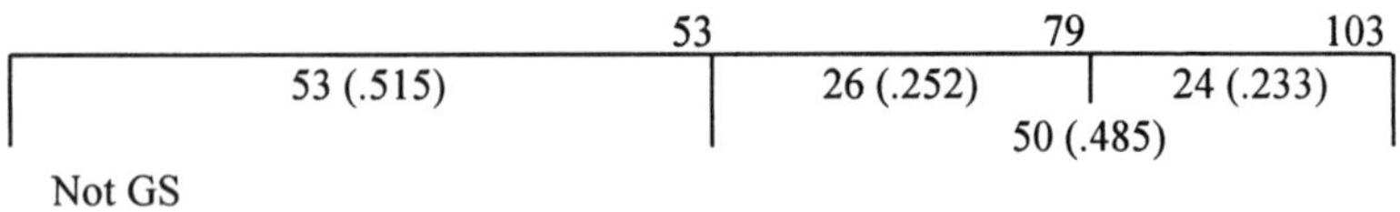

Motet no. 44 "Quam Pulcra Es" is one measure from bisected, as shown in figure 3.24.

Figure 3.24. Dunstable's *Motet no. 44* "Quam Pulcra Es"

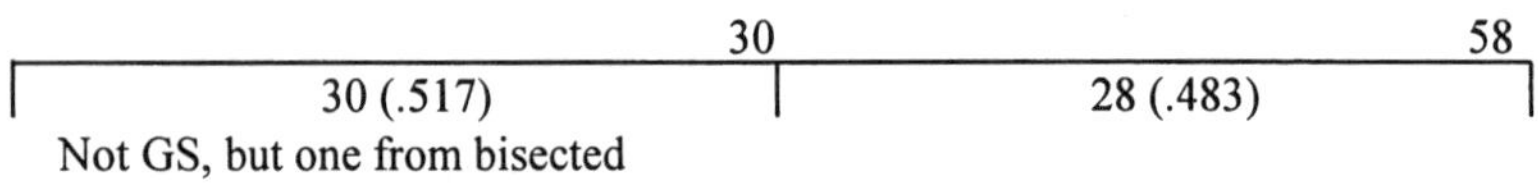

Figure 3.25. Dunstable's *Motet no. 45* "Salve Regina Mater Mire"

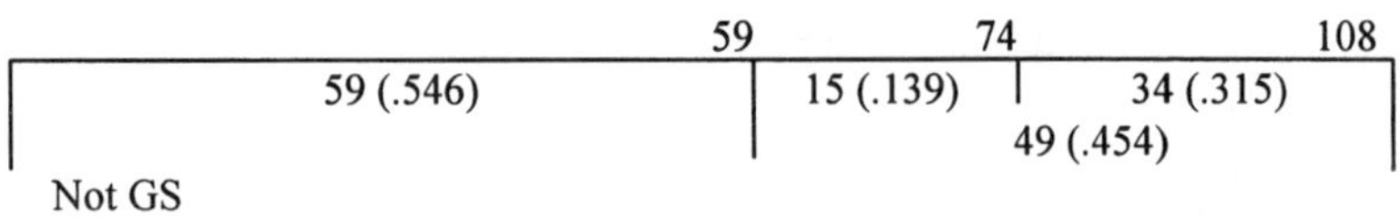

Figure 3.26. Dunstable's *Motet no. 46* "Salve Regina Misericordie"

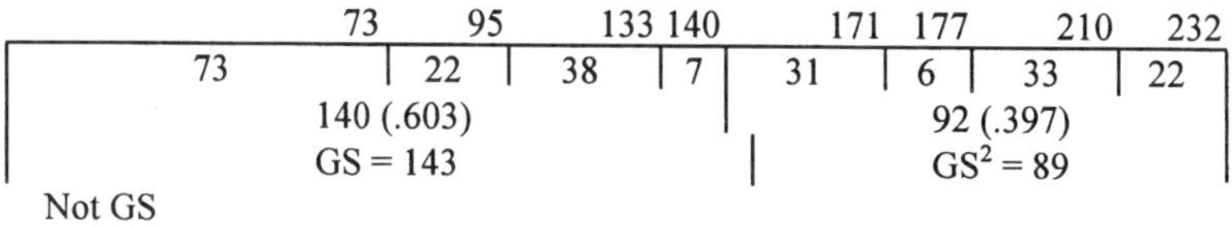

Counting the fermatas in *Motet no. 47* does not change it to GS.

Figure 3.27. Dunstable's *Motet no. 47* "Sancta Dei Genitrix"

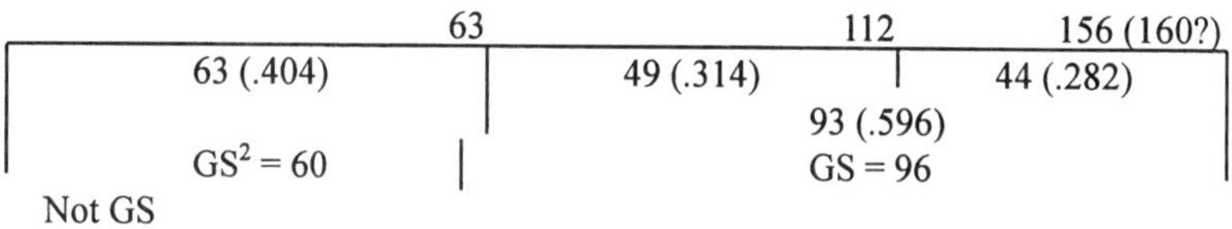

Motet no. 48 "Sancta Maria," is divided clearly at GS^2.[10] Its structure is shown in figure 3.28.

Figure 3.28. Dunstable's *Motet no. 48* "Sancta Maria"

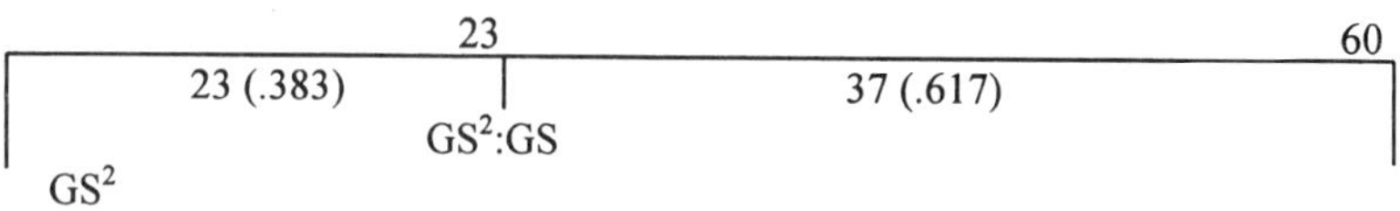

Figure 3.29. Dunstable's *Motet no. 49* "Sancta Maria Succurre Miseris"

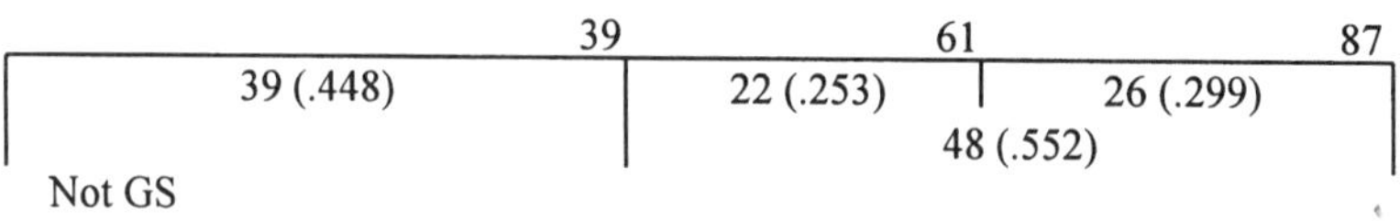

Motet no. 50 "Speciosa Facta Es" illustrated in figure 3.30, divides at one from GS and GS^2 in both short-long and long-short configurations.

Figure 3.30. Dunstable's *Motet no. 50* "Speciosa Facta Es" (Beats)

41 56 62 74 86 103 124 165

41 | 15 | 6 | 12 | 12 | 17 | 21 | 41

62 (.376) 103 (.624)

$GS^2 = 63$ GS = 102

103 (.624) 62 (.376)

GS = 102 $GS^2 = 63$

.001 from 3:5 and 5:3 and one measure from GS and GS^2

Motet No. 51 "Sub Tuam Protectionem" appears in figure 3.31. There is a structural point near 63 measures and a much stronger one at 53. Thus, this piece is marginally phi.

Figure 3.31. Dunstable's *Motet no. 51* "Sub Tuam Protectionem"

Strong Weak

53 63 78 101

53 (.525) | 10 | 15 | 23

63 (.624) 38 (.376)

GS = 62 $GS^2 = 39$

One from GS and .001 from 5:3

Table 3.2 summarizes the Dunstable study.

Table 3.2. Summary of the Dunstable study

Number	At GS or GS^2	One unit	Too far	Other
Polyphonic setting				
35	.618			Rogers/Roth.; disputed
			.491	Madden
36			.368	Ad hoc
37			.398	
38			.602	
39			.493	Nearly bisected
Motet				
40			.452	
41	.382			LN 47:76
42			.425	Close to 3:2:2

Number	At GS or GS^2	One unit	Too far	Other
43			.515	Nearly bisected
44			.517	Nearly bisected
45			.546	
46			.603	
47			.404	
48	.383			
49			.448	
50		.376/.624		.001 from 3:5 and 5:3
51		.624		.001 from 5:3

Only three of the seventeen motets and polyphonic settings have phi proportions, one of which is in dispute, but two others are acceptably close. Thus, Dunstable was an infrequent user of phi in his compositions.

Gilles Binchois

According to Rogers, James Rothwell's count of Gilles Binchois' *Rondeau* "De plus en plus" is in beats.[11] His result, as shown in figure 3.32a, is 5GN = 37:60 derived from the sequence 1, 4, 5, 9, 14, 23, 37, 60, etc.

Figure 3.32a. Binchois' *Rondeau* "De Plus en plus" (Rogers/Rothwell) (Beats)

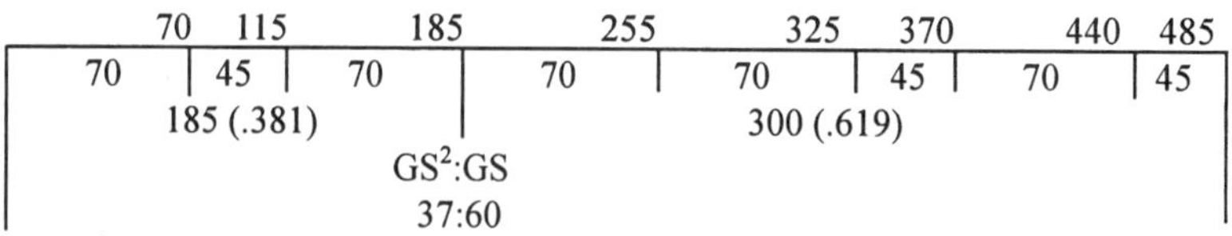

GS^2 and 5GN 14:9:14:14:14:9:14:9 or 5GN 37:60

As printed in *HAM*, however, this piece is structured slightly differently.[12] Counted in measures, it is still GS, but with a different FN sequence, as shown in figure 3.32b. If HAM is correct, Rogers and Rothwell omitted rests at the ends of the sections.

Since there is a dispute, all Binchois' rondeaux should be analyzed, but they are not available in transcription. The main point to note is that the rondeau form is standard, as discussed elsewhere. Hence, all of Binchois' rondeaux probably approach phi.

Figure 3.32b. Binchois' *rondeau* "De Plus en plus" (Madden) (Measures)

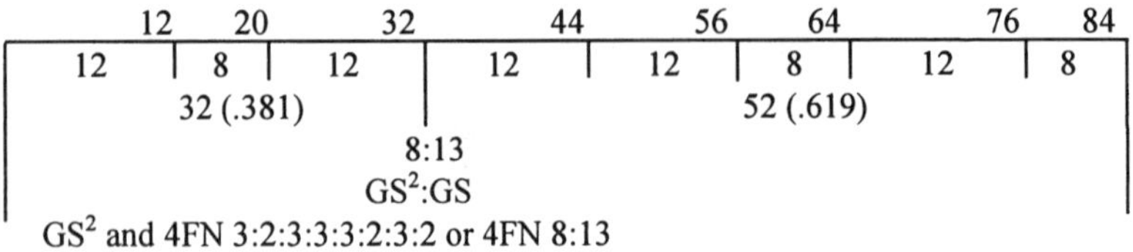

Arnold and Hugo de Lantins

Arnold de Lantins' *Rondeau* "Puisque je voy" shows evidence of fib and phi, as shown in figure 3.33.[13]

Figure 3.33. Arnold de Lantins' *Rondeau* "Puisque je voy"

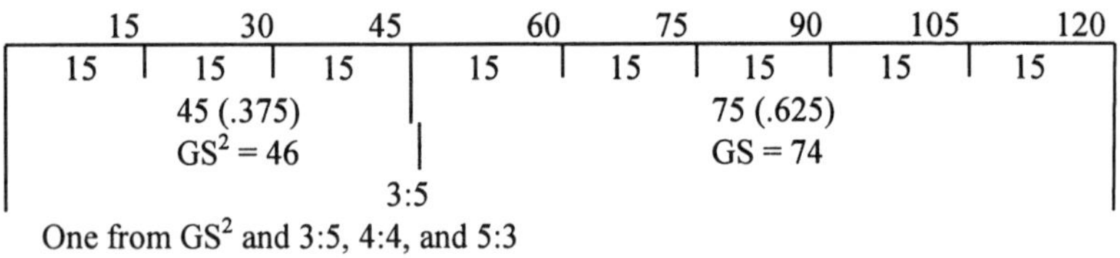

Hugo de Lantin's *Rondeau* "Ce ieusse fait" is not in phi proportion as presented because Davison and Apel have not shown all the verses.[14] But, this is a typical rondeau with an eight unit structure, as shown in figure 3.34. It is not GS, but divides at 3:5.

Figure 3.34. Hugo de Lantin's *Rondeau* "Ce ieusse fait"

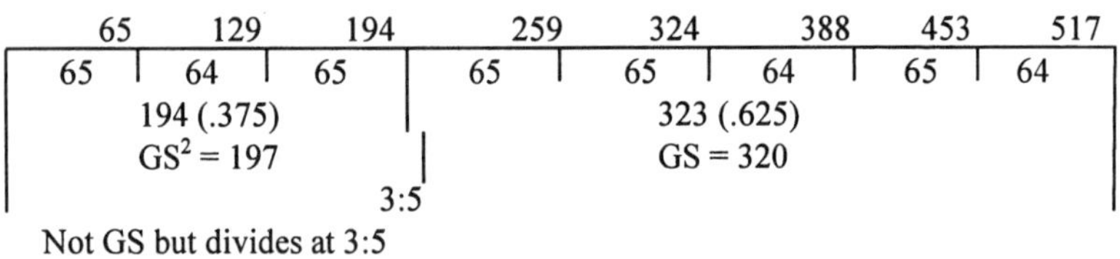

Johannes Ockeghem

Ockeghem's *Virelai* "Ma maîtresse" divides at one measure from GS^2 and at 3:2 and 17FN, as shown in figure 3.35.[15]

Figure 3.35. Ockeghem's *Virelai* "Ma maîtresse"

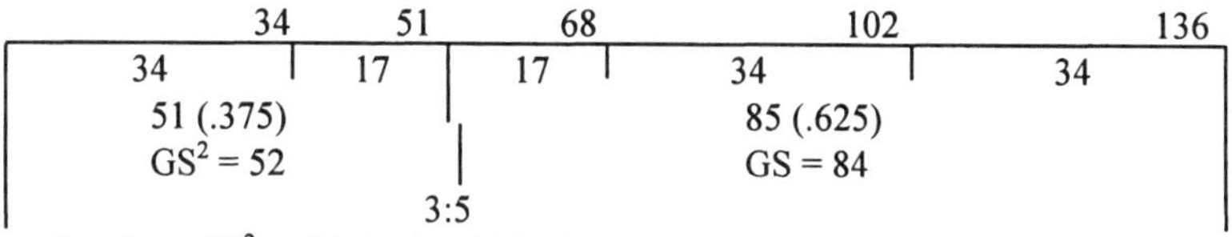

One from GS^2 and 3:5; also 17FN 2:1:1:2:2

The structure of Ockeghem's *Virelai* "Ma bouche rit" is shown in figure 3.36. It divides at GS^2, but not FN.[16]

Figure 3.36. Ockeghem's *Virelai* "Ma bouche rit"

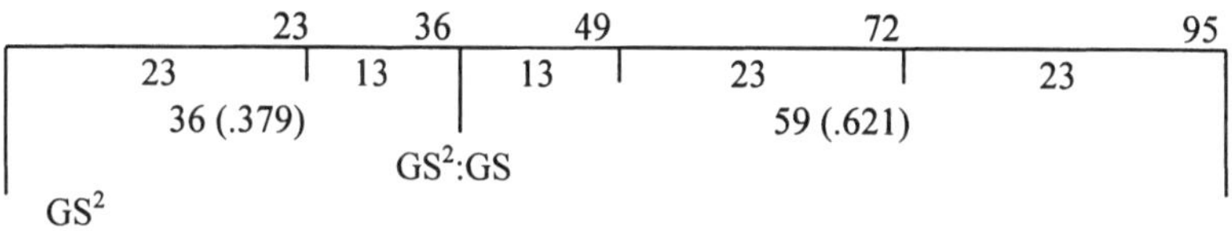

Jacob Obrecht

Jacob Obrecht's *Motet* "O vos omnes" is shown in figure 3.37. It divides strongly at GS.[17]

Figure 3.37. Obrecht's *Motet* "O vos omnes"

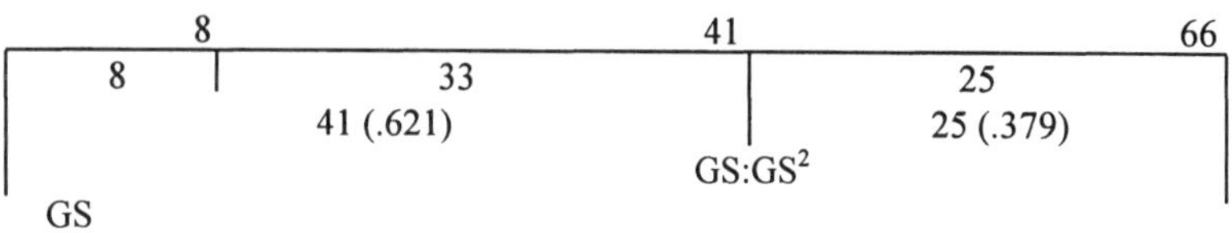

Obrecht's *Instrumental canzona* "Tsaat een meskin," seen in figure 3.38, is GS—due to an extended cadence from beat 122 to 129—but not FN.[18] Rothwell's proportions were slightly off, due to rounding.[19] He used 5/8 = .625 for GS, whereas the correct figure is 129/208 = .620. Since this is only .8% error, his work does not have to be rejected, but it should be remembered that approximations and careless rounding can lead to misunderstandings.

Figure 3.38. Obrecht's *Instrumental canzona* "Tsaat een meskin" (Quarter notes)

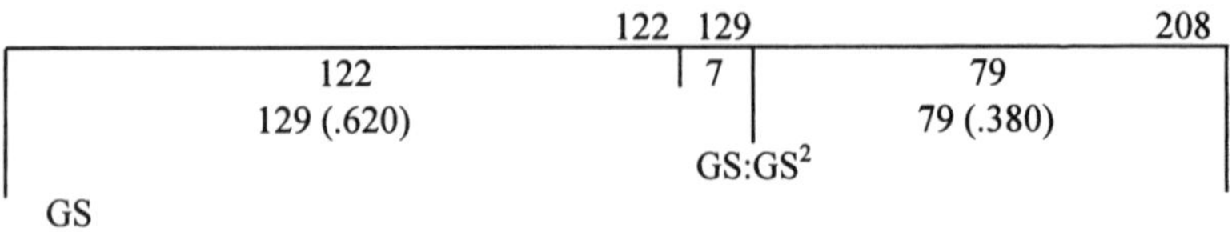

Loyset Compère

Loyset Compère's *Rondeau* "Royne du ciel" is shown in figure 3.39.[20] Unlike many rondeaux studied here, this is several points from either GS or FN.

Figure 3.39. Compère's *Rondeau* "Royne du ciel" (Quarter notes)

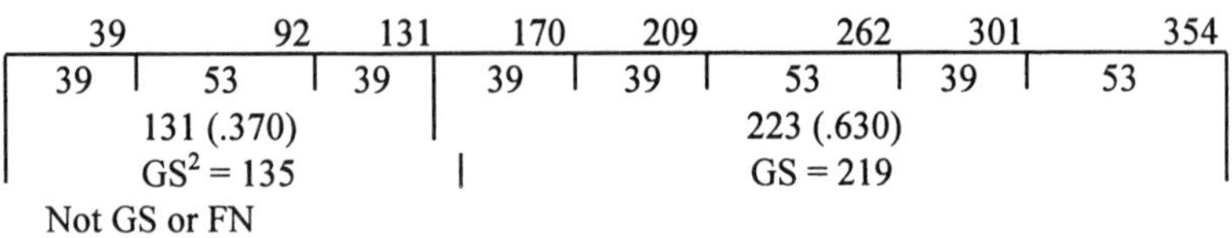

Conrad Paumann

Conrad Paumann's *Mit ganczem Willen* from *Fundamentum Organisandi* is weakly GS^2, but the stronger cadence at measure 26 is not phi.[21]

Figure 3.40. Paumann's instrumental piece *Mit ganczem Willen* (Measures)

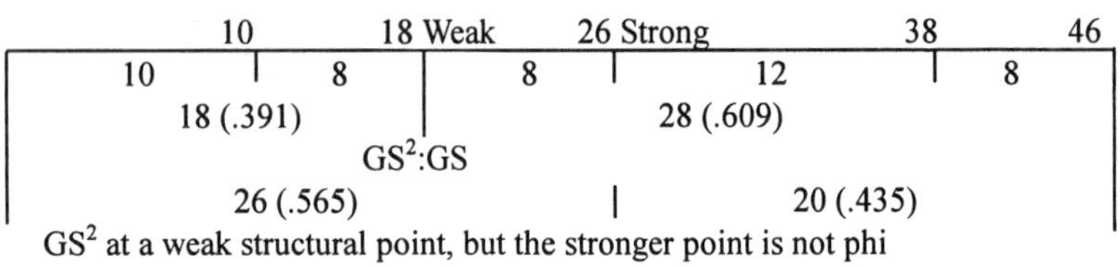

Glogauer Liederbuch

An instrumental piece from the *Glogauer Liederbuch* contains Fibonacci numbers in an unusual arrangement, as shown in figure 3.41.[22]

Figure 3.41. Instrumental piece from the *Glogauer Liederbuch* (Half notes)

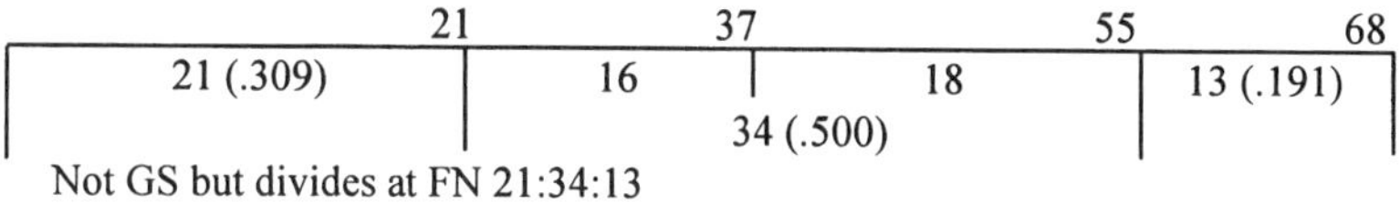

Arnolt Schlick

Arnolt Schlick's *Organ hymn* "Maria zart," seen in figure 3.42, may divide in GS^2 and GS, although the cadence at 21 is weak.[23] Reversing the division is more sensible because measure 35 has a stronger cadence.

Figure 3.42. Schlick's *Organ hymn* "Maria zart" (Measures)

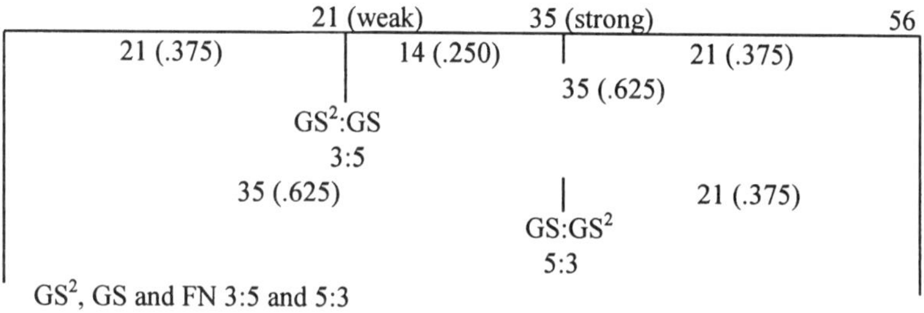

Paulus Hofhaimer

Paulus Hofhaimer's *German part song* "Mein's traurens ist" is divided at GS^2.[24]

Figure 3.43. Hofhaimer's *German part song* "Mein's traurens ist" (Half notes)

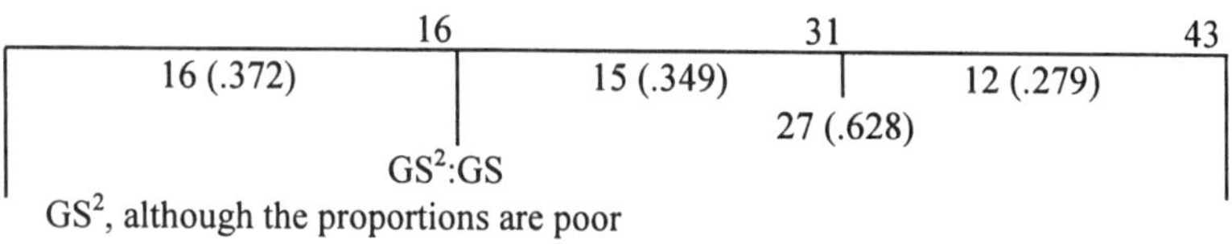

Orlando di Lasso

Orlando di Lasso's *Chanson* "Bon jour, mon Coeur" is GS^2 and BS.[25] The tonic cadence at beat 24 is not supported by the poetic rhythm.

Figure 3.44. Lasso's *Chanson* "Bon Jour, mon coeur" (Half notes)

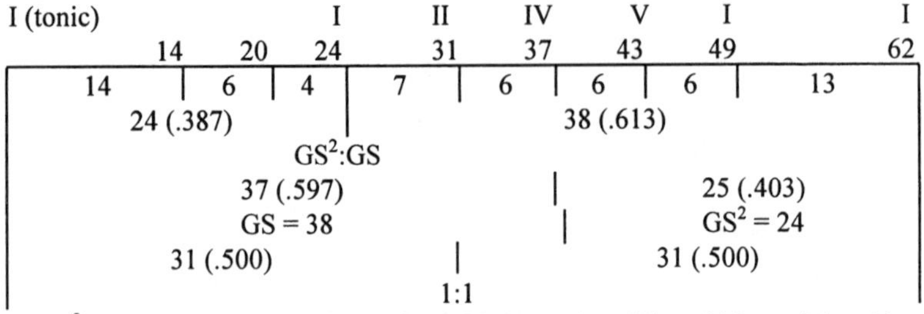

GS^2 (strong) on I, one half-note from GS (strong) on IV, and bisected (weak)

Hans Leo Hassler

Hans Leo Hassler's *Motet* "Quia vidisti me" divides at GS and GS^2.[26]

Figure 3.45. Hassler's *Motet* "Quia vidisti me" (Half notes)

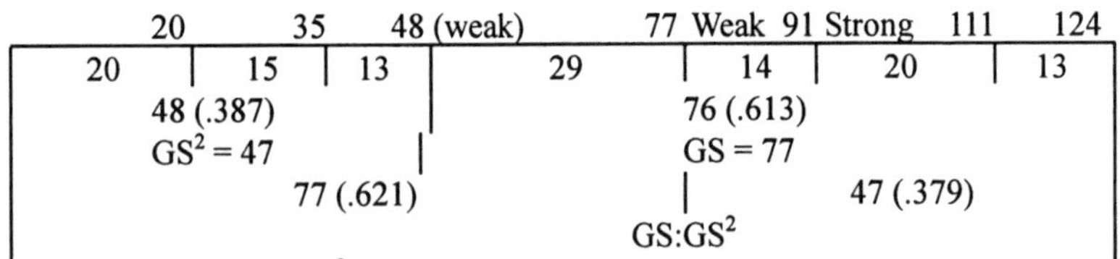

One from GS and GS^2 on weak cadences

Hassler's *Lied* "Ach Schatz" divides at GS^2 with each section bisected.[27]

Figure 3.46. Hassler's *Lied* "Ach Schatz" (Half notes)

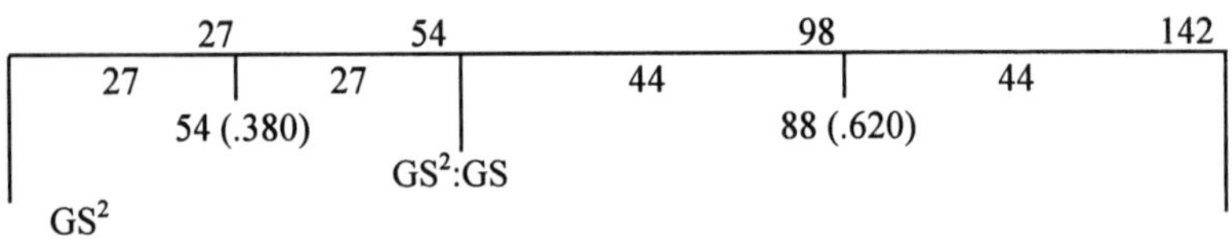

Melchior Franck

Melchior Franck's *Lied* "So wünsch ich ihr" divides at FN 13:13:21.[28]

Figure 3.47. Franck's *Lied* "So wünsch ich ihr" (Measures)

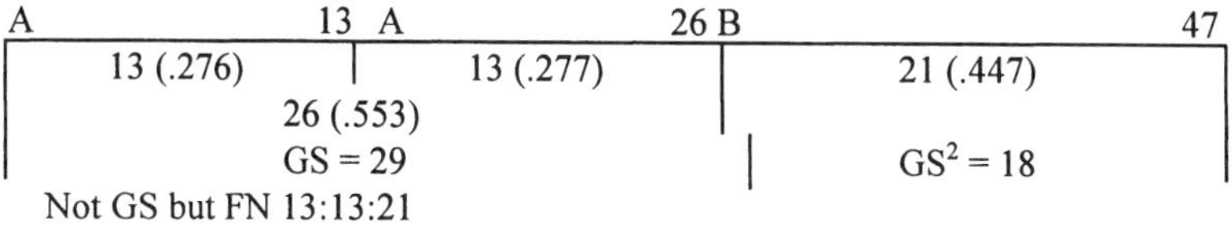

Orlando Gibbons

Orlando Gibbons' *Anthem* "O Lord, increase my faith" divides at GS.[29]

Figure 3.48. Gibbons' *Anthem* "O Lord, increase my faith" (Half notes)

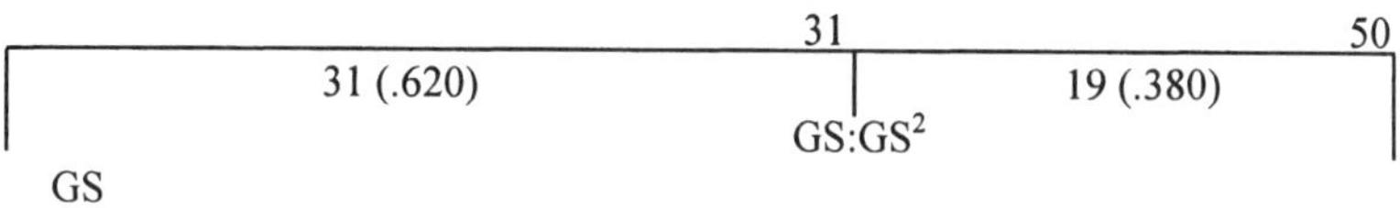

Figure 3.49 is Gibbons' *Verse anthem* "This is the record of John."[30] Counted in half notes, this piece is divided at one measure from GS, which is then bisected.

Figure 3.49. Gibbons' *Verse anthem* "This is the record of John" (Half notes)

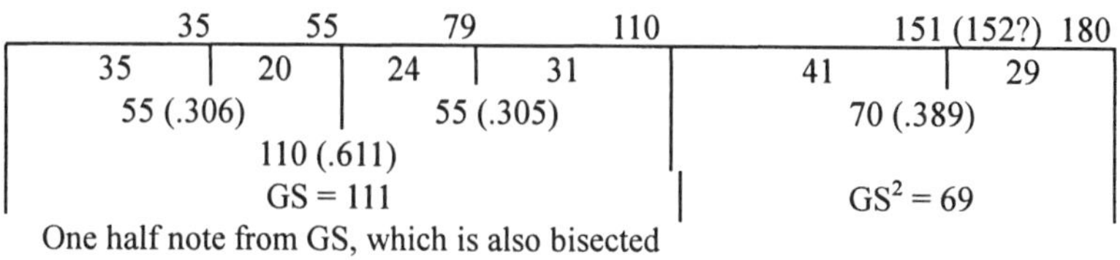

Figure 3.50 is Gibbons' *Pavane Lord Salisbury.*[31] It divides at GS and FN 34:21.

The Norton Scores printed Gibbons' *The Silver Swan.*[32] Shown in figure 3.51, it is one quarter note from GS.

Figure 3.50. Gibbons' *Pavane Lord Salisbury* (Measures)

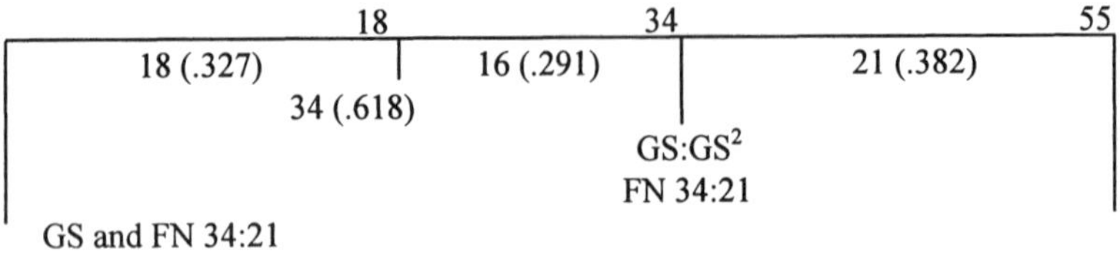

Figure 3.51. Gibbons' work *The Silver Swan* (Quarter notes)

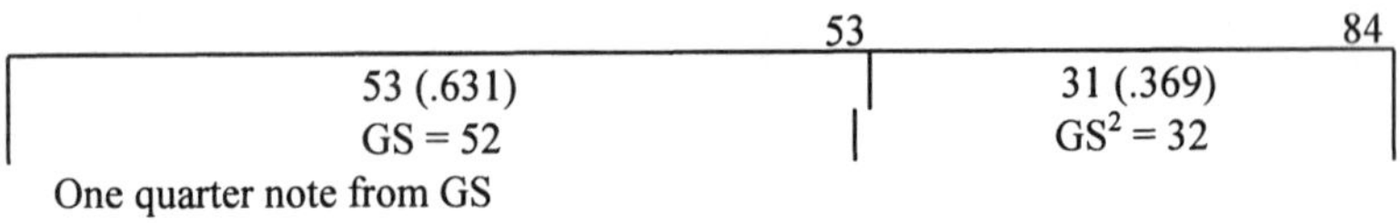

Rothwell's dissertation contained a score and analysis of Gibbons' *Fantasia,* shown in figure 3.52.[33] It divides at GS and 13:8.

Figure 3.52. Gibbons' keyboard work *Fantasia* (Rothwell) (Quarter notes)

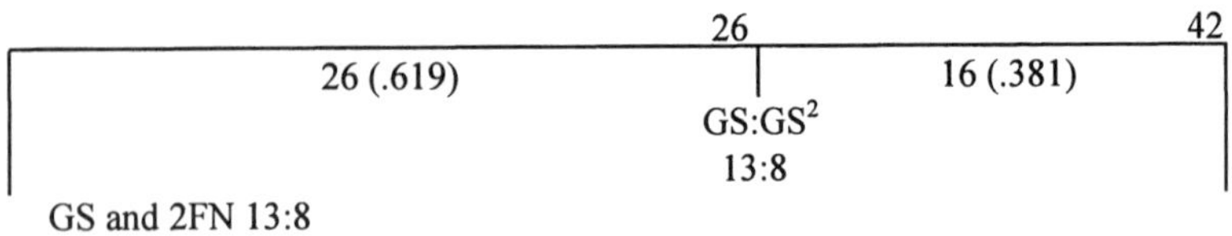

Josquin des Prés

The Norton Scores printed a piece by Josquin des Prés entitled *Ave Maria*, shown in figure 3.53.[34] It divides within one measure of GS.

Figure 3.53. Josquin's *Ave Maria* (Measures)

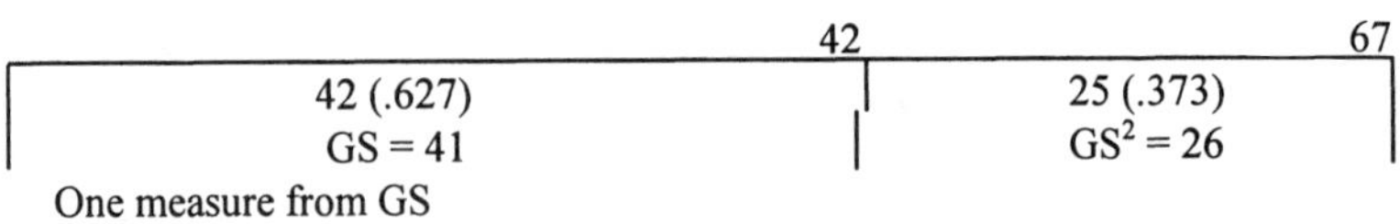

Luca Marenzio

Luca Marenzio's *S'io parto, io moro* is shown in figure 3.54.[35] This piece divides at one measure from GS.

Figure 3.54. Marenzio's *S'io parto, io moro* (Measures)

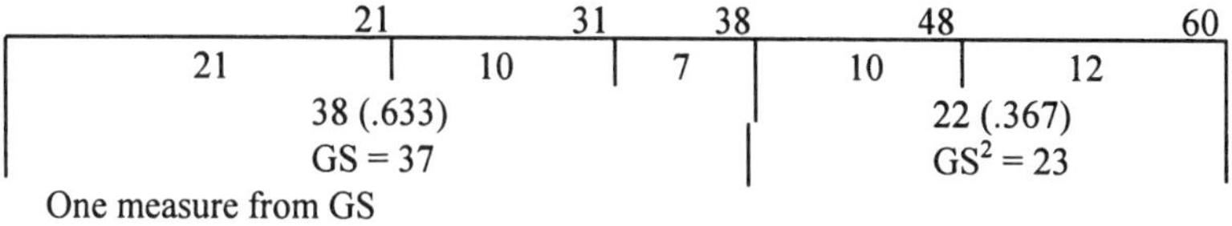

Claudin de Sermisy

Rothwell gave an example by Claudin de Sermisy from the *Maîtres Musiciens de la Renaissance Française*. The structure of *Au ioly boys* is shown in figure 3.55a.

Figure 3.55a. Sermisy's *Au ioly boys* (Rothwell) (Half notes) (Proportions corr.)

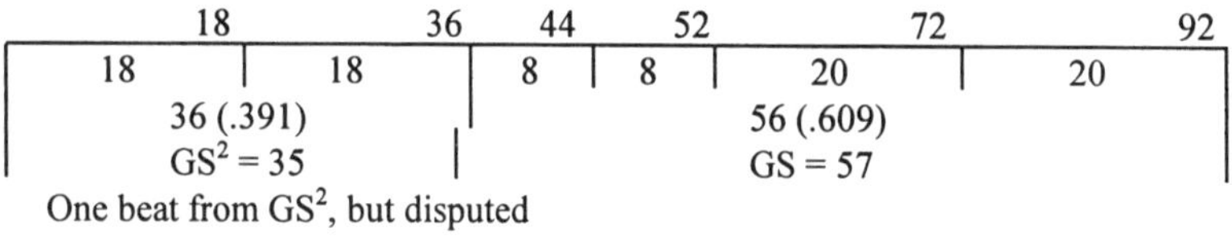

This is rather unlike the *Opera omnia* of Claudin de Sermisy, Vol. 3 no. 12, shown in figure 3.55b.[36]

Figure 3.55b. Sermisy's *Au ioly boys* (Madden) (Half notes)

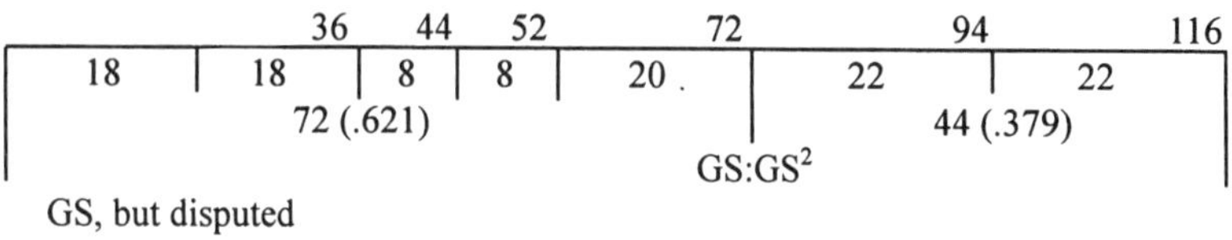

However, it is possible that the repeat sign on the last section of the *Opera omnia* is in error. Recasting this as in figure 3.55c gives a result very close to Rothwell's but at GS^2. The longer final section is due to an elongated cadence.

Figure 3.55c. Sermisy's *Au ioly boys* (Madden) (Beats)

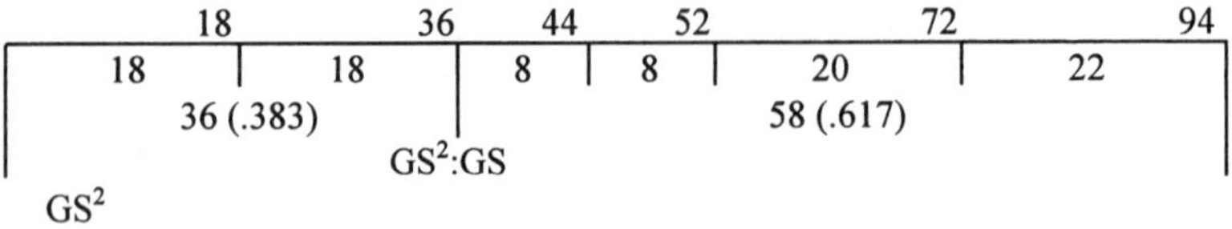

Clement Janequin

Rothwell also gave an example by Clement Janequin from the *Chansons Polyphoniques.*[37] *Guillot ung jour* is shown in figure 3.56. Checked against the score, it is seen to be miscounted at 74. Still it is accurate to one half-note.

Figure 3.56. Janequin's *Guillot ung jour* (Rothwell, Madden) (Corrected) (Half notes)

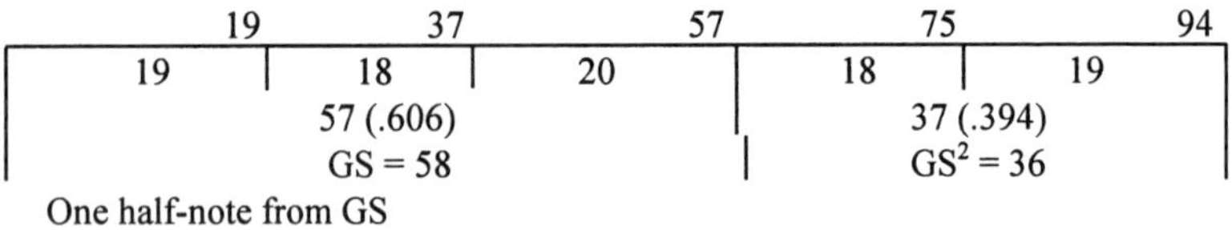

Rogers quoted from Rothwell's unpublished article "A Little Past Half" an analysis of a polyphonic chanson by Janequin entitled *Qui veult d'amour.*[38] Its structure is shown in figure 3.57a.

Figure 3.57a. Janequin's *Qui veult d'amour* (Rogers, Rothwell) (Quarter notes)

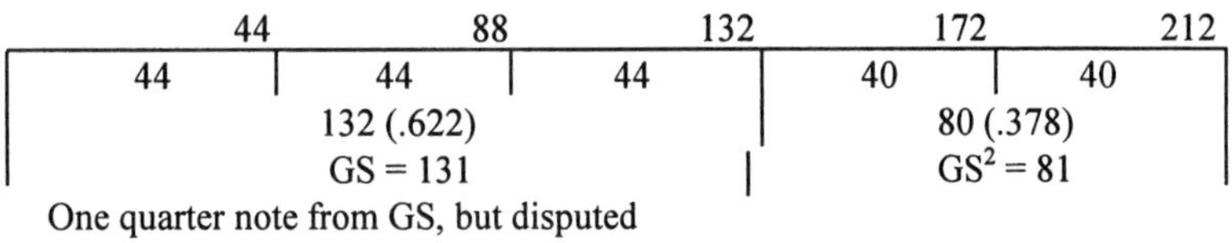

This appears to be accurate to one quarter note but checked against the *Chansons Polyphoniques* Vol. 3 no. 106, it proves to be exactly GS, as shown in figure 3.57b.

Figure 3.57b. Janequin's *Qui veult d'amour* (Madden) (Quarter notes)

44 | 88 | 131 | 171 | 212

44 | 44 | 43 | 40 | 41

131 (.618) | 81 (.382)

GS:GS^2

GS

Hieronimo Maffoni

James Rothwell gave an example by Hieronimo Maffoni from the *Italia Sacra Musica.*[39] *Quam pulchri sunt gressus tui* is shown in figure 3.58. Rothwell rightly approached this piece cautiously because the proportions do not support phi without fudging. He wanted to divide the first 98 beats or the last 96 beats. Neither is acceptable for this book. The piece as a whole is far from phi.

Figure 3.58. Maffoni's *Quam pulchri sunt gressus tui* (Rothwell) (Beats)

36 | 60 | 98 | 132

36 | 24 | 38 | 34

60 (.455) | 72 (.545)

Not GS

Giovanni Pierluigi da Palestrina

Rogers quoted from Rothwell once more to show an example from Giovanni Pierluigi da Palestrina's *Motet* "Manus Tuae."[40] This is illustrated in figure 3.59a. It appears to be accurate to one-half measure.

Figure 3.59a. Palestrina's *Motet* "Manus Tuae" (Rogers, Rothwell) (Measures)

22.5 | 57

22.5 (.395) | 34.5 (.605)

GS^2 = 22 | GS = 35

One-half measure from GS^2, but disputed

However, checking against the score, we do not find a convincing structural point at 22.5.[41] We say "maybe" at 27.5, 30, or 33.5, and "definitely" at 45 and 50.5, as shown in figure 33.19b. Of course, these are not phi. Measure 27.5 is one measure from bisected.

Figure 3.59b. Palestrina's *Motet* "Manus Tuae" (Madden) (Measures)

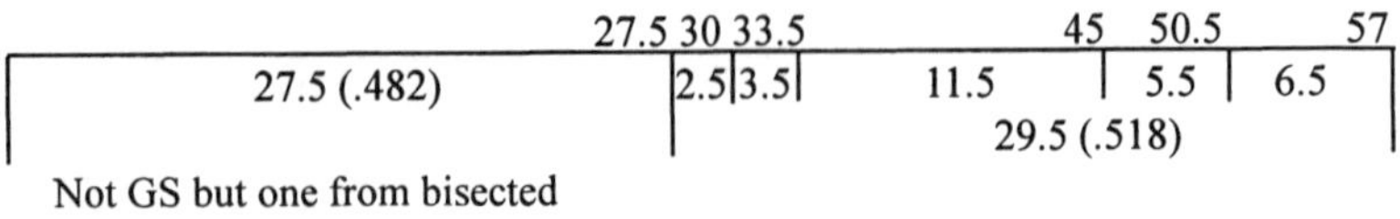

Since there is a dispute about Palestrina, three of his works from *HAM* will be analyzed, beginning with the "Agnus Dei I" from the *Missa Papae Marcelli.*[42] It is accurate to one measure. See figure 3.60.

Figure 3.60. Palestrina's "Agnus Dei I" from the *Missa Papae Marcelli* (Meas.)

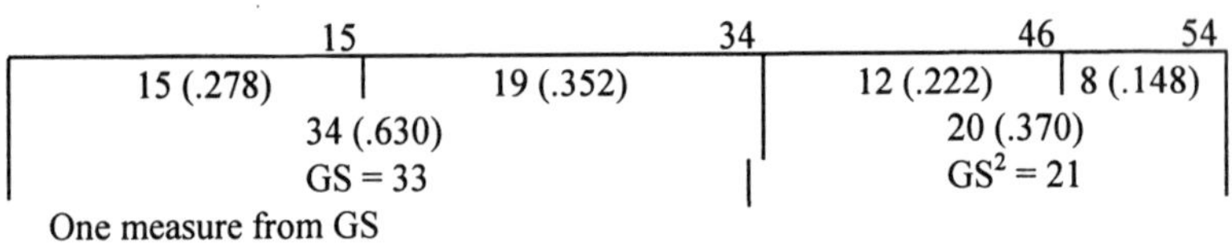

There follows *Motet* "Sicut Cervus" in figure 3.61.[43] It does not divide at fib or phi, but rather, is one half-measure from bisected.

Figure 3.61. Palestrina's *Motet* "Sicut Cervus" (Measures)

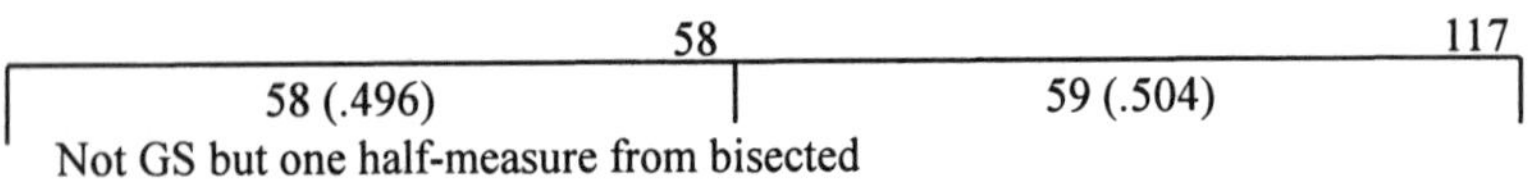

Finally, *Madrigal* "Alla riva del Tebro" is illustrated in figure 3.62.[44] This is two measures from GS^2.

Figure 3.62. Palestrina's *Madrigal* "Alla riva del Tebro"

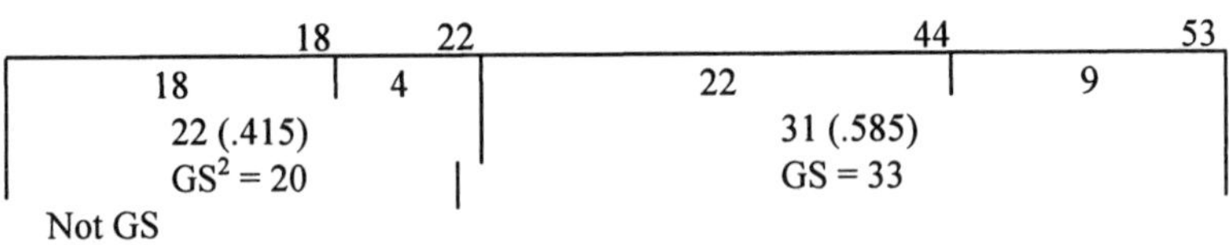

Table 3.3 summarizes the composers in this part of the chapter.

Table 3.3. Summary of the third part of the chapter

Composer	At GS or GS^2	One unit	Too far	Other
Binchois *De plus en plus*	.381			5GN or 4FN
Arnold de Lantins *Puisque*		.375		3:5, 4:4, & 5:3
Hugo de Lantins *Ce ieusse*		.375		3:5
Ockeghem *Ma maîtresse*		.375		3:5 & 17FN
Ma bouche rit	.379			
Obrecht *O vos omnes*	.621			
Tsaat Een meskin	.620			
Compère *Royne de ciel*			.370	
Paumann *Mit ganczem*	.391			Weak
Glogauer Liederbuch				21:34:13
Schlick *Maria zart*	.375			Also 3:5 and 5:3
Hofhaimer *Mein's traurens*	.372			Poor
di Lasso *Bon jour, mon*	.387			Also bisected
Hassler *Quia vidisti me*	.387			Weak. Also .621
Ach Schatz	.380			
Franck *So wünsch ich ihr*			.553	13:13:21
Gibbons *O Lord, increase*	.620			
This is the Record of John		.611		
Pavane Lord Salisbury	.618			FN 34:21
The Silver Swan		.631		
Fantasia	.619			2FN 13:8
des Prés *Ave Maria*		.627		
Marenzio *S'io parto*		.633		
De Sermisy *Au ioly boys*		.391		Rothwell, disputed
	.621			Madden, disputed
	.383			No repeat
Janequin *Guillot ung jour*		.606		
Qui veult d'amour		.622		Disputed
	.618			Madden
Maffoni *Quam pulchri*			.455	Unverified
Palestrina *Manus Tuae*	.395			Rog/Roth Disputed
			.482	Madden
Agnus Dei		.630		
Sicut Cervus			.496	Nearly bisected
Alla riva del Tebro			.415	

Conclusion

Apparently, many composers of the Renaissance had some connection with the golden proportion. It would seem to be a fertile field for research in the use of fib and phi by musicians. Rondeaux are a dependable source of these proportions, as are virelais. Less dependable are polyphonic settings and motets.

Powell and Sandresky hoped that Guillaume Dufay would be a fruitful source of phi, but their studies were flawed. Dufay was found to have used phi only rarely.

Similarly, Rogers and Rothwell hoped that John Dunstable would be a frequent user of phi. In this they were mistaken, but they found that he too used phi occasionally.

Disagreeing with Rogers and Rothwell about the counting of a rondeau by Binchois, we found that the rondeau was divided at GS and FN even after recounting it. Binchois wrote many rondeaux that have not been transcribed to modern notation, and we can be reasonably sure that they approach phi very closely.

Maffoni was studied hesitantly by Rothwell because the piece exhibited phi internally only when portions were excluded. This procedure is discouraged because it sheds more confusion than light onto the analysis.

All of the composers studied here used GS or FN with some accuracy and some frequency except Dufay and Dunstable, who used phi only rarely, and Compère and Maffoni who used it not at all.

Chapter Notes

[1] [San p. 292] A long is three measures.

[2] Is this error inadvertent, a difference in editions, or a fudge to fulfill a predetermined objective? This may seem trivial, but miscounting is a fault of many of the analyses discussed in this book.

[3] [San p. 296]

[4] [Pow p. 258–267]

[5] [Pow p. 260] Powell felt, as Joseph Schillinger did, that even this small application of Fibonacci numbers in rhythm is important.

[6] Similar geometric constructions are often found in literature about medieval architecture.

[7] [HAM1 p. 71]

[8] [Rog p. 40]

[9] [Dunst Vol. 8 p. 95]

[10] [HAM1 p. 66] [Dunst Vol. 8 p. 121]

[11] [Rog p.38] [Rothwell p. 57]

[12] [HAM1 p. 74]

[13] [HAM1 p. 75]

[14] [HAM1 p. 76] If the piece were described as alternating 66 and 63 units, as is possible, the proportions would be changed little.

[15] [HAM1 p. 78]

[16] [HAM1 p. 79] Most virelais prove to be Fibonaccian. They are built on a five-part form that can be described as AB BAA.

[17] [HAM1 p. 80] My ancient class notes say that this piece is by Loyset Compère, not Obrecht. I no longer know the source of that information or whether it is accurate.

[18] [HAM1 p. 82]

[19] [Rothwell p. 61]

[20] [HAM1 p. 83]

[21] [HAM1 p. 85]

[22] [HAM1 p. 87]

[23] [HAM1 p. 102]

[24] [HAM1 p. 96]

[25] [HAM1 p. 159]

[26] [HAM1 p. 186]

[27] [HAM1 p. 187]

[28] [HAM1 p. 190]

---

[29] [HAM1 p. 194]
[30] [HAM1 p. 195]
[31] [HAM1 p. 206]
[32] [Nort Std. Ed. p. 21]
[33] [Rothwell p. 36, 54]
[34] [Nort Std. Ed. p. 3]
[35] [Nort Std. Ed. p. 11]
[36] [Rothwell p. 59] [Sermisy]
[37] [Rothwell p. 64] [Janequin]
[38] [Rogers p.41] [Janequin]
[39] [Rothwell p. 66] I was unable to verify this music.
[40] [Rogers p. 43]
[41] [Pal Vol. V no. 2]
[42] [HAM1 p. 152]
[43] [HAM1 p. 153]
[44] [HAM1 p. 155]

Chapter 4
Baroque

Introduction

In chapters 2 and 3, phrase durations were used to mark structural points. Gregorian chant phrases were considered to end at breath marks and rondeau phrases at double bars. Though Renaissance polyphony was somewhat more complicated, cadences and their extensions were usually clear. Cadences became the de facto standard of structure.

On the other hand, in Baroque polyphony, cadences may be elided or extended, and phrases may overlap. The question then arises, Just where are the structural points?

This question has been asked and answered in different ways. Do highest and lowest notes, changes in texture or dynamics, entrances of fugal subjects, or other criteria suffice? How strongly must they be heard?

If cadences mark the ends of sections, should they be counted only up to—or all the way through—the note or chord that resolves the dominant? Should sections that are lengthened by extended cadences be counted to the end of the extension?

If subject entrances are used to determine structural points, how important are cadences? What if subject entrances and cadences do not coincide? Should cadences and subject entries be clearly audible, or is their appearance on the written score sufficient?

These questions have plagued this chapter, and in the hope of coming to the right conclusions, the following guidelines have been determined.

1. A judgment must be made as to whether themes or cadences dominate. If cadences are obscured by elisions and constant movement, thematic entries dominate. They are counted at the last full beat or measure prior to the entry.

2. If cadences are particularly strong and not mitigated by subject or theme entries, they dominate. The beat or measure of resolution should be counted as the structural point except in the case of clearly extended cadences, which are counted to the end of the extension and explained.

3. If cadence and entry are both strong, a choice has to be made, with an explanation of the alternatives.

4. Highest and lowest notes will be disregarded and textures will be used only as aids in finding cadences and entrances.

Although these principles cannot be used in every circumstance, they will be used often in fugues that have elisions. For example, *The Art of the Fugue* has had to be analyzed primarily on the basis of subject entries, whereas the *Two Part Inventions* were analyzed primarily on the basis of cadences.

These concepts have escaped several earlier commentators, resulting in confusing analyses. Corrections will be noted as appropriate.

As always, good judgment is needed, especially in difficult analyses. The rules may sometimes have to be broken, but overgeneralization and gross approximation should be avoided. Careful counting and notation is especially important.

Chapter 1 gave a standard of accuracy of one measure for phi in music. In some cases, this standard has to be changed to one beat. This has already been encountered with Gregorian chant, and will often be the case in the Baroque. Sometimes it will be necessary to say that something is "too far in beats, but close enough in measures." Even so, much of this music is close enough at one beat.

Johann Sebastian Bach

J. S. Bach provided a profound conclusion to the Baroque period. The following layout proceeds in approximately chronological order of studies referring to phi in Bach's music, beginning with Hugo Norden's reference to *The Art of the Fugue,* followed by three of the *Two Part Inventions* taken up by James Rothwell, and a Prelude from the *Well-Tempered Clavier* studied by Michael Richard Rogers. Some clear references to phi are found in these works. Because there are points of dispute, however, several additional pieces are studied from each collection.

The Art of the Fugue

Hugo Norden's article "Proportions in Music," presented an example from Bach's *The Art of the Fugue* that shows clear use of phi proportions. Results from his study of *Fuga I a 4 voci* (also known as *Contrapunctus I*) are shown in figure 4.1. Counting in measures, Norden relied upon subject entries to determine the structure. His result was 6FN 8:5. Rogers and Dorfman followed Norden in their discussions of this piece.[1]

Figure 4.1. Bach's *Fuga I a 4 voci* (*Contrapunctus I*) (Norden) (Measures)

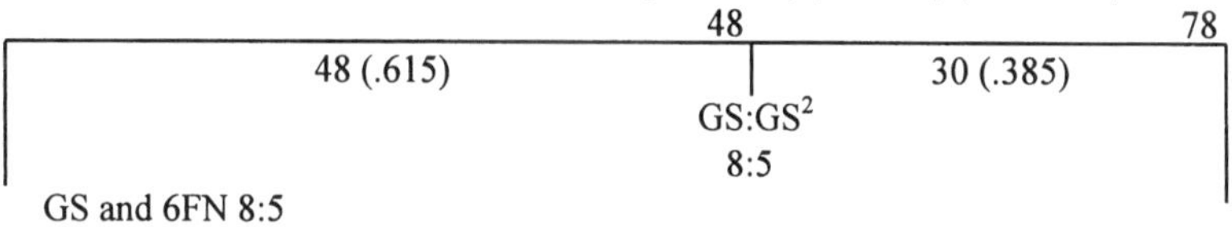

Rothwell showed an analysis of *Contrapunctus 2* (*II*). He did not state unequivocally whether the piece contains phi proportions, although he must have thought so, since it is in his paper. With Rothwell, it is often unclear as to whether measure numbers were meant to mark the duration of a section, or the beginning of the next. For example, did a structural point marked at measure 45 mean that the first section was 45 measures long, or that the new section began at measure 45?

Rothwell's analysis is shown with measure numbers corrected in figure 4.2a. It is roughly bisected. Our analysis, which consistently counted out to one less than his, is shown in figure 4.2b with more detail and counted in beats (half notes). Rothwell was correct, but his format was misleading and hard to read.[2]

Figure 4.2a. Bach's *Contrapunctus 2* (Rothwell) (Measures) (Corrected)

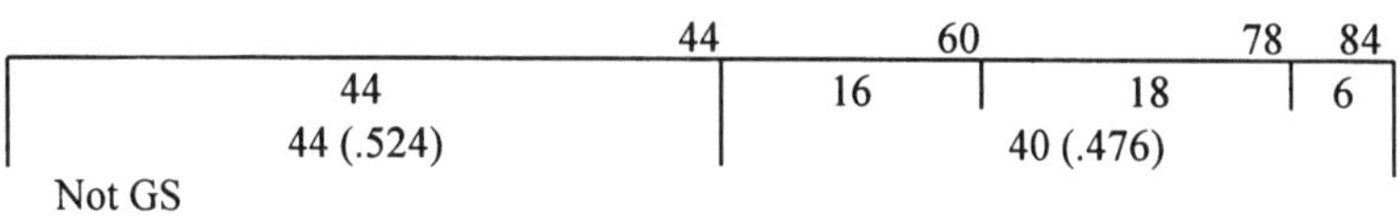

Figure 4.2b. Bach's *Contrapunctus II* (Madden) (Beats)

45 60 74 88 120 137 156 168
45 (.268) | 15 | 14 | 14 | 32 | 17 | 19 | 12
88 (.524) | 80 (.476)
Not GS

Contrapunctus III, in quarter notes, is seen in figure 4.3a–c. Ambiguous, it may be phi because of a weak subject entry near GS^2 and a weak cadence near GS, so we grudgingly say "close enough." A strong subject entry after beat 168, however, makes a ratio of 7:5, which is not phi.

Figure 4.3a. Bach's *Contrapunctus III* (Madden) (Quarter notes)

(Weak) (Strong)
73 89 113 168 182 200 248 288
73 | 16 | 24 | 55 | 14 | 18 | 48 | 40
113 (.392) | 175 (.608)
$GS^2 = 110$ | GS = 178
Three quarter notes from GS^2 (less than one measure)

Figure 4.3b. Bach's *Contrapunctus III* (Madden)

(Weak)
73 89 113 168 182 200 248 288
73 | 16 | 24 | 55 | 14 | 18 | 48 | 40
182 (.632) | 106 (.368)
GS = 178 | $GS^2 = 110$
Four quarter notes from GS (one measure)

Figure 4.3c. Bach's *Contrapunctus III* (Madden)

(Strong)
73 89 113 168 182 200 248 288
73 | 16 | 24 | 55 | 14 | 18 | 48 | 40
168 (.583) | 120 (.417)
7:5
Not GS

Contrapunctus IV, measured in half notes, is shown in figure 4.4. Also ambiguous, there is a cadence near GS^2 (half note 106), and a weak one near GS (174), but this is unsupported by subject entries. Therefore, the piece must be declared close to GS^2 based on cadence structure. The subject entries after 120 and 144 do not define the structure.

Figure 4.4. Bach's *Contrapunctus IV* (Madden) (Half notes)

52 106 120 144 174 212 256 276

52	54	14	24	30	38	44	20

106 (.384) | 170 (.616)

GS^2 = 105 | GS = 171

One from GS^2 (less than one measure) and percentage is very good

Contrapunctus V, measured in beats (quarter notes), is shown in figure 4.5. There are cadences at beats 210 and 258, which organist Glenn Gould emphasizes strongly.[3] These are not phi, so should they be emphasized? A weak subject entry after 224 qualifies this piece as weakly phi, correct to two beats, less than one measure. Strong entries are found after beats 64 and 128, which are not phi.

Figure 4.5. Bach's *Contrapunctus V* (Madden) (Beats)

(Weak)

64 128 210 224 258 278 304 340 360

64	64	82	14	34	20	26	36	20

224 (.622) | 136 (.378)

GS = 222 | GS^2 = 138

Two beats from GS, on a weak subject entry (less than one measure)

Contrapunctus VI, measured in beats (quarter notes), is shown in figure 4.6. This piece seems quite shapeless, but may be phi at one beat from GS^2, marked by a weak subject entry.

Figure 4.6. Bach's *Contrapunctus VI* (Madden) (Beats)

(Weak)

56 120 136 164 224 250 289 294 316

56	64	16	28	60	26	39	5	22

120 (.380) | 196 (.620)

GS^2 = 121 | GS = 195

One beat from GS^2 (less than one measure) and percentage is very good

Contrapunctus VII, measured in beats (quarter notes), is shown in figure 4.7. This piece contains another example of cadences and subject entries not agreeing. Beat 136 marks a strong entry, while beat 141 has a strong cadence extended to 149. With a very audible subject entry at 149, it must be judged close to GS.

Figure 4.7. Bach's *Contrapunctus VII* (Madden) (Beats)

50 88 136 149 (Strong) 196 244

50 | 38 | 48 | 13 | 47 | 48

149 (.611) | 95 (.389)

GS = 151 | $GS^2 = 93$

Two beats from GS (less than one measure)

Contrapunctus VIII, measured in quarter notes, is shown in figure 4.8. This piece is not near phi, but is nearly bisected at a strong subject entry.

Figure 4.8. Bach's *Contrapunctus VIII* (Madden) (Quarter notes)

138 158 194 242 268 366 373 (Strong) 494 586 678 752

138 | 20 | 36 | 48 | 26 | 98 | 7 | 121 | 92 | 92 | 74

373 (.496) | 379 (.504)

Three quarter notes from bisected

Figure 4.9. Bach's *Contrapunctus IX* (Madden) (Beats)

136 232 288 352 392 472 520

136 (.262) | 96 | 56 | 64 | 40 | 80 | 48

232 (.446) | 288 (.554)

288 (.554) | 232 (.446)

352 (.677) | 168 (.323)

Not GS

Contrapunctus X, is shown in figure 4.10. This piece is accurate to one beat from GS, but the proportion is almost exact. It is tempting to round down the GS value 296.64 instead of up, so as to fudge it.

Figure 4.10. Bach's *Contrapunctus X* (Madden) (Beats)

88 172 204 260 296 336 409 456 480

88 | 84 | 32 | 56 | 36 | 40 | 73 | 47 | 24

296 (.617) | 184 (.383)

GS = 297 | $GS^2 = 183$

One beat from GS (less than one measure) and percentage is excellent

Table 4.1 summarizes the findings for the ten fugues of this study. Four fugues lie within three points of GS and another three are within one measure. If Bach used phi consciously, he did not do so consistently, but it is unlikely that results this close could have occurred by accident. Meanwhile, Bach clearly did not use fib to any extent here.

Table 4.1. Summary of the study of Bach's *The Art of the Fugue*

Fugue	At GS or GS^2	One unit	One meas.	Too far	FN	Other
I	.615				6FN 8:5	Norden
II				.524		
III			.392			Close enough?*
			.632			Close enough?*
				.583		7:5
IV		.384				Close enough
V			.622			Close enough
VI		.380				Close enough
VII			.611			Weak
VIII				.496		Nearly bisected
IX				.446		Not close
X		.617				Close enough

* Percentages are distant

Two-Part Inventions

Rothwell analyzed numbers 3, 8, and 10 of Bach's *Two Part Inventions*. Since his numbers always need review, all fifteen of these pieces are analyzed below, using the Hans Bischoff edition.[4]

Analyzing the *Inventions* in the same manner as *The Art of the Fugue* feels wrong because the cadences are dominant. Therefore, the analyses usually are based on cadences, with comments to compare the use of motive entries.[5] *Invention I* is shown in figure 4.11, counted in beats. This piece is within one measure of GS. Analyzing by motives would bring it to within two beats of GS at 56.

Figure 4.11. Bach's *Invention I* (Madden) (Beats)

25 | 57 | 88

25 (.284) | 32 (.364) | 31 (.352)

57 (.648)

GS = 54 | GS^2 = 34

Three beats from GS (less than one measure)

Invention II is more complicated than the first piece. Its three prominent motives, a, b, and c, are shown in figure 4.12. Counted in beats, it is divided at phi if the cadence at beat 41 is considered structural, where motive *a* returns. To the ear, this seems a little weak. Stronger cadences are found at 25, 49, 73, and 89, which have nothing to do with phi. Analyzing by motives would move it one beat away from GS^2 to 40.

Figure 4.12. Bach's *Invention II* (Madden) (Beats)

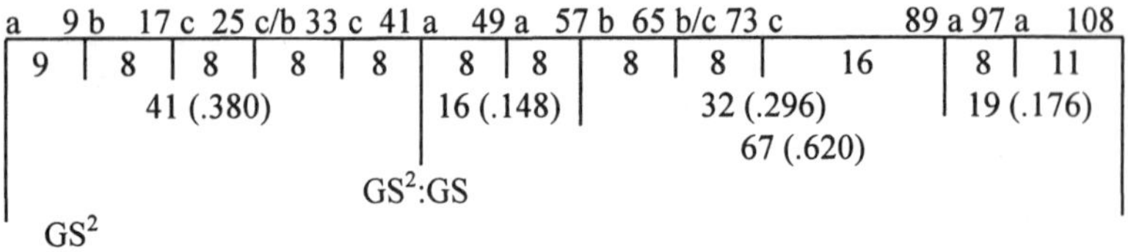

Rothwell discussed *Invention III* and found that phi proportions divide the subsections.[6] This is illustrated in figure 4.13a. The high B at measure 16 does divide the first 42 measures into Fibonacci ratio 8:13 (.381), but the claim that the 17 measures that complete the piece are also divided into phi proportion by a repeat of this B is not supported by the actual ratio 10:7 (.588) or the harmonic structure.

Figure 4.13a. Bach's *Invention 3* (Rothwell) (Measures)

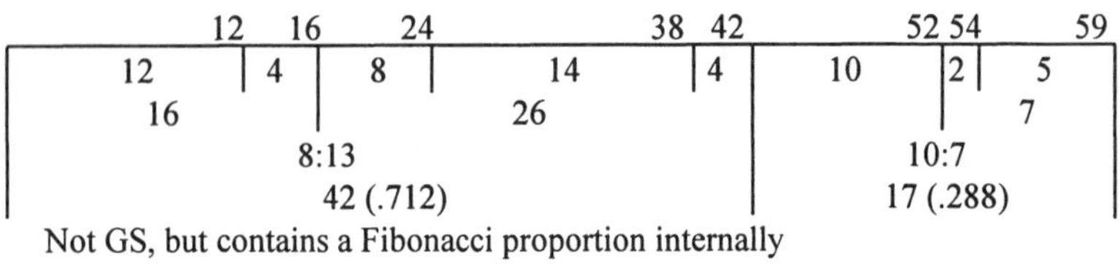

Our interpretation gives no importance to the high Bs and counts the piece in beats. It is three beats away from GS and GS^2, which is less than one measure. Keys are given at the cadences, as shown in figure 4.13b.

Invention IV is illustrated in figure 4.14. It is exactly trisected by a cadence at beat 52 and nearly so by a motive entry after 51.

Figure 4.13b. Bach's *Invention III* (Madden) (Beats)

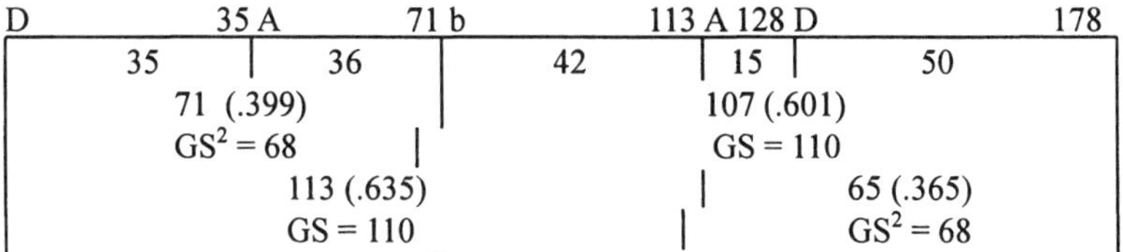

Three beats from GS and GS^2 (one measure) and .001 from 2:3

Figure 4.14. Bach's *Invention IV* (Madden) (Beats)

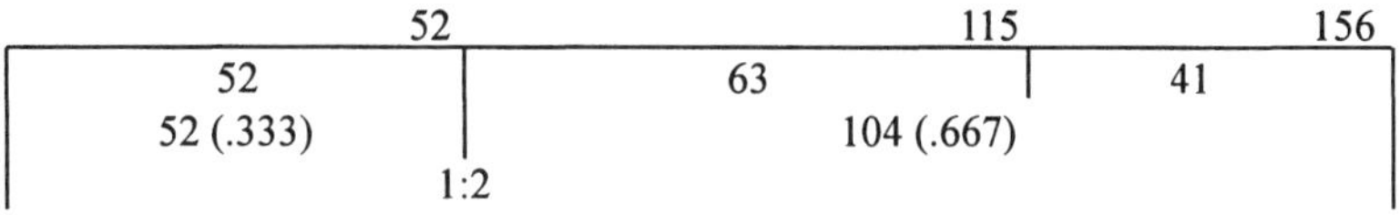

Trisected

Invention V is counted in beats. See figure 4.15. There is a strong motive entry after beat 44, which is five beats from GS^2. Weak entries after 32 and 76 do not seem complete enough to be structural, which is unfortunate because they are only two and three beats away from GS^3 and GS. The last 31 beats are only one away from GS^3. Based on cadences, this piece is one measure from GS^2, with a strong cadence at 45.

Figure 4.15. Bach's *Invention V* (Madden) (Beats)

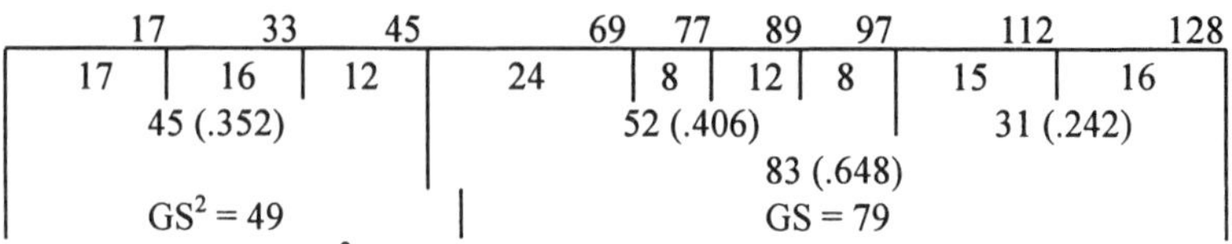

Four beats from GS^2 (one measure)

Invention VI comprises two large sections repeated verbatim with motives entering cleanly after full cadences and double bars. Illustrated in figure 4.16, it is tripartite AABA'BA' with approximately equal sections.

Figure 4.16. Bach's *Invention VI* (Madden) (Measures)

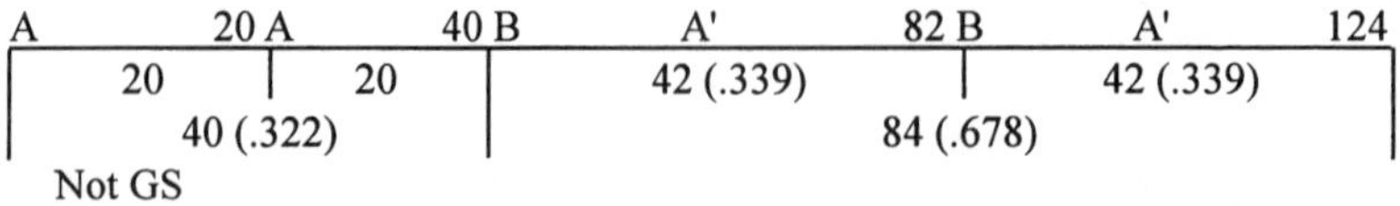

Invention VII follows in figure 4.17, counted in beats. It is less than one measure from GS. The percentages, however, are very poor. In this piece, there are no real subjects, and cadences dominate.

Figure 4.17. Bach's *Invention VII* (Madden) (Beats)

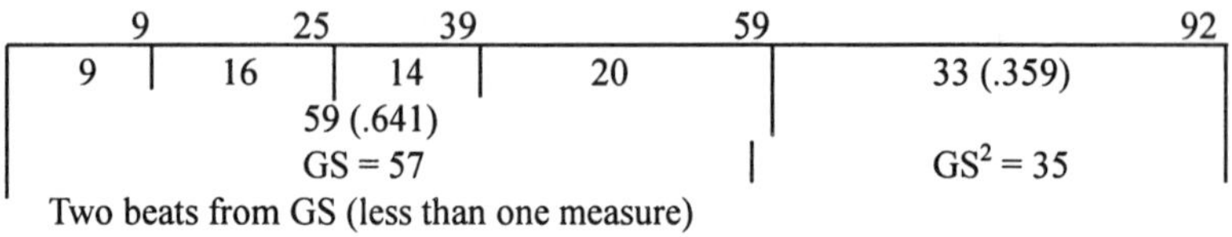

Rothwell's analysis of *Invention VIII* follows in figure 4.18a.[7] Several of his statements must be disputed. One of these is that the cadence in measure 12 "conforms closely with a 5/8 golden ratio." This should be 5:8 = 5/13. But where did this number come from, since 12/34 = .353 does not reduce to 5/13 = .385?

Figure 4.18a. Bach's *Invention 8* (Rothwell) (Measures)

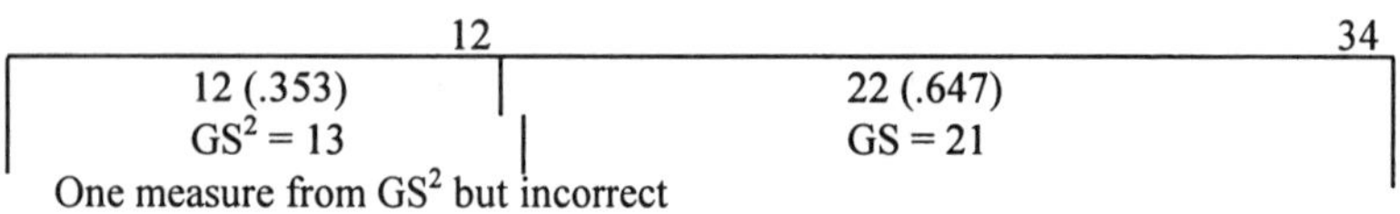

Another problem statement is, "Assuming the actual duration were equivalent to 33 measures [deleting the opening downbeat and the final two beats of measure 34], the smaller section in 5/8 [*sic*] golden ratio would be 12 measures, to the nearest whole number." But 33 × .382 = 12.606 (13), and the actual duration is 34 full measures, whose GS^2 is also 13.

Our interpretation, counted in beats, follows in figure 4.18b. This finds the piece to be exactly trisected based on a cadence at beat 34. Basing the structure on motive entries would move it to 33:69 = .324.

Figure 4.18b. Bach's *Invention VIII* (Madden) (Beats)

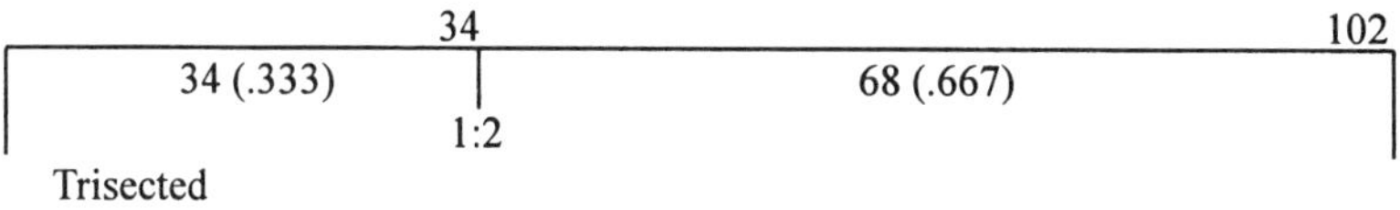

Invention IX is illustrated in figure 4.19. Cadences appear at beats 25 and 49, while motives are after 24 and 48.

Figure 4.19. Bach's *Invention IX* (Madden) (Beats)

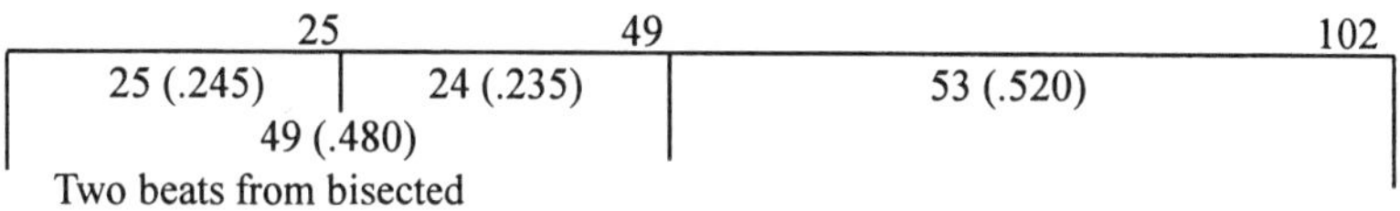

Rothwell's placement of the golden ratio at measure 20 in *Invention X*, shown in figure 4.20a, makes it seem that the piece is in phi proportion.[8] However, the trill at measure 20 marks only a pedal point, with at best a fleeting cadence.

Figure 4.20a. Bach's *Invention 10* (Rothwell) (Measures)

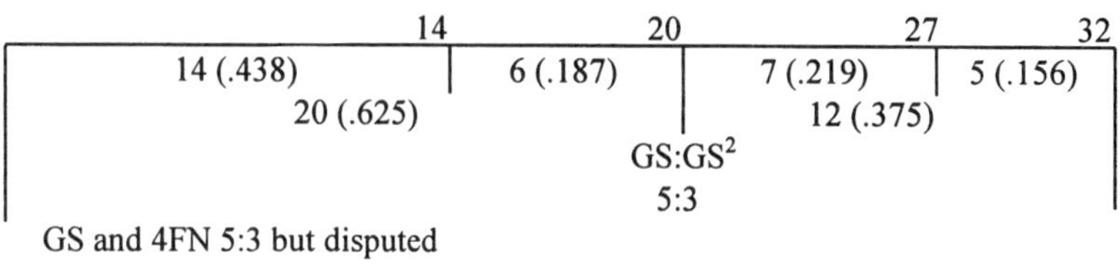

Our analysis places the structural points at cadences on beats 40 and 79, as shown in figure 4.20b.

Figure 4.20b. Bach's *Invention X* (Madden) (Beats)

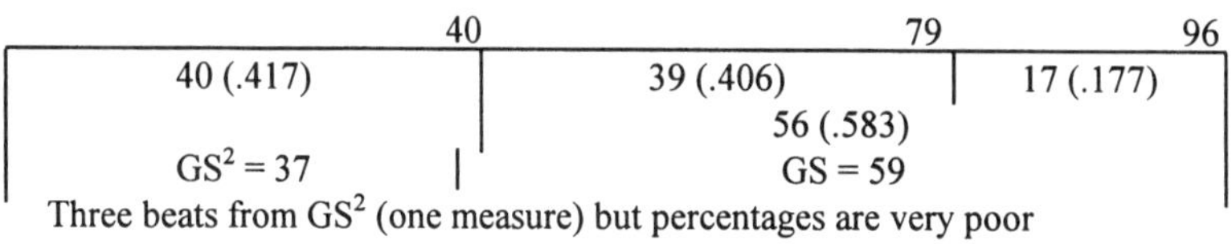

Invention XI has cadences at 41 and 63. It is not phi. See figure 4.21.

Figure 4.21. Bach's *Invention XI* (Madden) (Beats)

41	63	92
41 (.446)	22 (.239)	29 (.315)

Not GS, even though the second section is GS^3

Invention XII is shown in figure 4.22. Using cadences to determine structure, it is nearly bisected. Using a motive entry after 40 would not change the structure appreciably.

Figure 4.22. Bach's *Invention XII* (Madden) (Beats)

25	41	63	84
25	16	22	21
41 (.489)		43 (.511)	

One beat from bisected

Invention XIII is shown in figure 4.23. Based on cadences, it is three beats from bisected. Based on a motive entry after 52, it would be two beats from bisected.

Figure 4.23. Bach's *Invention XIII* (Madden) (Beats)

23	53	69	100
23 (.230)	30 (.300)	16 (.160)	31 (.310)
53 (.530)		47 (.470)	

Three beats from bisected

Invention XIV is tricky because of numerous elided cadences. Use of cadences as structural points is shown in figure 4.24a and motive entries in figure 4.24b.

Figure 4.24a. Bach's *Invention XIV* (Madden) (Beats)

20	33	45	53	67	80
20	13	12	8	14	13
33 (.413)		47 (.587)			
$GS^2 = 31$		GS = 49			

Two beats from GS^2 (less than one measure)

There are numerous references to GS in the sub-sections and the major structural points fall one beat from phi and 16FN 2:3.

Figure 4.24b. Bach's *Invention XIV* (Madden) (Beats)

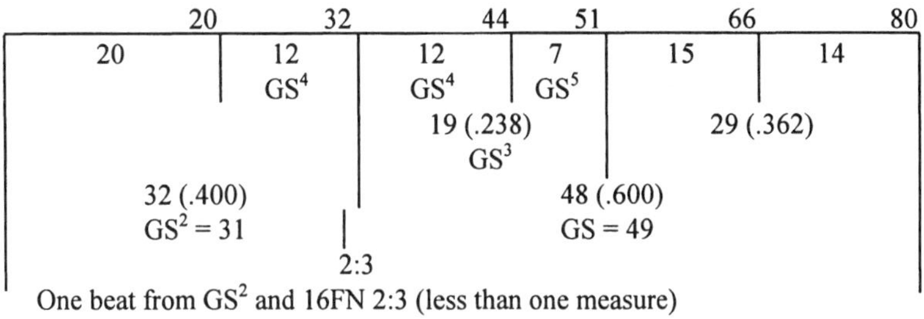

Invention XV is shown in figure 4.25. It is bisected. The first half is 44 beats long on the basis of motive entries, or 45 beats on the basis of cadences. Motive entries have been chosen as structural for the illustration, except for the cadence at measure 29, which has no motive entry.

Figure 4.25. Bach's *Invention XV* (Madden) (Beats)

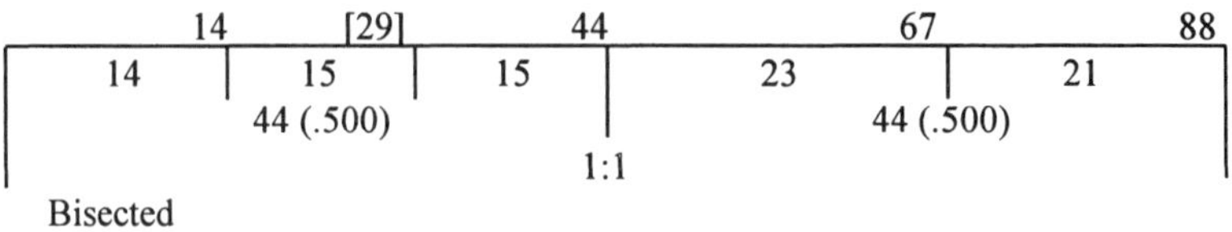

Table 4.2 summarizes the study of the *Two Part Inventions*.

Exact phi proportions are poorly represented in the *Inventions*, since only one of them is indisputably phi in terms of beats. Although seven are close enough in terms of measures, the percentages are poor, ranging from .017/.618 = 2.8% (III) to .035/.382 = 9.2% error (X). Judging from the low proportion of structures occurring exactly at important ratios, Bach was probably only using intuition to structure these works.

Table 4.2. Summary of the Bach *Inventions*

Invention	At GS or GS^2	One meas.	Too far	Bi- or Tri-sected	FN	Other
I		.648				
II	.380					Weak
III					2FN 8:13	Rothwell*
		.399			.001 fm 2:3	Madden
IV				.333		
V		.352				
VI			.322			
VII		.641				
VIII		.353				Rothwell*
				.333		Madden
IX			.480			
X	.625				4FN 5:3	Rothwell*
		.417				Madden (Far)
XI			.446			Elem. of GS
XII			.489			
XIII			.530			
XIV		.413				Far
		.400			16FN 2:3	Elem. of GS
XV				.500		

*Doubtful

The Well-Tempered Clavier

Several pieces in *The Well-Tempered Clavier* have been analyzed before now with respect to the use of fib or phi. Since some of the analyses are disputed, others of the collection will be analyzed.

Preludes

Rogers analyzed *Prelude no. 1* from Book I of *The Well-Tempered Clavier* and found phi somewhere in measure 22 or 23, where a sort of secondary dominant occurs, as shown in figure 4.26a.[9] As noted before, this is too vague for this book. It would be more accurate to say that the structure is one measure from GS and to make a reference to the Fibonacci proportion between the first and second sections.

Figure 4.26a. Bach's *Prelude 1, WTC I* (Rogers) (Measures)

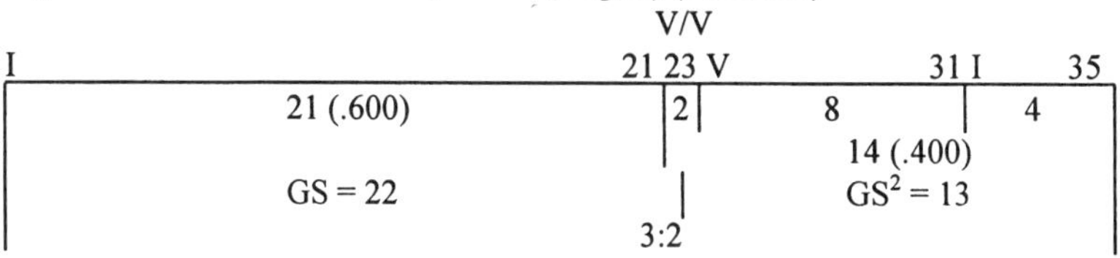

One measure from GS, and 7FN 3:2, but vague

Nevertheless, our analysis, as shown in figure 4.26b, has an even worse percentage and is off one measure to the other side of GS.[10]

Figure 4.26b. Bach's *Prelude 1, WTC I* (Madden) (Measures)

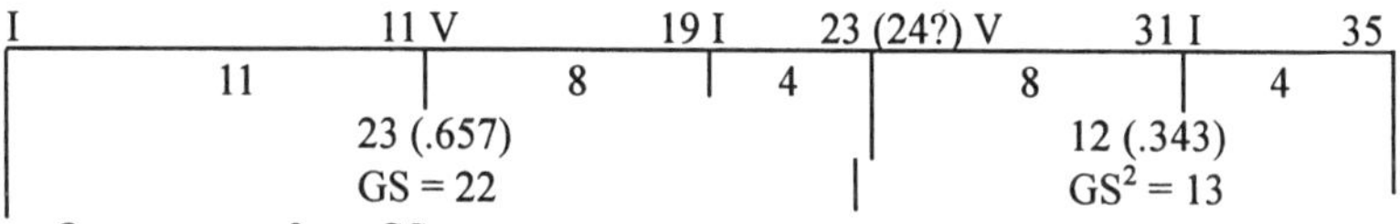

One measure from GS

Webster was of the opinion that *Prelude No. 9* from Book I represented phi exactly, as shown in figure 4.27a.[11]

Figure 4.27a. Bach's *Prelude 9, WTC I* (Webster) (Measures)

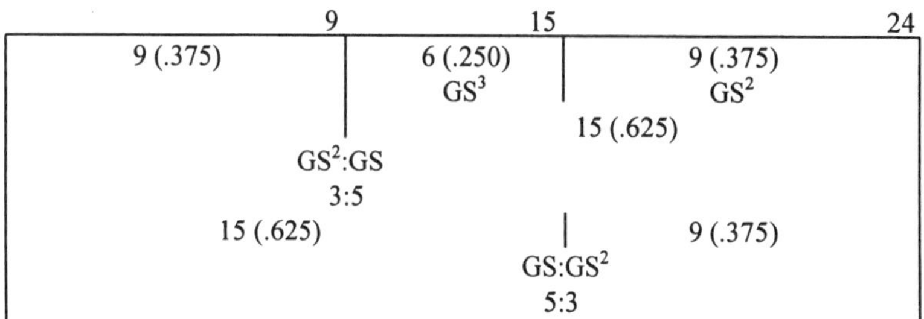

GS and FN, but incorrect

If Webster's numbers were correct, it would indeed be phi. However, he must have miscounted. Our analysis shows it to be trisected, as shown in figure 4.27b.

Figure 4.27b. Bach's *Prelude 9, WTC I* (Madden) (Measures)

8 14 24

8 (.333) | 6 (.250) | 10 (.417)

1:2

Trisected

Prelude 13 from Book I is shown in figure 4.28. It is GS^2 and GS based on cadences. On the basis of themes it would be off by one half-measure.

Figure 4.28. Bach's *Prelude 13, WTC I* (Madden) (Half-measures)

23 37 60

23 (.383) | 14 (.234) GS^3 | 23 (.383) GS^2

37 (.617)

GS^2:GS

37 (.617) | 23 (.383)

GS:GS^2

GS and GS^2 and symmetrical

Prelude 14, Book I is bisected, based on cadences. See figure 4.29.

Figure 4.29. Bach's *Prelude 14, WTC I* (Madden) (Half-measures)

24 48

24 (.500) | 24 (.500)

1:1

Bisected

Prelude 17 from Book I is shown in figure 4.30. Accurate to GS and GS^2, measure 17 is based on a theme entry and measure 27 on a cadence.

Figure 4.30. Bach's *Prelude 17, WTC I* (Madden) (Measures)

17 27 44

17 (.386) | 10 (.228) | 17 (.386)

27 (.614)

GS^2:GS

27 (.614) | 17 (.386)

GS:GS^2

GS and GS^2 and symmetrical

Prelude 21 from Book I is shown in figure 4.31. Based on cadences, it is bisected and accurate to GS and 5FN 5:3.

Figure 4.31. Bach's *Prelude 21, WTC I* (Madden) (Half-measures)

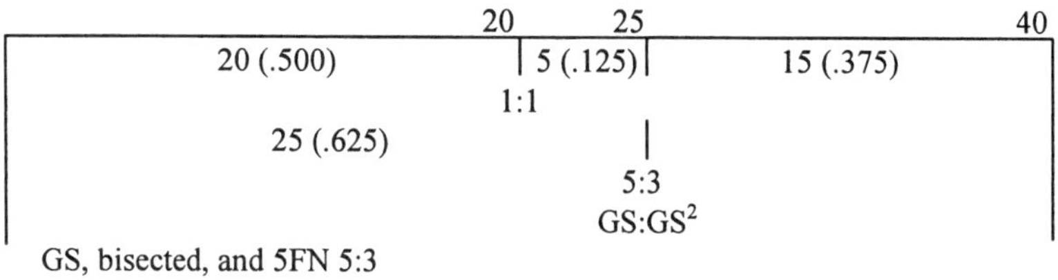

Prelude 3 from Book II is exactly bisected. See figure 4.32. Not everyone will like this interpretation because of the tempo change at bar 25.

Figure 4.32. Bach's *Prelude 3, WTC II* (Madden) (Measures)

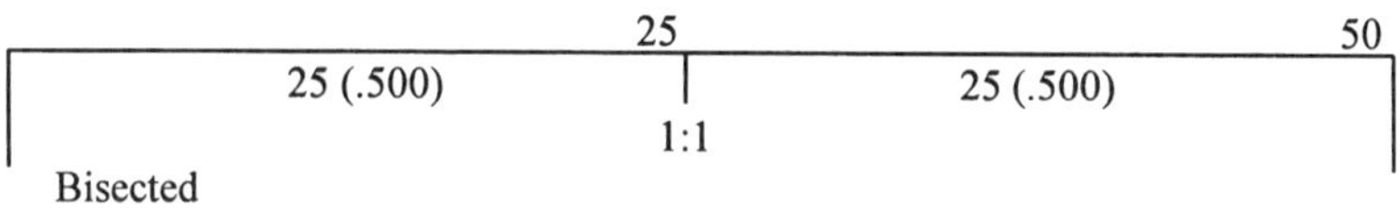

Prelude 5 from Book II is shown in figure 4.33. It is an awkward FN 2:5.

Figure 4.33. Bach's *Prelude 5, WTC II* (Madden) (Measures)

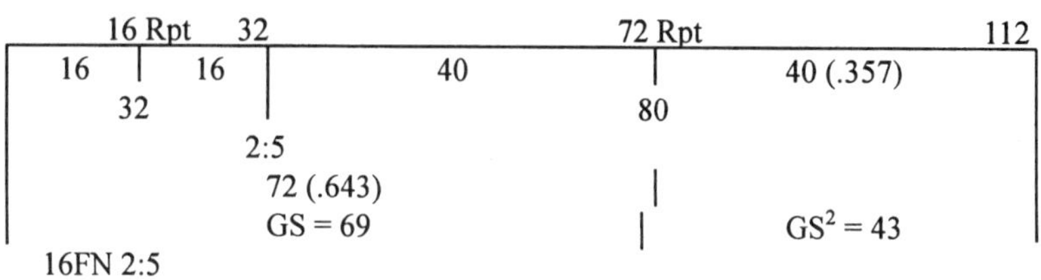

Webster claimed that *Prelude 6* from Book II is bisected.[12] Our analysis showed that it is only close. See figure 4.34.

Figure 4.34. Bach's *Prelude 6, WTC II* (Madden) (Measures)

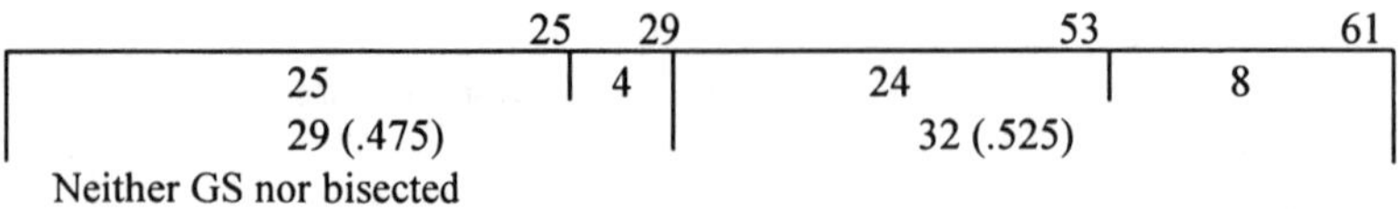

Prelude 12 from Book II is a clear example of FN 2:3. See figure 4.35.

Figure 4.35. Bach's *Prelude 12, WTC II* (Madden) (Measures)

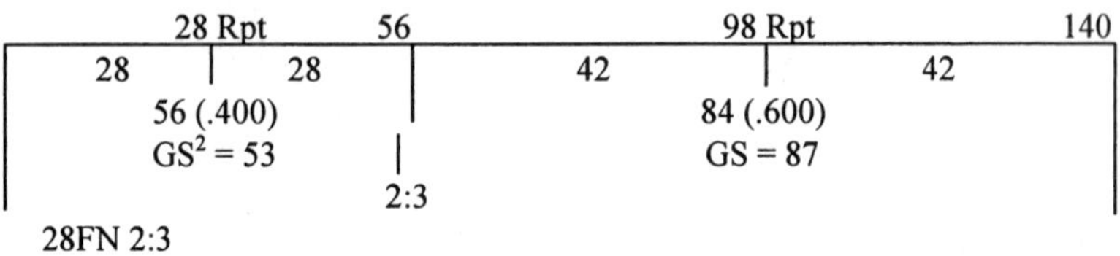

Webster's analysis of *Prelude 13* from Book II is shown in figure 4.36.[13] It is one measure from GS and accurate to 15FN 3:2.

Figure 4.36. Bach's *Prelude 13, WTC II* (Webster) (Measures)

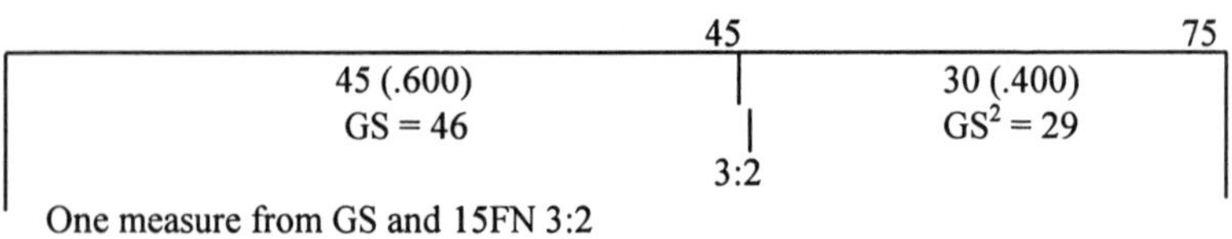

Prelude 15 from Book II is clearly in FN 1:2, as shown in figure 4.37.

Figure 4.37. Bach's *Prelude 15, WTC II* (Madden) (Measures)

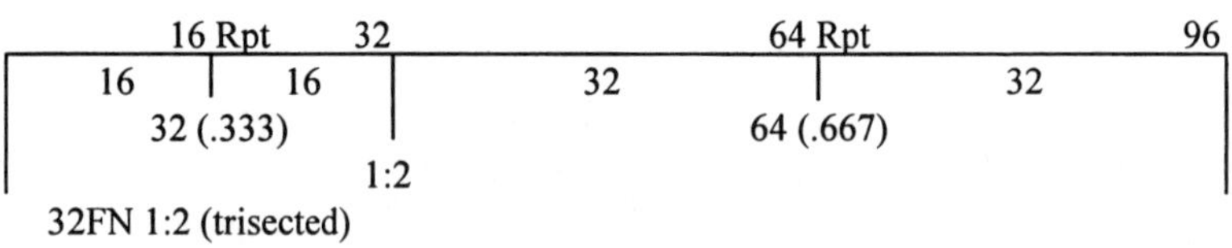

Prelude 16 from Book II is shown in figure 4.38. In half-measures, it is bisected on the basis of cadences.

Figure 4.38. Bach's *Prelude 16, WTC II* (Madden) (Half-measures)

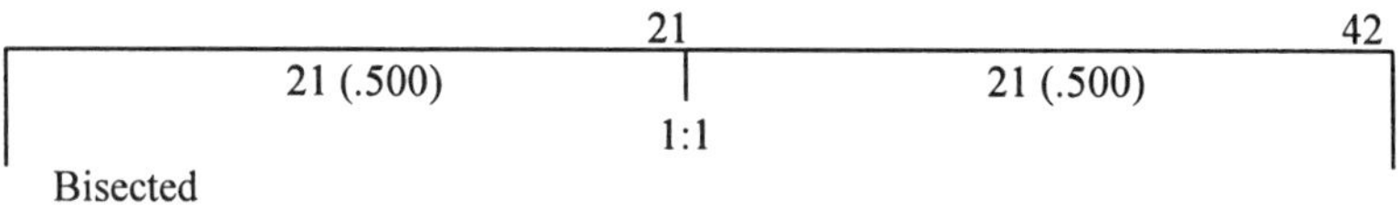

Prelude 19 from Book II, is shown in figure 4.39. On the basis of cadences, it is accurate to GS, but has a poor percentage.

Figure 4.39. Bach's *Prelude 19, WTC II* (Madden) (Measures)

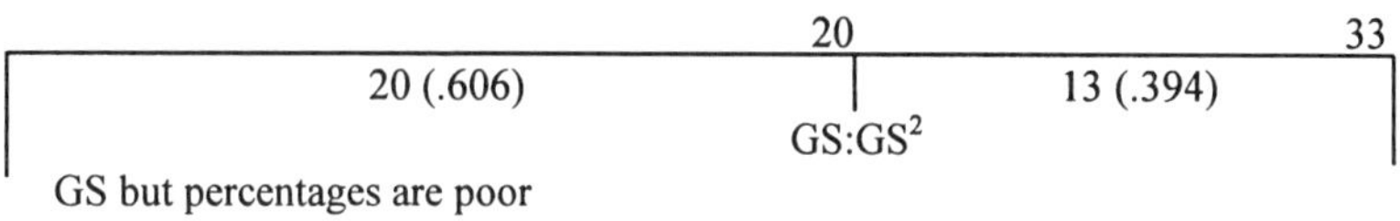

Prelude 20 is shown in figure 4.40. It is binary, with each section repeated and exactly bisected.

Figure 4.40. Bach's *Prelude 20, WTC II* (Madden) (Measures)

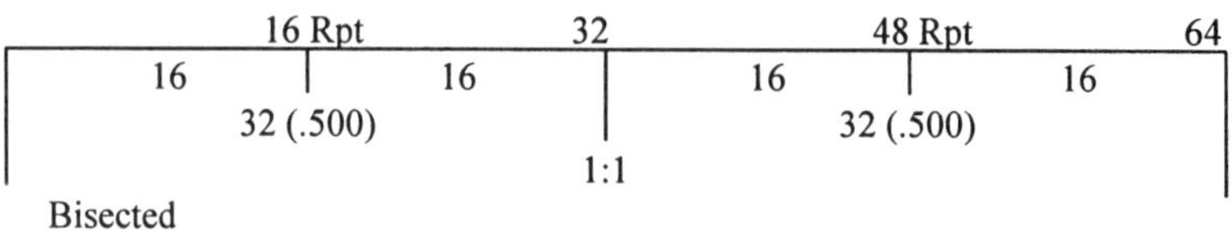

Prelude 21 from Book II is shown in figure 4.41a. Divided into two sections, each of which is repeated, it is accurate to GS based on cadences. An alternative interpretation is shown in figure 4.41b.

Figure 4.41a. Bach's *Prelude 21, WTC II* (Madden) (Measures)

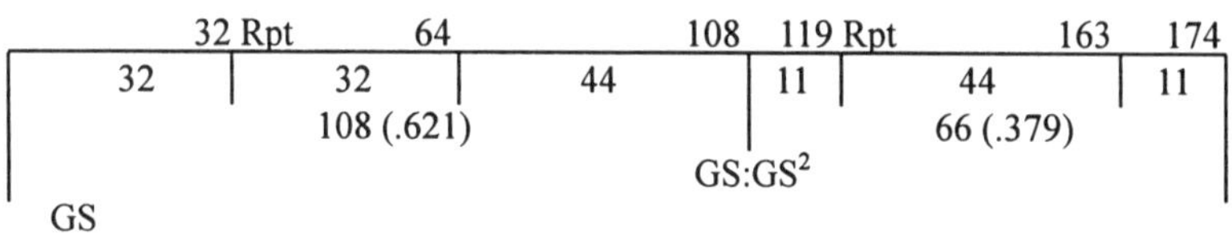

Figure 4.41b. Bach's *Prelude 21, WTC II* (Madden) (Measures)

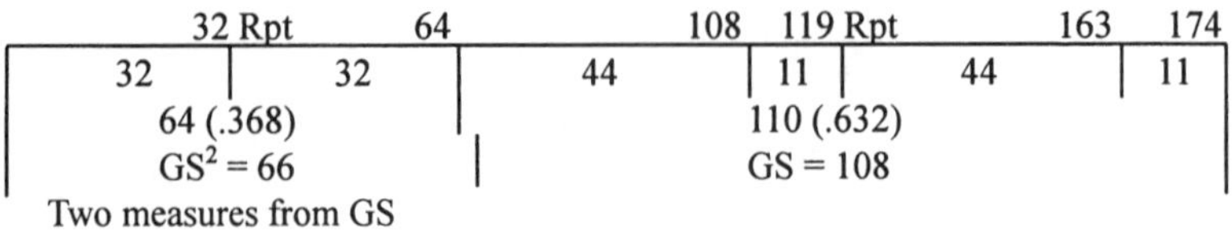

Fugues

Fugue 1 from Book I is bisected, based on cadences. See figure 4.42.

Figure 4.42. Bach's *Fugue 1, WTC I* (Madden) (Half-measures)

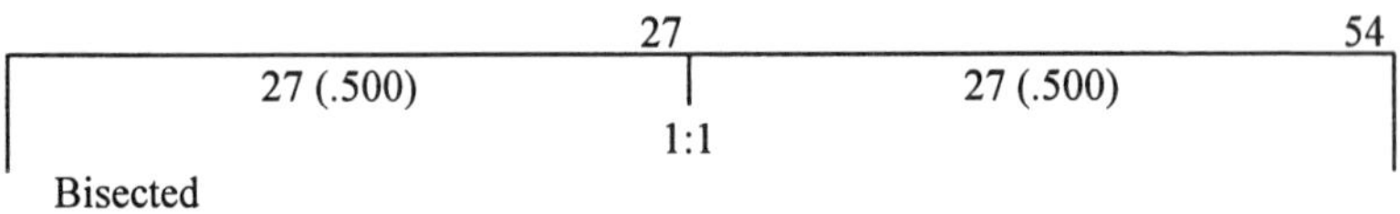

Fugue 2 from Book I is shown in figure 4.43. Based on cadences, it is accurate to GS, or on motive entries one beat away.

Figure 4.43. Bach's *Fugue 2, WTC I* (Madden) (Quarter notes)

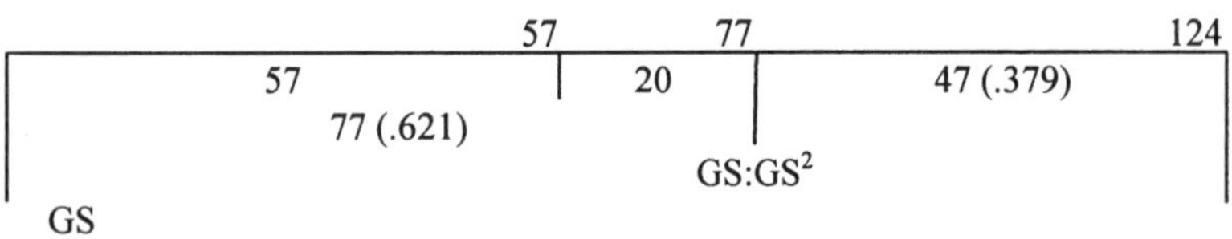

Fugue 3 from Book I is shown in figure 4.44. On the basis of cadences, the piece is divided exactly at GS and 2FN 34:21.

Figure 4.44. Bach's *Fugue 3, WTC I* (Madden) (Half-measures)

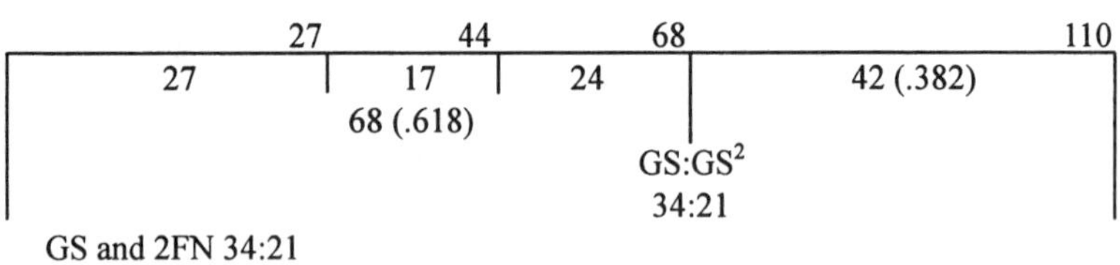

Fugue 8, Book I, is bisected based on subject entries. See figure 4.45.

Figure 4.45. Bach's *Fugue 8, WTC I* (Madden) (Half-measures)

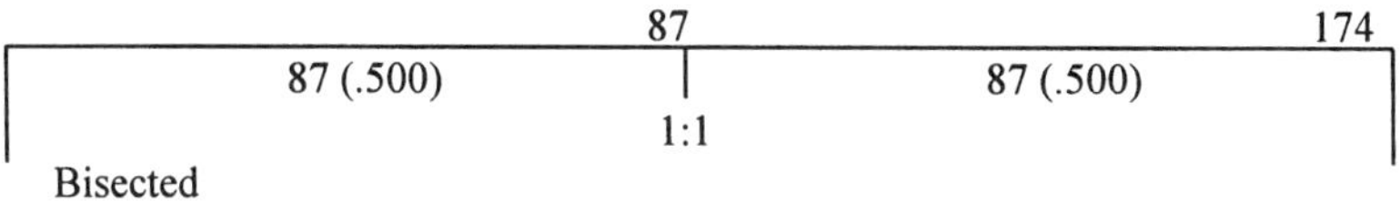

Fugue 9 from Book I is shown in figure 4.46. On the basis of subject entries, it is accurate to GS.

Figure 4.46. Bach's *Fugue 9, WTC I* (Madden) (Half-measures)

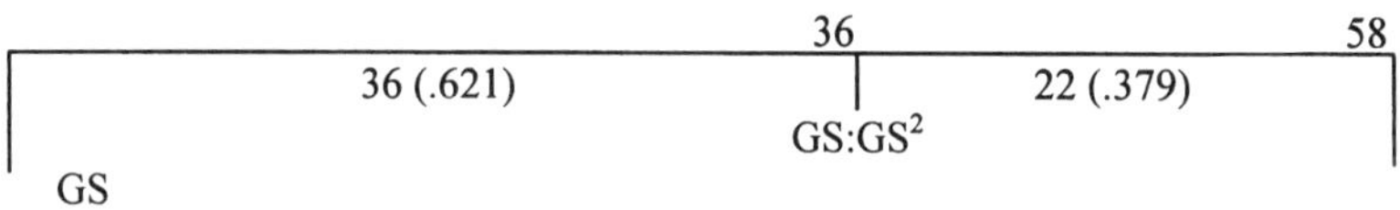

Fugue 10 from Book I is shown in figure 4.47. The bisecting entry is in the middle of an expository section that begins at 19.

Figure 4.47. Bach's *Fugue 10, WTC I* (Madden) (Measures)

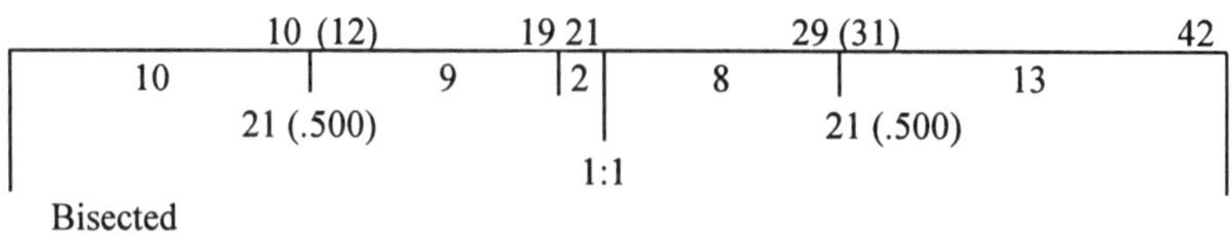

Fugue 11 from Book I is shown in figure 4.48. On the basis of both cadences and subject entries, this piece is bisected.

Figure 4.48. Bach's *Fugue 11, WTC I* (Madden) (Measures)

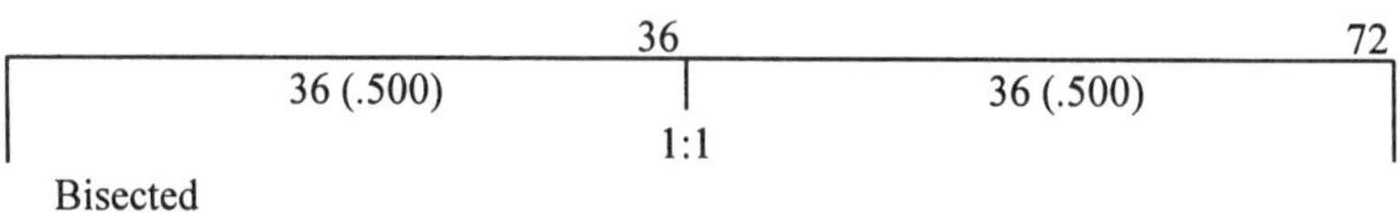

Fugue 17 from Book I is GS based on subject entries. See figure 4.49.

Figure 4.49. Bach's *Fugue 17, WTC I* (Madden) (Measures)

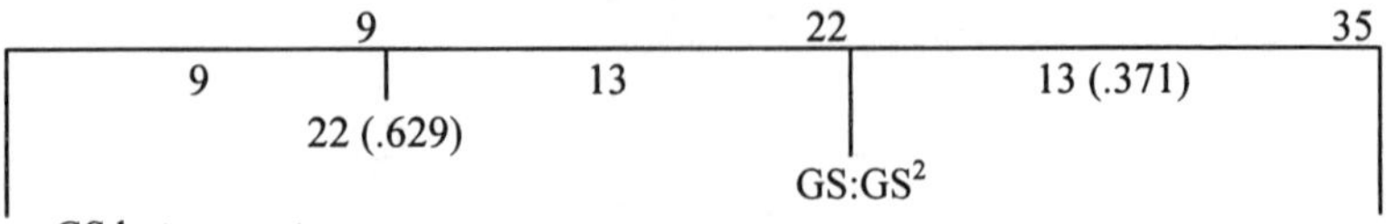

GS but percentages are poor

Fugue 8 from Book II is shown in figure 4.50. On the basis of subject entries, it is bisected.

Figure 4.50. Bach's *Fugue 8, WTC II* (Madden) (Measures)

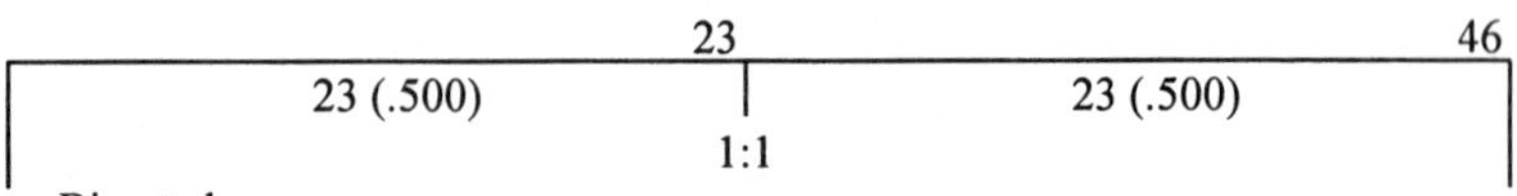

Bisected

Fugue 13 from Book II is shown figure 4.51. Based on cadences, it is accurate to both GS and GS^2.

Figure 4.51. Bach's *Fugue 13, WTC II* (Madden) (Beats)

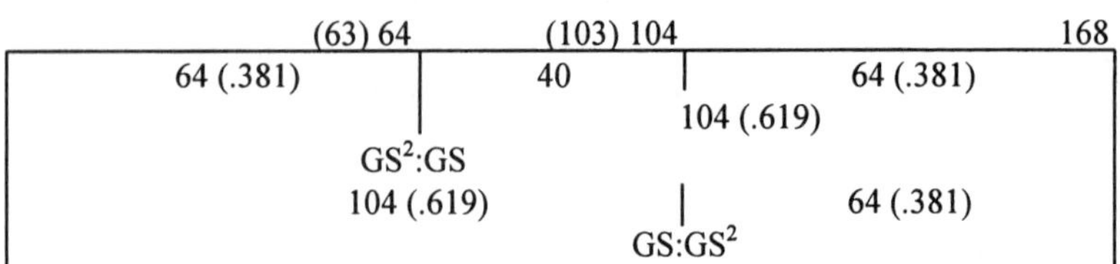

GS and GS^2 and symmetrical

Fugue 16 from book II is shown in figure 4.52. Judged to be divided at GS^2, it is tricky due to the entry coming in the middle of an exposition.

Figure 4.52. Bach's *Fugue 16, WTC II* (Madden) (Measures)

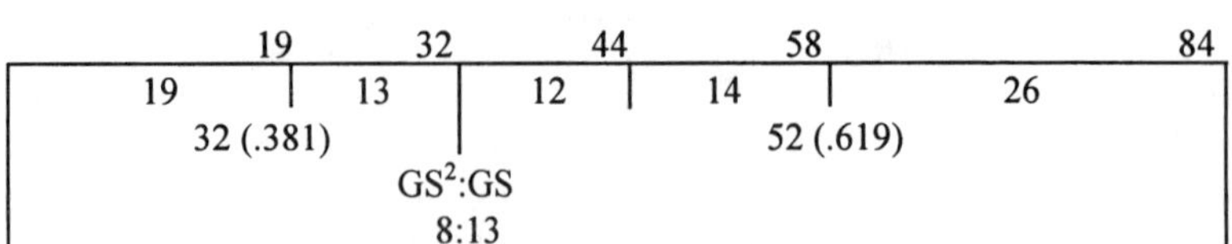

GS^2 and 4FN 8:13

Fugue 17 from Book II is shown in figure 4.53. Based on subject entries, it is accurate to GS; based on cadences, it is divided at 3:2 by a cadence at half-note 60.

Figure 4.53. Bach's *Fugue 17, WTC Book II* (Madden) (Half-notes)

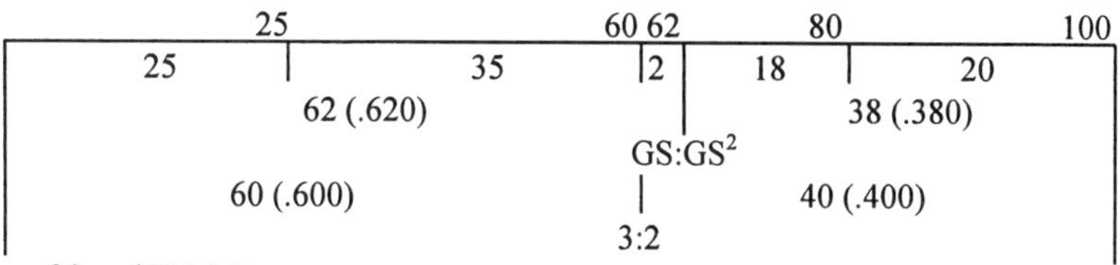

GS and FN 3:2

Fugue 18 from Book II is one measure from GS^2 based on subject entries, but it is exactly GS^2 based on cadences. See figure 4.54.

Figure 4.54. Bach's *Fugue 18, WTC II* (Madden) (Measures)

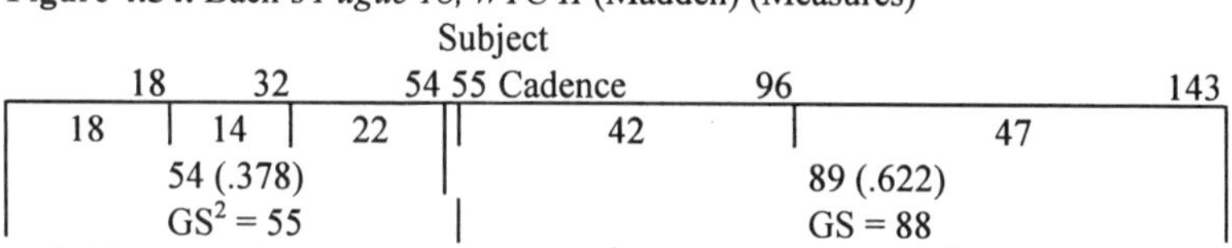

Subject entry is one measure from GS^2, but cadence is at GS^2

Table 4.4 shows the proportions for the Preludes and Fugues.

Table 4.4. Summary of the Bach *WTC*

Prelude/ Book	At GS or GS^2	One meas.	Bi- or trisected	FN	Other
1/I		.600		7FN 3:2	Rogers*
		.657			Madden
9/I	.375/.625			3FN 3:2:3	Webster*
			.333		Madden
13/I	.383/.617				
14/I			.500		
17/I	.386/.614				
21/I	.625		.500	5FN 5:3	Poor
3/II			.500		
5/II				16FN 2:5	

Prelude/ Book	At GS or GS^2	One meas.	Bi- or trisected	FN	Other
6/II					.475
12/II				28FN 2:3	.400
13/II		.600		15FN 3:2	Webster
15/II			.333	32FN 1:2	
16/II			.500		
19/II	.606				Poor
20/II			.500		
21/II	.621				
					.368
Fugue/ Book					
1/I			.500		
2/I	.621				
3/I	.618			2FN 34:21	
8/I			.500		
9/I	.621				
10/I			.500		
11/I			.500		
17/I	.629				Poor
8/II			.500		
13/II	.381/.619				
16/II	.381			4FN 8:13	
17/II	.620			20FN 3:2	
18/II		.378			Subject
	.385				Cadence

*Doubtful

Most of these items have been selected for their accuracy, while leaving out an equal number which are close or debatable. This sample shows that 14 of the 96 pieces in the collection may divide at GS (although two of these are disputed), while 10 are exactly bisected. These data, and the ratios shown in table 4.3 above, show that Bach was very careful about his proportions. Accuracy at this level could not have been accidental.

Other Music of Bach

Rogers mentioned John O'Donnell's article "And yet they are not Three Fugues; but One Fugue (Another look at Bach's Fugue in E♭)." Also called the *St. Anne Fugue*, it is *BWV 552* from the *Clavierübung.*[14]

O'Donnell gave a stimulating discussion of several relationships used by Bach to represent the numerology of his name, the divinity of God, and the structural points marking these relationships. O'Donnell's structure is shown in figure 4.55a and an alternative interpretation in 4.55b.

Figure 4.55a. Bach's *St. Anne Fugue BWV552* (O'Donnell)

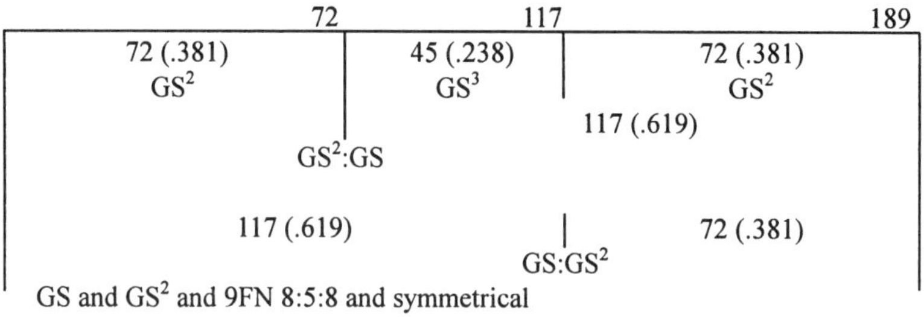

The beauty of this is the precision of the relationships. It gives GS accurate to 1 point (.16%), clear Fibonacci relations, and the number 45, which is the number of measures in the first movement of the *Credo* of the *B Minor Mass* and an important number for Bach. The correctness of O'Donnell's theory has been the subject of much debate.

The trouble with it is that the structural points are determined by the meter changes, which occur simultaneously with the cadences, eliding them. Then, do the sections end at 72 and 117, where the meters change, or at 73 and 118, where the cadences occur? The second interpretation would force the structure to be slightly off.

Figure 4.55b. Bach's *St. Anne Fugue BWV552* (Madden)

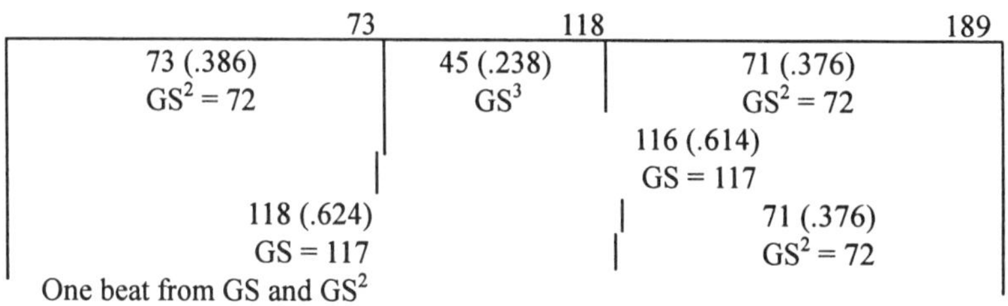

There is a certain beauty in fudging, but it can also be a source of consternation. Is it justified? What was Bach's intention? How should other works of his be interpreted? It is a problem of judgment.

Rothwell discussed Bach's organ work *Kleines harmonishes Labyrinth.* It is shown in figure 4.56a.[15]

Figure 4.56a. Bach's *Kleines harmonisches Labyrinth* (Rothwell/Madden) (Measures)

Introitus 20 Centrum 32 Exitus 52

20 (.385) 12 (.230) 20 (.385)

GS^2 GS^3 GS^2

32 (.615)

GS^2:GS

5:8

32 (.615) 20 (.385)

GS:GS^2

8:5

GS and GS^2 and 4FN 5:3:5 and symmetrical

This analysis consists of three pieces with durations of 20, 12, and 20 measures marked by double bars. It is divided in Fibonacci proportions 4FN 5:3:5 and GS and GS^2. Although the overall assessment is valid, whether the individual pieces are divided by Fibonacci numbers is less certain. Counting in beats makes a difference, as shown in figure 4.56b.

Figure 4.56b. Bach's *Kleines harmonisches Labyrinth* (Rothwell/Madden) (Beats)

Introitus *Centrum* *Exitus*

24 51 65 80 105 128 158 176 208

24 | 27 | 14 | 15 | 25 (24?) | 23 (24?) | 30 | 18 | 32

51 (50?) | 29 (30?) 50

51:29 (near 5:3) 25:23 (near 1:1) 30:50 (3:5)

48 | 32

(3:2)

80 (.385) 48 (.231) 80 (.385)

$GS^3 = 49$ $GS^2 = 79$

128 (.615)

$GS^2 = 79$ GS = 129

128 (.615) 80 (.385)

GS = 129 $GS^2 = 79$

105 (.505) | 103 (.495)

One beat from GS and GS^2, 16FN 5:3:5, and one beat from bisected

Here, the *Exitus* is clearly 3:5 and 3:2, as Rothwell claimed, but the best that can be said for the *Introitus* and the *Centrum* is that they are *near* 5:3 and 1:1, respectively. Based on cadences and melodies, there is little justification for fudging, and doing so would require fudging the *Exitus* as well, throwing off its perfect proportions.

Rothwell's analysis of the chorale prelude *Wachet auf: ruft uns die Stimme* is considerably more difficult to accept. Counted in measures, his analysis is shown in figure 4.57a.[16]

Figure 4.57a. Bach's *Wachet auf: ruft uns die Stimme* (Rothwell) (Measures)

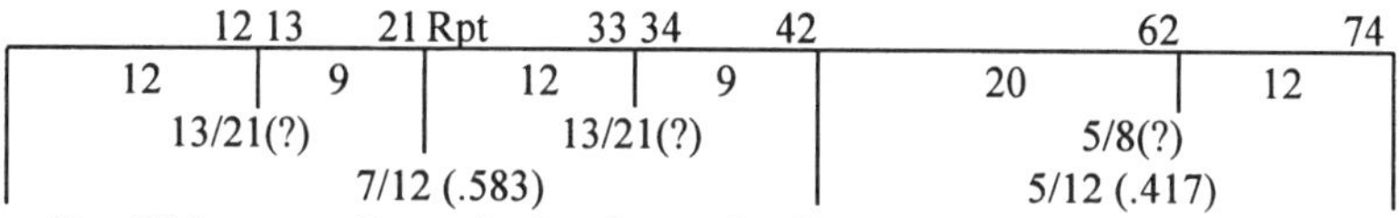

Not GS but according to Rothwell contains fib

Rothwell claimed that the first section is divided into 13:8 (13/21), but he should have said 12:9 (12/21). He also stated that "the large form of the work divides very closely into 7/12 ratio." This is 42/72 = .583, where he should have used 42/74 = .568. It is corrected in figure 4.57b.

Figure 4.57b. Bach's *Wachet auf: ruft uns die Stimme* (Madden) (Measures)

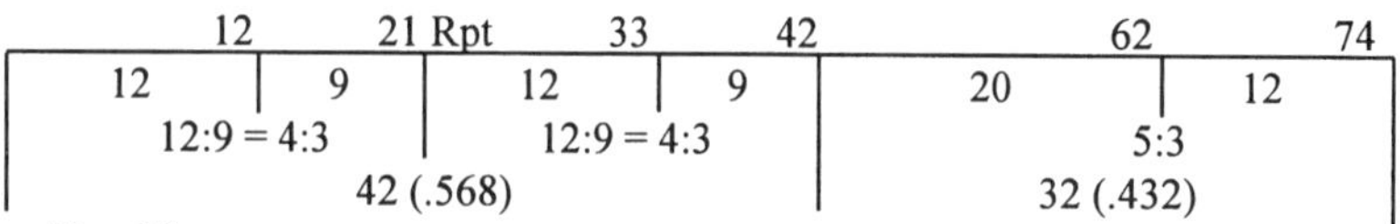

Not GS

Dorfman did not dispute Rothwell and seemed to support his inaccuracies, saying that differences of up to 4.2% are "so slight as to have practically no meaning with regard to musical form."[17] We oppose that attitude.

Rogers mentioned the analysis by Robert Cogan and Pozzi Escot of Bach's *Chaconne for Violin Unaccompanied* from *Partita no. 2* in their book *Sonic Design.* Although they mentioned phi only in passing, their analysis is worth comment.[18] An illustration follows in figure 4.58.

Figure 4.58. Bach's *Chaconne for Violin Unaccompanied* (Cogan and Escot) (Phrases)

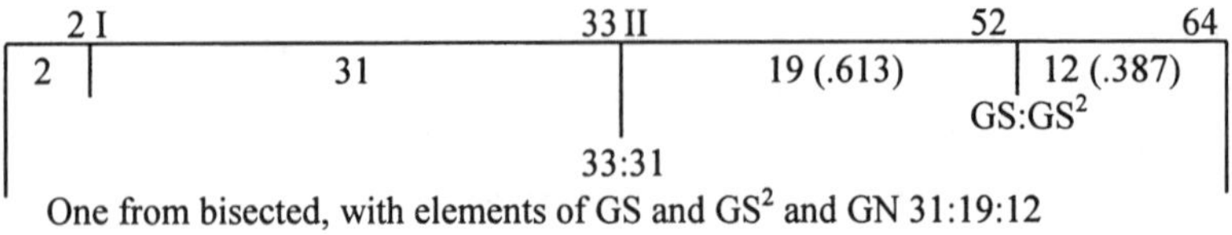

Cogan and Escot stated that "In section II, the smaller part . . . relates to the larger part . . . in the ratio of 12:19; similarly, the larger part relates to the whole section in the ratio of 19:31. Such a proportional relationship has been known historically as the 'Golden Section.'" In other words, 12:19::19:31. This is not strictly correct and could easily be misunderstood. If taken to be 12/19 = 19/31, it would not be correct because (12/19 = .632) ≠ (19/31 = .613), but if restated as 12/31 = .387 and 19/31 = .613, as in the illustration, the sections *are* phi related because GS and GS^2 are 19 and 12. This is a clear example of the need to avoid equating the two ratios unless they are both precisely phi.

Douglass M. Green's explanation of the *Passacaglia for organ* was incorrect also.[19] He stated, "The grouping of these variations manifests . . . the symmetry of the 'golden mean,' [i.e.,] 3 is to 5 as 5 is to 8 [and] 8 is to 13 as 13 is to 21," or 3:5::5:8 and 8:13::13:21. This is incorrect because (3/8 = .375) ≠ (5/13 = .385) and (8/21 = .381) ≠ (13/34 = .382), although the latter is admittedly close. Restating this as in figure 4.59, we see that overall it *is* GS since GS is 13 and GS^2 is 8.

Figure 4.59. Bach's *Passacaglia* (Green) (Sections)

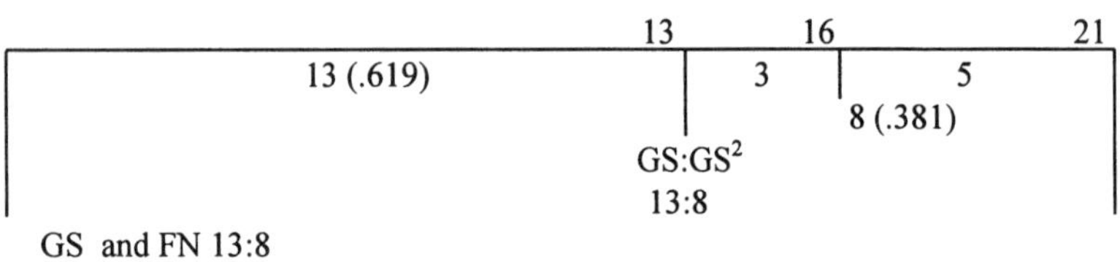

Pascoe's approach to fib and phi in music involves a sequence of proportions running from G1 to G14, with G7 and G8 equivalent to GS^2 and GS as used in this book.[20] With the values shown in table 4.5, subdivisions of GS structure can be identified in the music. Thus, for example, module G14 occurs .979 of the way along, while G8 occurs at .618.

Table 4.5. Pascoe's modules and their proportions

Module	GS Equivalent	Proportion
G1	GS^8	.021
G2	GS^7	.034
G3	GS^6	.056
G4	GS^5	.090
G5	GS^4	.146
G6	GS^3	.236
G7	$GS^2 (1-GS)$	.382
G8	$GS (1-GS^2)$	.618
G9	$1-GS^3$	.764
G10	$1-GS^4$	.854
G11	$1-GS^5$	.910
G12	$1-GS^6$	.944
G13	$1-GS^7$	.966
G14	$1-GS^8$	.979

This seems like an excellent approach, except that Pascoe allowed a 3% range or "area of influence," a tolerance too loose for this book.[21] In addition, he did not clearly identify these points in his detailed measure-by-measure analyses. It is difficult to tell where GS actually was meant to be in an analysis. Worse, our analyses of all four Bach movements analyzed by Pascoe show no connection with fib or phi, as we will see in the next four illustrations.

Pascoe's analysis of the first movement of the *Violin Concerto no. 2* is shown accurate to GS in figure 4.60a, but disputed. It is not phi because nothing structural happens at the points indicated. Our analysis is shown in figure 4.60b.[22]

Figure 4.60a. Bach's *Violin Concerto no. 2* Mvt I (Pascoe) (Measures)

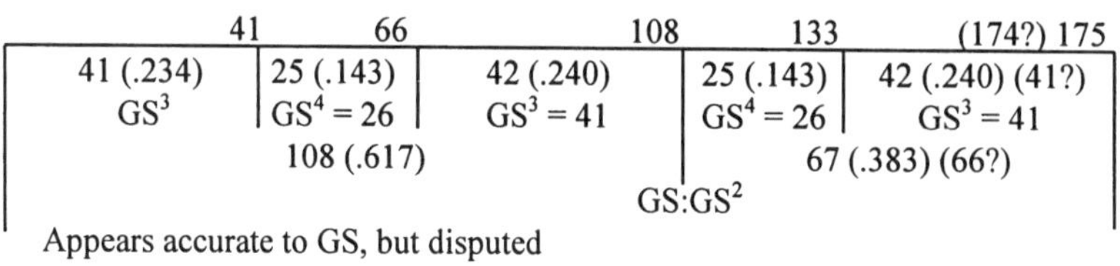

Figure 4.60b. Bach's *Violin Concerto no. 2* Mvt I (Madden) (Measures)

52 122 174

52 (.299) | 70 (.402) | 52 (.299)

122 (.701)

Not GS but symmetrical

The third movement, a rondo, is shown in figure 4.61a. Pascoe marked it as accurate to GS. Rather, it is bisected. See figure 4.61b.[23]

Figure 4.61a. Bach's *Violin Concerto no. 2* Mvt III (Pascoe) (Measures)

37 61 98 122 160

37 (.231) $GS^3 = 38$ | 24 (.150) $GS^4 = 23$ | 37 (.231) $GS^3 = 38$ | 24 (.150) $GS^4 = 23$ | 38 (.238) GS^3

61 (.381) | 99 (.619)

GS^2:GS

Appears to be GS^2, but disputed

Figure 4.61b. Bach's *Violin Concerto no. 2* Mvt III (Madden) (Measures)

A 16 B 32 A 48 B 64 A 80 C 96 A 112 D 144 A 160

16 | 16 | 16 | 16 | 16 | 16 | 16 | 32 | 16

80 (.500) | 80 (.500)

1:1

64 (.400) | 96 (.600)

96 (.600) | 64 (.400)

GS = 99 $GS^2 = 61$

Bisected and FN 2:3 and FN 3:2

Pascoe's analysis of the first movement of the *English Suite no. 3* is illustrated in figure 4.62a.[24] Although he acknowledged that the movement as a whole is not divided at phi, he claimed that Part B is. This would be true only to very loose tolerances, as can be seen in the illustration.

Figure 4.62a. Bach's *English Suite no. 3* "Prelude" (Pascoe) (Measures)

A 33 67 B 99 125 161 213

33 | 34 | 32 | 26 | 36 | 52

67 | 58 (.397) GS^2 of 146 = 56 | 88 (.603) GS of 146 = 90

No elements of GS

Dorfman showed measure 108 as a return to part A, dividing the piece nearly in half, as shown in figure 4.62b.[25]

Figure 4.62b. Bach's *English Suite no. 3* "Prelude" (Dorfman) (Measures)

A 33 B 67 A′ 99 B 108A 125 B′ 161 C 180 A 213

33 | 34 | 32 | 9 | 17 | 36 | 19 | 33

108 (.507) | 105 (.493)

Nearly 1:1

Nearly bisected

Dorfman was unimpressed with Pascoe's module analysis, saying that the events identified as appearing at G modules "are . . . rather unspectacular when viewed relative to the whole of the prelude." Although we agree with Dorfman's opinion, we disagree with both analyses. Ours is based on strong cadences. It omits Dorfman's division at 108 because there is no cadence there. Our analysis is shown in figure 4.62c.[26] Of course, it is not phi.

Figure 4.62c. Bach's *English Suite no. 3* "Prelude" (Madden) (Measures)

67 99 125 213

67 (.315) | 32 | 26 | 88

146 (.685)

99 (.465) | 114 (.535)

125 (. 587) | 88 (.413)

Not GS

The "Gigue" is shown in figure 4.63a. It is not phi. The two sections were analyzed on different bases and are misleading. Our analysis appears in figure 4.63b.[27] It, too, is not phi.

Figure 4.63a. Bach's *English Suite no. 3* "Gigue" (Pascoe) (Measures)

A 13 20 B 29 35 44

13 | 7 | 9 | 6 | 9

GS of 20 = 12 | GS^2 of 20 = 8 | 15

20 (.455) | $GS:GS^2$

Elements of GS?

Figure 4.63b. Bach's *English Suite No. 3* "Gigue" (Madden) (Measures)

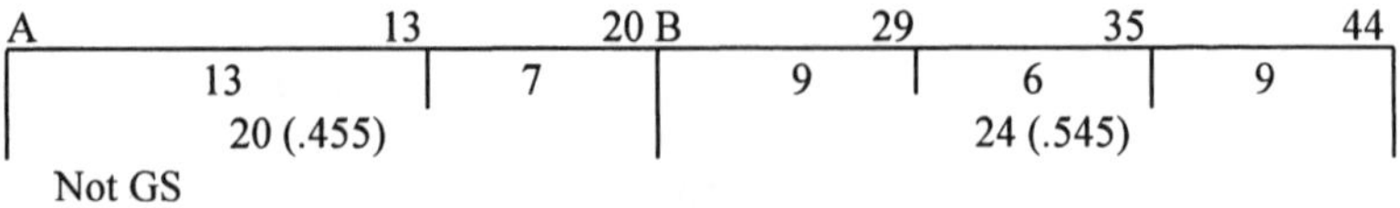

Other Music of the Baroque

Pascoe, Rothwell, and Rogers discussed music of Handel, Scarlatti, and Monteverdi with regard to the use of fib and phi in music. After looking at these, we will draw a few works from anthologies. Usage of phi by composers other than Bach is surprisingly uncommon.

Handel

Rothwell studied the "Hallelujah Chorus" from George Frideric Handel's *Messiah* claiming to have found phi.[28] See figure 4.64a.

Figure 4.64a. Handel's *Messsiah* "Hallelujah Chorus" (Rothwell) (Measures)

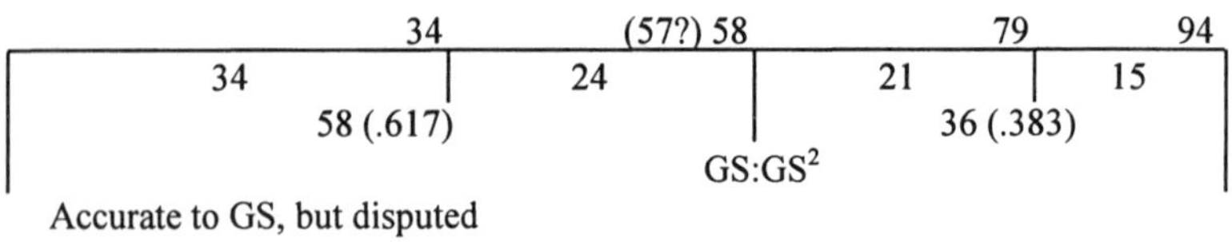

Rothwell's positioning of the structural point is spread over measures 57 and 58, which gives too much leeway. Placing it after 57 gives one measure from GS, while 58 would give GS, but this is not a structural point. Further subdivision at 79 is too weak to be worth consideration, regardless of the prominence of the trumpet solo claimed by Rothwell.

Rothwell's fractions are also wrongly rounded and distort the relationships. The correct fraction 57/94 is 12 points, or 1.9% from GS. The best that can be said is that this piece is divided at one measure from GS, as shown in figure 4.64b.[29]

Figure 4.64b. Handel's *Messiah* "Hallelujah Chorus" (Madden) (Measures)

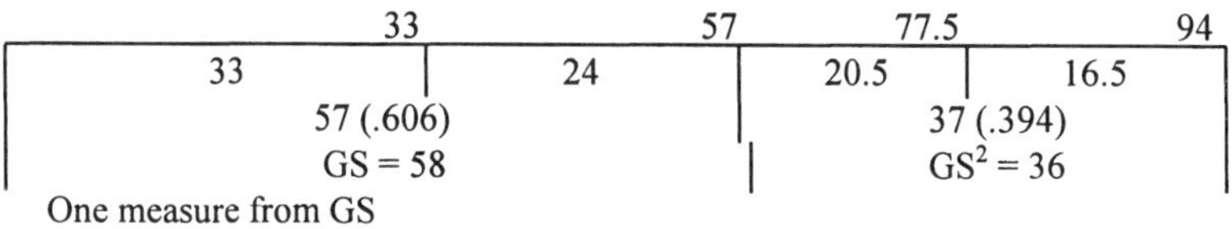

Pascoe discussed the second movement of the *Oboe Concerto op. 3 no. 3* of Handel. The main points are illustrated in figures 4.65a.

Figure 4.65a. Handel's *Oboe Concerto op. 3 no. 3* Mvt II (Pascoe) (Measures)

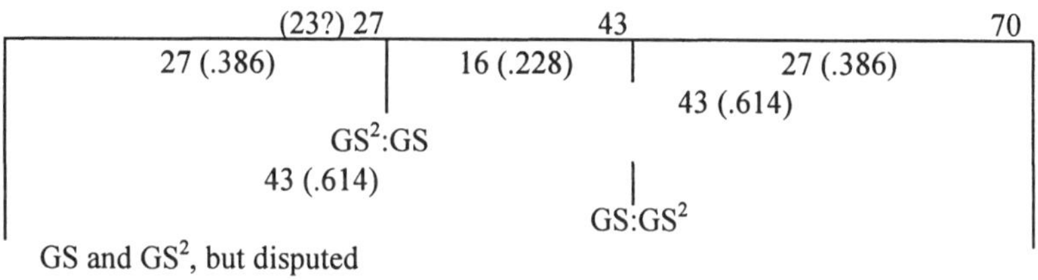

We are more comfortable with the divisions shown in figure 4.64b.[30]

Figure 4.65b. Handel's *Oboe Concerto op. 3 no. 3* Mvt II (Madden) (Half-measures)

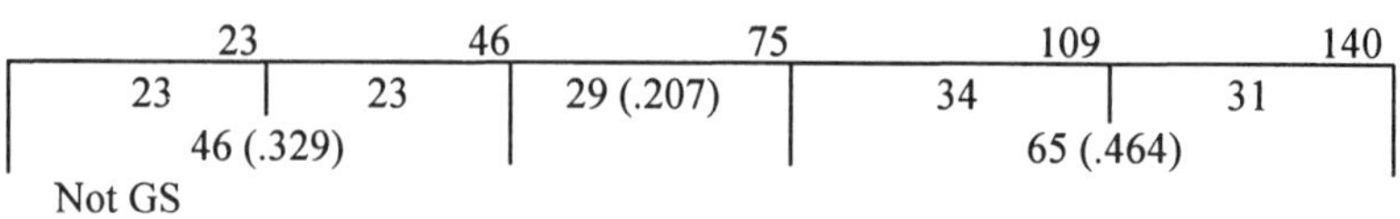

We accept Pascoe's analysis of the fourth movement with the exception of a probable division at 75 instead of 77 (one measure from GS). See figure 4.66.[31]

Pascoe applied his modules to the score with descriptions of the events, but without clearly identifying structural points such as cadences and theme entries. Thus, something may happen at each point of GS (such as a change in dynamic), but not necessarily something structural. This makes the analyses difficult to evaluate.

Figure 4.66. Handel's *Oboe Concerto op. 3 no. 3* Mvt IV (Pascoe)

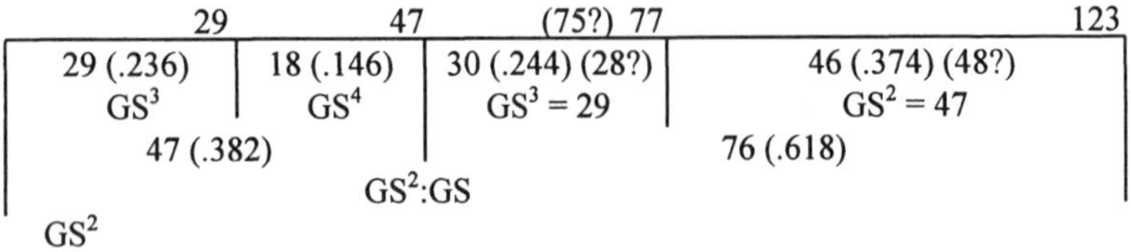

Pascoe's analyses of Handel's *Concerto Grosso op. 6 no. 4* are also problematic. Nothing happens at phi in either piece. We show them here to point out their errors. Pascoe's construction of the first movement is shown in figure 4.67a. Although it appears accurate to GS, the actual structural points are at measures 6, 12, and 20, as shown in figure 4.67b.[32]

Figure 4.67a. Handel's *Concerto Grosso op. 6 no. 4* Mvt I (Pascoe)

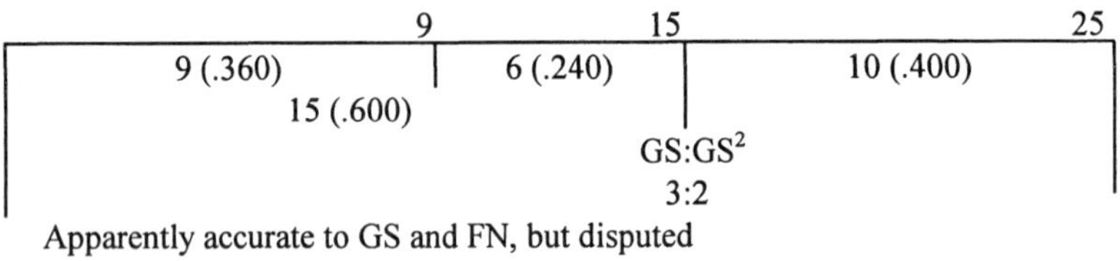

Figure 4.67b. Handel's *Concerto Grosso op. 6 no. 4* Mvt I (Madden)

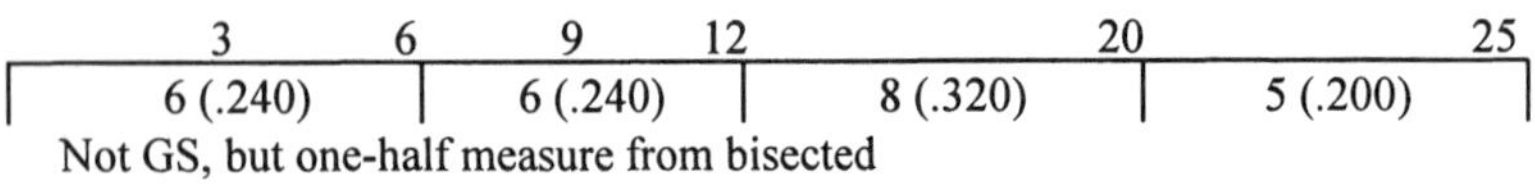

Pascoe's analysis of Movement IV is shown in figure 4.68a, apparently accurate to GS. But, again, nothing happens at or near measures 47 or 77, as shown in figure 4.68b.

Figure 4.68a. Handel's *Concerto Grosso op. 6 no. 4* Mvt IV (Pascoe)

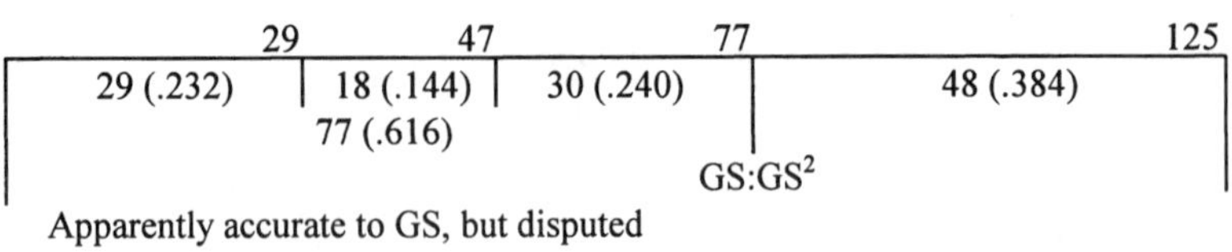

Figure 4.68b. Handel's *Concerto Grosso op. 6 no. 4* Mvt IV (Madden)

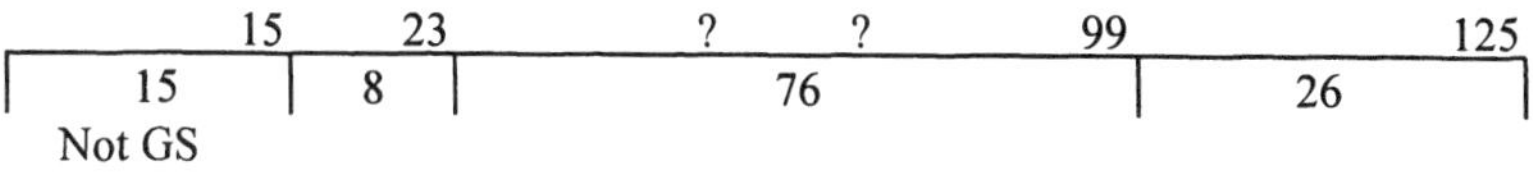

Scarlatti

Rogers found phi in a phrase of Domenico Scarlatti's *Sonata in D minor*, Longo number 413. See figure 4.69a.

Figure 4.69a. Scarlatti's *Sonata in D Minor L. 413* (Rogers) (Eighth notes)

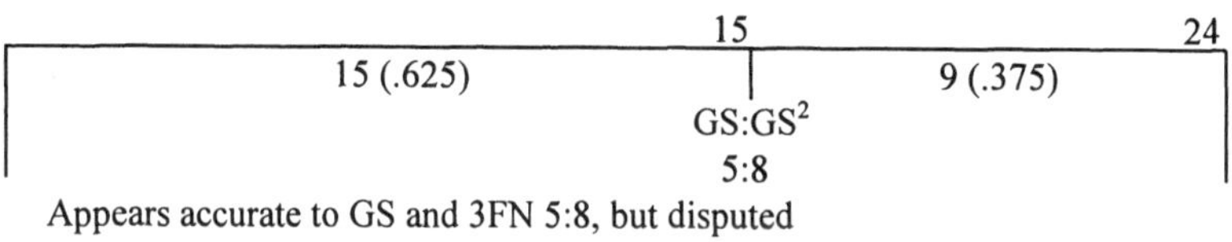

But the first section is only 14 eighth notes long. It is corrected in figure 4.68b.[33]

Figure 4.69b. Scarlatti's *Sonata in D Minor L. 413* (Madden) (Eighth notes)

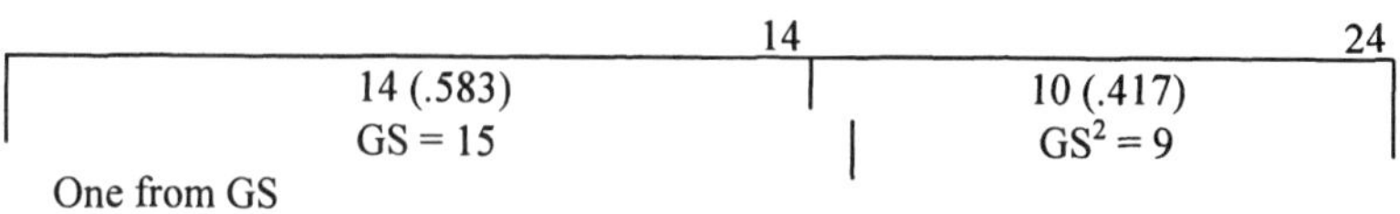

The "Pastorale" that this example is taken from is also not divided at fib or phi, nor are any other pieces of Scarlatti examined.

Monteverdi

Rogers examined the aria "Tu se morta" from the opera *Orfeo* of Claudio Monteverdi. See figure 4.70a.

Rogers expected something to happen at each of Pascoe's modules from G3 to G8, but there is nothing at G3 (count 5), G4 (8), G5 (13), or G7 (34). There may be something at G6 (21) and G8 (55), but the total count should be 90 half notes, so that G8 is 55/90 = .611, not .618. Our analysis, shown in figure 4.70b with the revised count, puts the structural point after measure 55.[34]

Figure 4.70a. Monteverdi's *Orfeo* "Tu se morte" [*sic*] (Rogers) (Half notes)

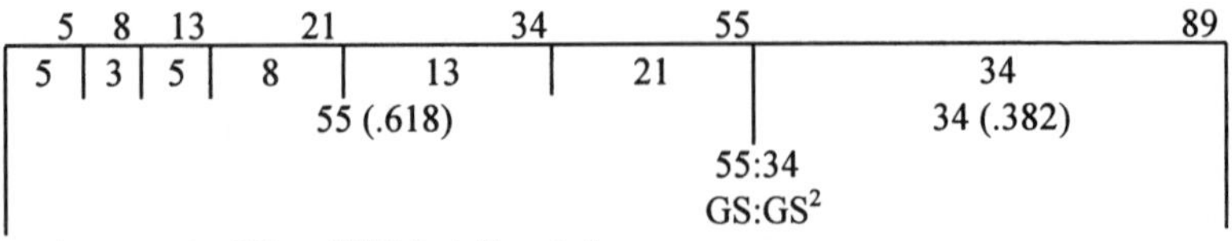

Accurate to GS and FN, but disputed

Figure 4.70b. Monteverdi's *Orfeo* "Tu sei morta" [*sic*] (Madden) (Half notes)

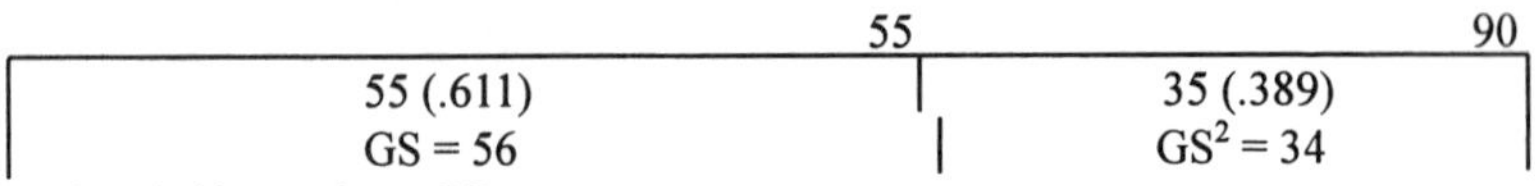

One half-note from GS

Monteverdi's recitative "Ma che temi" from *Orfeo* is illustrated in figure 4.71.[35] It is divided at GS^2, if we count measures. It should be noted, however, that *HAM* shorts the 69th measure one half note.

Figure 4.71. Monteverdi's "Ma che temi" from *Orfeo* (Madden) (Measures)

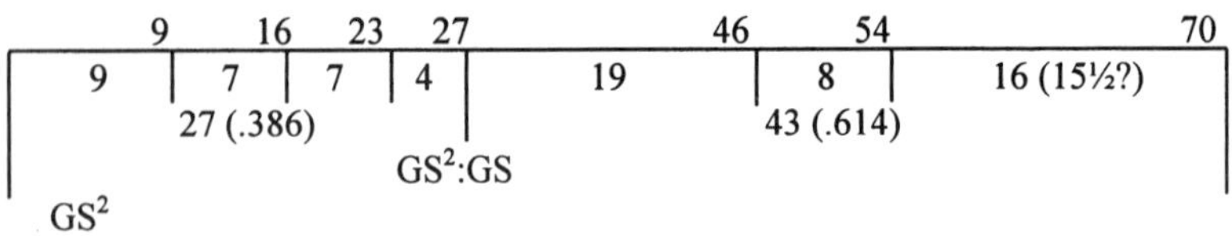

Monteverdi's *Madrigal* "Ohimè se tanto amate" also is divided at GS.[36] See figure 4.72.

Figure 4.72. Monteverdi's *Madrigal* "Ohimè se tanto amate" (Madden) (Measures)

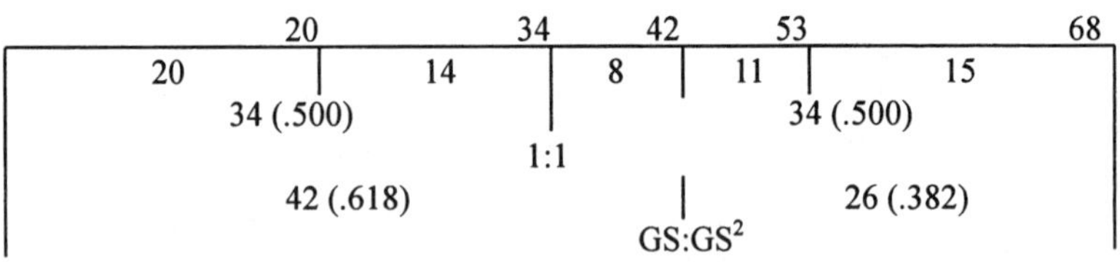

GS and bisected

Music Selected from *HAM, Volume 2*

Stephano Landi wrote a *Sinfonia* in his opera *Il San Alessio*.[37] Shown in figure 4.73, it is one measure from GS^2. It also further subdivides into GS^3 and GS^4.

Figure 4.73. Landi's *Sinfonia* from *Il San Alessio* (Madden) (Measures)

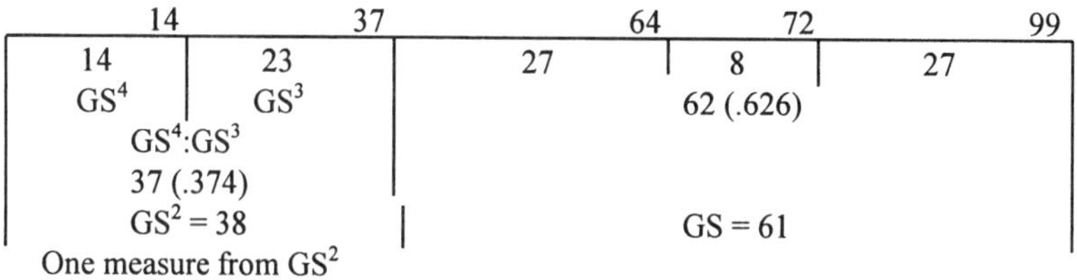

Jacques Champion de Chambonnières' *Chaconne* is shown in figure 4.74.[38] It is a rondeau in eleven parts ABA'CA"DA'EAFA, divided at GS and 4FN 3:2:3:3:3:2:3:2:3:2:3.

Figure 4.74. Chambonnières' *Chaconne* (Madden) (Measures)

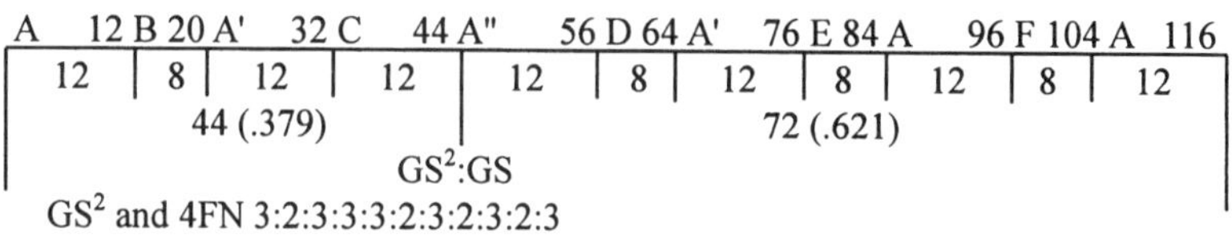

Antonio Vivaldi's *Concerto Grosso op. 3 no. 6,* first movement, is shown in figure 4.75.[39]

Figure 4.75. Vivaldi's *Concerto Grosso op. 3 no. 6* Mvt I (Madden)

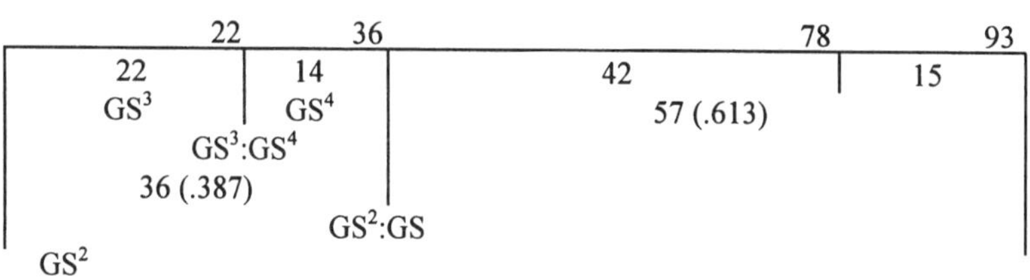

It is accurate to GS^2 and further subdivides into GS^3 and GS^4. Who would have predicted this for Vivaldi? It might be worthwhile to examine other works of this composer.

Wilhelm Friedemann Bach's *Fugue IV* has clear entries after measure 24, and a clear cadence at measure 34.[40] Shown in figure 4.76, it is one measure from GS, and accurate to 8FN 3:2.

Figure 4.76. W.F. Bach's *Fugue IV* (Madden) (Measures)

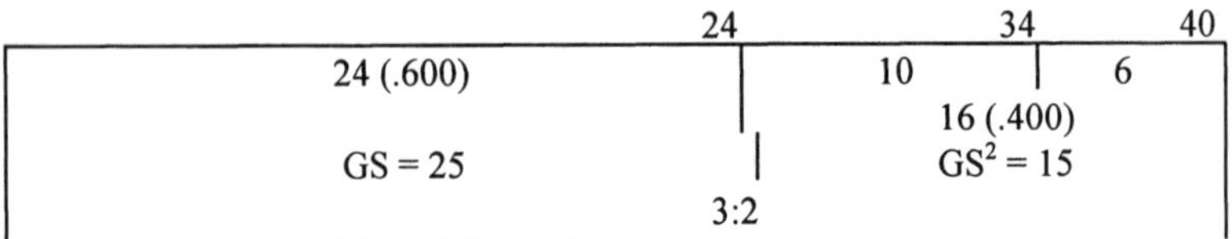

One measure from GS, and 8FN 3:2

Table 4.6 summarizes the music analyzed since table 4.4.

Table 4.6. Summary of the "other music" study

Composer/ Piece	GS or GS^2	One unit	Bi/ tri- sected	FN or elem. GS	Too far	Other
Bach/						
St. Anne	.381/.619					O'Donnell
	.386	.624				Madden
Labyrinth	.385	.615		5:3:5		Rothwell
	.385	.615		5:3:5		Madden
Wachet auf				13:8	.583	Rothwell*
					.568	Madden
Chaconne				GS		Cogan
Passacaglia	.619					Green
Vn Conc, I	.617					Pascoe*
					.701	Madden
Vn Conc, III	.381					Pascoe*
			.500	2:3, 3:2		Madden
Eng Ste, I				GS?		Pascoe**
			.507			Dorfman*
					.315	Madden
Eng Ste, VI				GS?	.455	Pascoe**
					.455	Madden

Composer/ Piece	GS or GS^2	One unit	Bi/ tri-sected	FN or elem. GS	Too far	Other
Handel/						
Messiah	.617					Rothwell*
		.606				Madden
Oboe, II	.386/.614					Pascoe*
					.329	Madden
Oboe, IV	.382					Pascoe
Grosso, I	.600			3:2		Pascoe*
					.480	Madden
Grosso, IV	.616					Pascoe*
						Madden†
Scarlatti/						
Sonata	.625			5:8		Rogers*
		.583				Madden
Monteverdi/						
Orfeo, aria	.618					Rogers*
		.611				Madden
Orfeo, recit.	.386					Madden
Ohimè	.618		.500			
Landi/						
Sinfonia		.374				
Chambonnières	.379					
Vivaldi/						
Grosso, I	.387					
W.F. Bach/						
Fugue		.600		3:2		

*Doubtful **Part B divided in GS? †No clear structure

Conclusion

After the problem of how to determine an imitative composition's structure was solved, it became clear that some composers used phi in the Baroque. J. S. Bach used phi in at least 14.6% of the pieces in *The Well-Tempered Clavier*, 40% of the fugues in *The Art of the Fugue* (70% if an error of one measure is allowed), and to a lesser extent in the *Inventions*.

Only a few other composers used phi, including Monteverdi and Vivaldi. This was a pleasant surprise. Scarlatti cannot be said to have used phi, despite Pascoe's and Rogers' claims. Handel was marginal.

Many errors were encountered in the previous studies. Researchers should remember that the elements of a ratio are the durations of sections, whether determined by theme entries or cadences. A common error is to use the number that follows—instead of the one that ends—the section.

Another common error is careless reduction of fractions. Today, the exact value of a proportion is easily derived with a pocket calculator by dividing the length of a section by the total number of measures or beats and rounding to three decimal places.

Chapter Notes

[1] [Nord64 p. 219] [Rog p. 32] [Dor p. 97] [BachA] The title is seen in several forms, with and without "the". The form used here is given by Alfred Dörfel and Wolfgang Graeser, editors for the Breitkopf & Härtel Complete Works Edition.

[2] [Rothwell p. 78] [BachA]

[3] [Gou]

[4] [BachI]

[5] Kent Kennan and Theodore Baker called subject entries "motives" in their discussions of the *Inventions*, whereas Douglass M. Green and Thomas Benjamin called them "themes". [Kennan p. 60] [Bak95 p. 123] [Gre p. 283] [Benj p. 145]

[6] [Rothwell p. 81]

[7] [Rothwell p. 84]

[8] [Rothwell p. 86]

[9] [Rog p. 60]

[10] [BachW]

[11] [Webster p. 240]

[12] [Webster p. 240]

[13] [Webster p. 240]

[14] [Rog p. 33] [O'Don p. 11] [BachF p. 105] Humphries [Hum p. 27] has this as 36|45|36 due to counting breves rather than whole notes in sections 1 and 3.

[15] [Rothwell p. 42–47, 67–76]

[16] [Rothwell p. 79–80]

[17] [Dor p. 125–31]

[18] [Rog p. 33] [CogEsc p. 262] [BachC]

[19] [Gre p. 124]

[20] [Pas p. 77]

[21] [Pas p. 86]

[22] [Pas p. 114] [BachV p. 35]

[23] [Pas p. 118] [BachV p. 59]

[24] [Pas p. 119–21]

[25] [Dor p. 121–22]

[26] [BachE p. 50]

[27] [Pas p. 122–23] [BachE p. 62]

[28] [Rothwell p. 89]

[29] [Nort Std. Ed. p. 57–67]

[30] [Pas p. 165–7] [Hand Vol. 11 p. 49]

[31] [Pas p. 167–9]

---

[32] [Pas p. 170–1] [Hand Vol. 14 p. 73]

[33] [Rog p. 58] [Scar p. 44]

[34] [Rog p. 92] [Nort Std. Ed. p. 18] This is spelled "Tu se morte" by Rogers, "Tu sei morta" in the Norton Scores, and "Tu se' morta" in the text of the music.

[35] [HAM2 p. 7]

[36] [HAM2 p. 9]

[37] [HAM2 p. 47]

[38] [HAM2 p. 54, 283]

[39] [HAM2 p. 183]

[40] [HAM2 p. 233]

Chapter 5

Classic

Introduction

So far, several dissertations have been used as sources of information on the use of fib and phi in music, including Pascoe, Rogers, Rothwell, and Dorfman. Several more will be added in this chapter, including Adkins, Camp, Demaree, Hoffman, and Sidorowicz.

The outline of this chapter will be to summarize the enormous body of previous work. The first task, of course, is to verify that the data were recorded correctly. The second is to determine to what extent phi is present, and to find some way to make the results visible.

All of the symphonies, string quartets, and piano sonatas of Beethoven, Haydn, and Mozart have been analyzed. Pulling this work into a manageable form is a huge challenge but handling it adequately should provide a firm basis on which to conclude whether these composers knew about fib and phi.

But first, let us clear up a misconception that has gotten into the common wisdom, that the first movement of the *Fifth Symphony* of Beethoven is divided exactly into a Fibonacci sequence. It will be shown here that not even the measure count is correct in those analyses, and that neither fib nor phi is present in the movement.

Chapter 5

Hoffman

We begin with Michael Hoffman's dissertation "Fibonacci Proportions in Composition." It uncritically quoted Derek W. Haylock's article "The golden section and Beethoven's fifth" and two of Trudi Garland's books.[1]

Haylock began by discarding the final twenty measures of the first movement as "musically insignificant," along with measure 511 as "apparently a later addition," and the final motto statement (da-da-da-dah), in order to get an even 600 measures to divide up. Claiming that GS and GS^2 were at measures 372 and 228, he showed the piece divided into phi proportions, framed and separated by motto statements.[2] See figure 5.1a.

Figure 5.1a. Haylock's interpretation of Beethoven's *Fifth Symphony* Mvt I

		377			610	615
5	(5)	372 One from GS of 600	5	(5)	228 One from GS^2 of 600	5

One problem with this was that the 600 initial measures already contained the motto statements, as shown by the numbers in parentheses; thus the "framing" added two redundant motto statements. (Only the final one is valid as a frame.) Framing also gave him 615 measures instead of the 600 that he began with.

Garland saw that 5 + 372 = FN 377 and 228 + 5 = FN 233, as shown in figure 5.1b. This could almost have worked, as seen in the Fibonacci sequence 1, 1, 2, 3, 5, 8, 13, 21, 34, 55, 89, 144, 233, 377, 610, . . . , but there was still the problem of the extra motto statement between the sections.[3] In addition, there were still too many measures and the extra motto statements established by Haylock.

Figure 5.1b. Garland's interpretation of Haylock

		377			610	615
5	(5)	372 377 (FN)	5	(5)	228 233 (FN)	5

Although beautiful and imaginative, this structure has nothing to do with the music, which consists of 626 measures, none of which can be arbitrarily excised. The proportions are closer to FN 2:1:2 than 377:233, as shown in figure 5.2.[4]

Figure 5.2. Corrected analysis of Beethoven's *Fifth Symphony* Mvt I (Madden)

Expo	124 Rpt	248 Devel	371 Recap	601 Coda 626
124 (.198)	124 (.198)	123 (.197)	230 (.367)	25(.040)
248 (.396)			255 (.407)	

Neither fib nor phi

A few adjustments can be made without doing violence to this analysis. Sidorowicz placed the recapitulation after measure 376, forcing the motto to be interpreted as an introduction and/or transition.[5] This action would also require placing the exposition after measure 5 and the coda after 606. Although this may be a valid interpretation, it seems more realistic to consider the mottos as imbedded at the beginning of each section: the exposition, its repetition, the development, the recapitulation, and the coda.

In any case, placing the coda after measure 496, as Haylock, Sidorowicz, and others have done, cannot be correct. Nothing happens there to suggest a coda. Also, codas of the Classic period were short and usually integrated into the recapitulations, as we will see. Apparently these analysts wanted the recapitulation and coda each to be the same length as the unrepeated exposition, but the final motto statement after measure 601 clearly marks the beginning of the coda.

Because the analyses by Haylock and Garland showed a precedent for Hoffman's work, it was not in his interest to dispute them. All the same, not challenging them worsened an already bad situation. Beethoven will be shown to have been in no way beholden to fib or phi.

Sidorowicz

Alexander E. Sidorowicz's dissertation was entitled in part "The Proportional and Spatial Analysis of the First Movement Sonata Form of Mozart, Haydn, and Beethoven."

His printout of raw data alone consumed twenty-four pages. It came with three sets of values. The first recognized no repeats; the second, only the first repeat if there was one; and the third, the repeat of both sections, if marked. He made several sets of charts from this data to compare the results of taking, or not taking, the repeats.

In the following tables, each line contains two proportions: the smaller marking the end of the exposition, and the larger marking the end of the development. (They are not expected to be complements of one another, as they are in the figures.) Table 5.1, using figures from Sidorowicz's

Haydn printouts, shows that taking the first repeat drastically shifts the structural points away from phi as compared to taking no repeats. When the repeat is taken, there are only three values within .03 of phi.[6]

Table 5.2 shows Sidorowicz's data for taking both repeats. It shows no difference to table 5.1 for no repeats. But, how can no-repeats and both-repeats have the same proportions? Well, they can't! Sidorowicz assumed that in taking both repeats, the proportions would simply be doubled, so he cut each proportion of the second section in half, redistributed the results, and made the proportions the same as those where repeats were not taken.

This does not seem like a good idea. If, as with the Bach binary pieces, this music were measured as first part to second part, these structures would indeed be the same whether or not the repeats were taken. But a sonata is a three part form; when the second repeat is taken, not only is the development repeated, but also the recapitulation. Thus, where should the recapitulation be marked; at the first occurrence or the second?

The first occurrence would ordinarily not be far enough in; the second would be too far. Thus, it is a nonsensical question whose only answer can be that, for purposes of analysis, no repeats are taken, even if marked.[7] As seen in table 5.1, this gives the largest quantity of phi proportions while simultaneously providing the best structural logic of the three groupings. Exceptions to this rule will be reported clearly.

Sidorowicz's proportions have been tested and found generally reliable. His adjustments for slow introductions using a complicated formula have been listed in a few cases, but some seem too large. For example, in Mozart's *K. 543,* the introduction goes from 25 measures to the equivalent of 125 with the adjustment, making it 37% of the piece. This does not seem reasonable. Inflating the introductions to more than twice their notated lengths seems excessive and many adjustments must be rejected. In addition, the adjustments always move the proportions away from GS.

The sonata form is quite symmetrical, with exposition and recapitulation of approximately equal lengths. Since the developments are usually shorter than either of the other sections, they control the proportions more than any other factor.

In the following figures and tables, proportions rounded to two decimal places are Sidorowicz's. Figures rounded to three decimal places are ours.

Table 5.1. Summary of the first movements of some early Haydn symphonies as given by Sidorowicz.

No repeats taken

Symph No.	At GS or GS^2	Near GS or GS^2 (±.01)	Near GS or GS^2 (±.02)	Near GS or GS^2 (±.03)	Bi- or tri-sected	Too far
35				.59		.34
36			.36			.72
37			.40			.48
38		.39				.68
39						.43/.70
40		.37/.63				
41		.39		.65		
42			.36	.59		
43	.38	.63				
44		.39	.64			
45					.67	.34
46		.39				.69
47				.35		.71
48						.42/.68

First repeats taken

Symph No.	At GS or GS^2	Near GS or GS^2 (±.01)	Near GS or GS^2 (±.02)	Near GS or GS^2 (±.03)	Bi- or tri-sected	Too far
35						.51/.70
36						.53/.80
37		.63				.58
38						.56/.77
39		.61				.80
40						.54/.73
41						.56/.75
42						.53/.70
43						.56/.74
44						.56/.74
45						.51/.76
46						.56/.78
47						.52/.79
48				.59		.77

Table 5.2. Summary of the first movements of some early Haydn symphonies as given by Sidorowicz

Both repeats taken

Symph No.	At GS or GS^2	Near GS or GS^2 (±.01)	Near GS or GS^2 (±.02)	Near GS or GS^2 (±.03)	Bi- or tri-sected	Too far
35				.59		.34
36			.36			.72
37			.40			.48
38		.39				.68
39						.43/.70
40		.37/.63				
41		.39		.65		
42			.36	.59		
43	.38	.63				
44		.39	.64			
45					.67	.34
46		.39				.69
47				.35		.71
48						.42/.68

Figures 5.3 to 5.5 and tables 5.3 to 5.5 show summaries of three sets of music by Mozart. These show the proportions that result when repeats are not taken. Where an important proportion results by taking one or both repeats or deleting the introduction, the result is given in a note.

Figure 5.6 and table 5.6 show some of Haydn's symphonies. The earlier symphonies are not commercially available in score and, due to copyright restrictions, it is deemed unwise to use Sidorowicz. Most of the symphonies have slow introductions that have been adjusted by Sidorowicz with a formula. Although this may be the right thing to do, the adjustments are difficult to justify and move the proportions away from phi. We have chosen to count by measures without adjusting for tempos. Table 5.6b shows that removing the introductions makes phi rather more apparent.[8] This became even more clear when we put Sidorowicz's adjustments into this table.

Figures 5.7 and 5.8 and tables 5.7 and 5.8 show Haydn's string quartets and keyboard sonatas, and figures 5.9 to 5.11 and tables 5.9 to 5.11 show Beethoven's symphonies, string quartets, and piano sonatas.

Mozart

Figure 5.3. Mozart symphonies

K. 16, 19, 22, 43, 45, 74, 84, 181, 318, 444. Not in sonata form.

Symph. 2 K. 17

Expo	34 Devel	58 Recap 96
34 (.354)	24 (.250)	38 (.396)
$GS^2 = 37$	GS = 59	$GS^2 = 37$

One from GS

Symph. 3 K. 18

Expo	54 Devel	92 Recap 141
54 (.383)	38 (.269)	49 (.348)
$GS^2 = 54$	GS = 87	$GS^2 = 54$

GS^2

Symph. 8 K. 48

Expo	33 Devel	59 Recap 93
33 (.355)	26 (.279)	34 (.366)
$GS^2 = 36$	GS = 57	$GS^2 = 36$

Symph. 9 K. 73

Expo	45 Devel	58 Recap 105
45 (.429)	13 (.123)	47 (.448)
$GS^2 = 40$	GS = 65	$GS^2 = 40$

Symph. 12 K. 110

Expo	69 Devel	92 Recap 157
69 (.439)	23 (.147)	65 (.414)
$GS^2 = 60$	GS = 97	$GS^2 = 60$

Symph. 13 K. 112

Expo	54 Devel	70 Recap 124
54 (.435)	16 (.130)	54 (.435)
$GS^2 = 47$	GS = 77	$GS^2 = 47$

Exactly symmetrical

Symph. 14 K. 114

Expo	59 Devel	79 Recap 139
59 (.425)	20 (.143)	60 (.432)
$GS^2 = 53$	GS = 86	$GS^2 = 53$

Figure 5.3. Mozart symphonies (Cont)

Symph. 15 K. 124

Expo	36 Devel	72 Recap	111
36 (.324)	36 (.325)	39 (.351)	
$GS^2 = 42$	GS = 69	$GS^2 = 42$	

Development is unusually long

Symph. 16 K. 128

Expo	54 Devel	85 Recap	138
54 (.391)	31 (.225)	53 (.384)	
$GS^2 = 53$	GS = 85	$GS^2 = 53$	

GS and one from GS^2

Symph. 17 K. 129

Expo	46 Devel	67 Recap	114
46 (.404)	21 (.184)	47 (.412)	
$GS^2 = 44$	GS = 70	$GS^2 = 44$	

Symph. 18 K. 130

Expo	62 Devel	83 Recap	132
62 (.470)	21 (.159)	49 (.371)	
$GS^2 = 50$	GS = 82	$GS^2 = 50$	

One from GS

Symph. 19 K. 132

Expo	59 Devel	89 Recap	148
59 (.399)	30 (.202)	59 (.399)	
$GS^2 = 57$	GS = 91	$GS^2 = 57$	

Exactly symmetrical and .001 from both FN 2:3 and FN 3:2

Symph. 20 K. 133

Expo	78 Devel	100 Recap	160? 182
78 (.429)	22 (.120)	82 (.451)	
$GS^2 = 70$	GS = 112	$GS^2 = 70$	

At measure 101, key is right for a recapitulation but theme is not. Sidorowicz put it there. What might be called a false recapitulation appears at measure 160. This was composed in 1772, apparently prior to Haydn's use of such constructions, so who invented it?

Figure 5.3. Mozart symphonies (Cont)

Symph. 21 K. 134

Expo | 68 Devel | 95 Recap | 155 Coda 173

68 (.393) $GS^2 = 66$ | 27 (.156) $GS = 107$ | 78 (.451) $GS^2 = 66$

Sidorowicz miscounted. The exposition should end at .393 instead of .44

Symph. 22 K. 162

Intro 12 Expo | 67 Devel | 79 Recap | 135

67 (.496) $GS^2 = 52$ | 12 (.089) $GS = 83$ | 56 (.415) $GS^2 = 52$

The development is short, but the interpretation seems reasonable

Symph. 24 K. 182

Expo | 48 Devel | 75 Recap | 137 Coda 146

48 (.329) $GS^2 = 56$ | 27 (.185) $GS = 90$ | 71 (.486) $GS^2 = 56$

Ambiguous form. Sidorowicz miscounted both sections

Symph. 25 K. 183

Expo | 82 Devel | 116 Recap | 200 Coda 214

82 (.383) $GS^2 = 82$ | 34 (.159) $GS = 132$ | 98 (.458) $GS^2 = 82$

GS^2

Symph. 26 K. 184

Expo | 42 Devel | 67 Recap | 111 Coda 135

42 (.311) $GS^2 = 52$ | 25 (.185) $GS = 83$ | 68 (.504) $GS^2 = 52$

Omitted by Sidorowicz, but added. Without the coda, the proportions are improved to .378 and .604 but it seems like a bad idea to drop it

Symph. 27 K. 199

Expo | 58 Devel | 86 Recap | 146

58 (.397) $GS^2 = 56$ | 28 (.192) $GS = 90$ | 60 (.411) $GS^2 = 56$

Our proportions do not support those given by Sidorowicz

Figure 5.3. Mozart symphonies (Cont)

Symph. 28 K. 200

Expo	67 Devel	90 Recap	165 Coda 173
67 (.387)	23 (.133)		83 (.480)
$GS^2 = 66$		GS = 107	$GS^2 = 66$

One from GS^2

Symph. 29 K. 201

Expo	76 Devel	106 Recap	184 Coda 206
76 (.369)	30 (.146)		100 (.485)
$GS^2 = 79$		GS = 127	$GS^2 = 79$

Symph. 30 K. 202

Expo	78 Devel	112 Recap	207
78 (.377)	34 (.164)		95 (.459)
$GS^2 = 79$		GS = 128	$GS^2 = 79$

One from GS^2

Symph. 31 K. 297

Expo	128 Devel	163 Recap	295
128 (.434)	35 (.119)		132 (.447)
$GS^2 = 113$		GS = 182	$GS^2 = 113$

Symph. 33 K. 319

Expo	128 Devel	207 Recap	370
128 (.346)	79 (.213)		163 (.441)
$GS^2 = 141$		GS = 229	$GS^2 = 141$

The recapitulation should be at measure 207 instead of 244

Symph. 34 K. 338

Expo	100 Devel	157 Recap	264
100 (.379)	57 (.216)		107 (.405)
$GS^2 = 101$		GS = 163	$GS^2 = 101$

One from GS^2. The development should be at measure 100, rather than 111

Symph. 35 K. 385

Expo	94 Devel	128 Recap	204
94 (.461)	34 (.166)		76 (.373)
$GS^2 = 78$		GS = 126	$GS^2 = 78$

Figure 5.3. Mozart symphonies (Cont)

Symph. 36 K. 425

Intro 19 Expo	122 Devel	172 Recap	264 Coda 287
122 (.425)	50 (.174)		115 (.401)
GS^2 = 110		GS = 177	GS^2 = 110

.001 from FN 3:2. Without intro, expo ends at one measure from GS^2 (.384)

Symph. 38 K. 504

Intro 36 Expo	142 Devel	214 Recap	302
142 (.470)	72 (.238)		88 (.292)
GS^2 = 115	GS = 187		GS^2 = 115

Symph. 39 K. 543

Intro 25 Expo	142 Devel	183 Recap	309
142 (.460)	41 (.132)		126 (.408)
GS^2 = 118	GS = 191		GS^2 = 118

Adjustment for introduction seems too large (.63|.09|.28)

Symph. 40 K. 550

Expo	100 Devel	163 Recap	299
100 (.334)	63 (.211)		136 (.455)
GS^2 = 114	GS = 185		GS^2 = 114

The repeat would make this almost exactly bisected

Symph. 41 K. 551

Expo	120 Devel	188 Recap	276 Coda 313
120 (.383)	68 (.218)		125 (.399)
GS^2 = 120	GS = 193		GS^2 = 120

GS^2 and .001 from FN 3:2

Except for *K. 124,* the developments are shorter than the expositions or recapitulations

A quick survey reveals that many of the marks identifying the beginnings and endings of the developments lie between the marks representing GS^2 and GS. Thus, these pieces are highly symmetrical, but only two are exactly so. Also, is 29% of pieces within one measure of GS enough to qualify as knowing use? Three cases of .001 from GS^2 might be.

Table 5.3. Summary of the Mozart symphonies

K. No.	At GS or GS^2	1 meas.	Bi- or trisected	Too far	Comments
17		.604		.354	
18	.383			.652	
48				.355/.634	
73				.429/.552	
110				.439/.586	
112				.435/.565	Symmetrical
114				.425/.568	
124				.324/.649	Long development
128	.616	.391			
129				.404/.588	
130		.629		.470	
132				.399/.601	Sym., .001 from 2:3
133				.429/.549	False recapitulation?
134				.393/.549	
162				.496/.585	
182				.329/.514	
183	.383			.542	
184				.311/.496	
199				.397/.589	
200		.387		.520	
201				.369/.515	
202		.377		.541	
297				.434/.553	
319				.346/.559	
338		.379		.595	
385				.461/.627	
425				.425/.599	.001 from 3:2
504				.470/.708	
543				.460/.592	
550			.334	.545	
551	.383			.601	GS and .001 from 2:3

4/31 = 12.9% at GS or GS^2
9/31 = 29.0% within one measure
2/31 = 6.5% symmetrical
1/31 = 3.2% long development

Figure 5.4. Mozart string quartets

K. 170, 171. Not in sonata form

Qtet. 1 K. 80

Expo	28 Devel	39 Recap	67
28 (.418)	11 (.164)	28 (.418)	
$GS^2 = 26$	GS = 41	$GS^2 = 26$	

Exactly symmetrical

Qtet. 2 K. 155

Expo	49 Devel	71 Recap	119
49 (.412)	22 (.185)	48 (.403)	
$GS^2 = 45$	GS = 74	$GS^2 = 45$	

Qtet . 3 K. 156

Expo	71 Devel	109 Recap	180
71 (.394)	38 (.212)	71 (.394)	
$GS^2 = 69$	GS = 111	$GS^2 = 69$	

Exactly symmetrical

Qtet. 4 K. 157

Expo	52 Devel	74 Recap	126
52 (.413)	22 (.174)	52 (.413)	
$GS^2 = 48$	GS = 78	$GS^2 = 48$	

Exactly symmetrical

Qtet. 5 K. 158

Expo	45 Devel	74 Recap	127
45 (.354)	29 (.229)	53 (.417)	
$GS^2 = 49$	GS = 78	$GS^2 = 49$	

Qtet. 6 K. 159

Expo	29 Devel	44 Recap	71
29 (.408)	15 (.212)	27 (.380)	
$GS^2 = 27$	GS = 44	$GS^2 = 27$	

GS

Qtet. 7 K. 160

Expo	48 Devel	59 Recap	105
48 (.457)	11 (.105)	46 (.438)	
$GS^2 = 40$	GS = 65	$GS^2 = 40$	

Figure 5.4. Mozart string quartets (Cont)

Qtet. 8 K. 168

Expo	41 Devel	62 Recap 108
41 (.380)	21 (.194)	46 (.426)
$GS^2 = 41$	GS = 67	$GS^2 = 41$

GS^2

Qtet. 9 K. 169

Expo	36 Devel	72 Recap 117
36 (.308)	36 (.307)	45 (.385)
$GS^2 = 45$	GS = 72	$GS^2 = 45$

GS. Development is unusually long

Qtet. 12 K. 172

Expo	52 Devel	71 Recap 127
52 (.409)	19 (.150)	56 (.441)
$GS^2 = 49$	GS = 78	$GS^2 = 49$

Qtet. 13 K. 173

Expo	45 Devel	64 Recap 118 Coda 135
45 (.333)	19 (.141)	71 (.526)
$GS^2 = 52$	GS = 83	$GS^2 = 52$

Without the coda, divides at .381

Qtet. 14 K. 387

Expo	56 Devel	108 Recap 171
56 (.327)	52 (.305)	63 (.368)
$GS^2 = 65$	GS = 106	$GS^2 = 65$

Qtet. 15 K. 421

Expo	41 Devel	69 Recap 117
41 (.350)	28 (.240)	48 (.410)
$GS^2 = 45$	GS = 72	$GS^2 = 45$

Qtet. 16 K. 428

Expo	68 Devel	100 Recap 164
68 (.415)	32 (.195)	64 (.390)
$GS^2 = 63$	GS = 101	$GS^2 = 63$

One from GS

Figure 5.4. Mozart string quartets (Cont)

Qtet. 17 K. 458

Expo 90 Devel 137 Recap 230 Coda 278

90 (.324) | 47 (.169) | 141 (.507)

$GS^2 = 106$ | GS = 172 | $GS^2 = 106$

Qtet. 18 K. 464

Expo 87 Devel 161 Recap 270

87 (.322) | 74 (.274) | 109 (.404)

$GS^2 = 103$ | GS = 167 | $GS^2 = 103$

Qtet. 19 K. 465

Intro 22 Expo 106 Devel 154 Recap 226 Coda 246

106 (.431) | 48 (.195) | 92 (.374)

$GS^2 = 94$ | GS = 152 | $GS^2 = 94$

.001 from FN 5:3. Without the coda, divides at .375/.589.

Qtet. 20 K. 499

Expo 98 Devel 141 Recap 242 Coda 266

98 (.368) | 43 (.162) | 125 (.470)

$GS^2 = 102$ | GS = 164 | $GS^2 = 102$

Qtet. 21 K. 575

Expo 77 Devel 116 Recap 193

77 (.399) | 39 (.202) | 77 (.399)

$GS^2 = 74$ | GS = 119 | $GS^2 = 74$

Exactly symmetrical and .001 from both FN 2:3 and FN 3:2

Qtet. 22 K. 589

Expo 71 Devel 130 Recap 208

71 (.341) | 59 (.284) | 78 (.375)

$GS^2 = 79$ | GS = 129 | $GS^2 = 79$

5:3

One from GS and FN 5:3

Qtet. 23 K. 590

Expo 74 Devel 111 Recap 186 Coda 198

74 (.374) | 37 (.187) | 87 (.439)

$GS^2 = 76$ | GS = 122 | $GS^2 = 76$

.001 from FN 3:5

Table 5.4. Summary of the Mozart string quartets

K. No.	At GS or GS^2	1 meas.	Bi- or tri-sected	Too far	Comments
80				.418/.582	Symmetrical
155				.412/.597	
156				.394/.606	Symmetrical
157				.413/.587	Symmetrical
158				.354/.583	
159	.620			.408	
160				.457/.562	
168	.380			.574	
169	.615			.308	Long development
172				.409/.559	
173			.333	.474	
387				.327/.632	
421				.350/.590	
428		.610		.415	
458				.324/.493	
464				.322/.596	
465				.431/.626	.001 from 5:3
499				.368/.530	
575				.399/.601	Sym. & .001 from 2:3
589		.625		.341	5:3
590				.374/.561	.001 from 3:5

3/21 = 14.3% at GS or GS^2
5/21 = 23.8% within one measure
4/21 = 19.4% exactly symmetrical

There is more symmetry in the quartets than in the symphonies; the developments are shorter than the expositions or recapitulations except for *K. 169.*

Figure 5.5. Mozart piano sonatas

K. 282, 331. Not in sonata form.

Sonata 1 K. 279

Expo | 38 Devel | 57 Recap | 100

38 (.380)	19 (.190)	43 (.430)
$GS^2 = 38$	GS = 62	$GS^2 = 38$

GS^2

Figure 5.5. Mozart piano sonatas (Cont)

Sonata 2 K. 280

Expo	56 Devel	82 Recap 144
56 (.389)	26 (.180)	62 (.431)
$GS^2 = 55$	GS = 89	$GS^2 = 55$

One from GS^2

Sonata 3 K. 281

Expo	40 Devel	69 Recap 109
40 (.367)	29 (.266)	40 (.367)
$GS^2 = 42$	GS = 67	$GS^2 = 42$

Exactly symmetrical. With repeats, the recapitulation is at .500

Sonata 5 K. 283

Expo	53 Devel	71 Recap 120
53 (.442)	18 (.150)	49 (.408)
$GS^2 = 46$	GS = 74	$GS^2 = 46$

Sonata 6 K. 284

Expo	51 Devel	71 Recap 127
51 (.402)	20 (.157)	56 (.441)
$GS^2 = 49$	GS = 78	$GS^2 = 49$

Sonata 7 K. 309

Expo	58 Devel	93 Recap 155
58 (.374)	35 (.226)	62 (.400)
$GS^2 = 59$	GS = 96	$GS^2 = 59$

One from GS^2, FN 3:2, and .001 from 3:5

Sonata 8 K. 310

Expo	49 Devel	79 Recap 133
49 (.368)	30 (.226)	54 (.406)
$GS^2 = 51$	GS = 82	$GS^2 = 51$

Sonata 9 K. 311

Expo	39 Devel	56? 59? Recap 98? 112
39 (.348)	20 (.179)	53 (.473)
$GS^2 = 43$	GS = 69	$GS^2 = 43$

False recap after measure 98. Sidorowicz has recap at measure 56

Figure 5.5. Mozart piano sonatas (Cont)

Sonata 10 K. 330

Expo	58 Devel	87 Recap	150
58 (.387)	29 (.193)	63 (.420)	
$GS^2 = 57$	GS = 93	$GS^2 = 57$	

One from GS^2

Sonata 12 K. 332

Expo	93 Devel	132 Recap	229
93(.406)	39 (.170)	97 (.424)	
$GS^2 = 87$	GS = 142	$GS^2 = 87$	

Sonata 13 K. 333

Expo	63 Devel	93 Recap	165
63 (.382)	30 (.182)	72 (.436)	
$GS^2 = 63$	GS = 102	$GS^2 = 63$	

GS^2

Sonata 14 K. 457

Expo	74 Devel	99 Recap	167 Coda 185
74 (.400)	25 (.135)	86 (.465)	
$GS^2 = 71$	GS = 114	$GS^2 = 71$	

FN 2:3

Sonata 15 K. 533

Expo	102 Devel	145 Recap	239
102 (.427)	43 (.180)	94 (.393)	
$GS^2 = 91$	GS = 148	$GS^2 = 91$	

Sonata 16 K. 545

Expo	28 Devel	41 Recap	73
28 (.384)	13 (.178)	32 (.438)	
$GS^2 = 28$	GS = 45	$GS^2 = 28$	

GS^2. Recapitulation is in the wrong key. False recap again?

Sonata 17 K. 547a

Expo	78 Devel	126 Recap	196
78 (.398)	48 (.245)	70 (.357)	
$GS^2 = 75$	GS = 121	$GS^2 = 75$	

Figure 5.5. Mozart piano sonatas (Cont)

Sonata 18 K. 570

Expo	79 Devel	132 Recap	209
	79 (.378)	53 (.254)	77 (.368)
	$GS^2 = 80$	$GS = 129$	$GS^2 = 80$

One from GS^2

Sonata 19 K. 576

Expo	58 Devel	98 Recap	160
	58 (.363)	40 (.250)	62 (.387)
	$GS^2 = 61$	$GS = 99$	$GS^2 = 61$

One from GS

Table 5.5. Summary of the Mozart piano sonatas

K. No.	At GS or GS^2	1 meas.	Bi- or trisected	Too far	Comments
279	.380			.570	
280		.389		.569	
281				.367/.633	Symmetrical
283				.442/.592	
284				.402/.559	
309		.374		.600	3:2, .001 from 3:5
310				.368/.594	
311				.348/.527	False recapitulation?
330		.387		.580	
332				.406/.576	
333	.382			.564	
457				.400/.535	2:3
533				.427/.607	
545	.384			.562	False recapitulation?
547a				.398/.643	
570		.378		.632	
576		.613		.363	

3/17 = 17.6% at GS^2
8/17 = 47.1% within one measure
1/17 = 5.8% exactly symmetrical
These proportions fit very tightly into the symmetry of GS and GS^2
Is this a high enough percentage to suggest that he knew?

Haydn

Figure 5.6. Late Haydn symphonies (With introductions)

Symph. 88

Intro 16 Expo — 103 Devel — 179 Recap — 266

103 (.387)	76 (.286)	87 (.327)
GS^2 = 102	GS = 164	GS^2 = 102

One from GS^2. Symmetrical without introduction

Symph. 89

Expo — 58 Devel — 110 Recap — 172

58 (.337)	52 (.303)	62 (.360)
GS^2 = 66	GS = 106	GS^2 = 66

Recapitulation is at measure 110, with right key but wrong melody. False recapitulation?

Symph. 90

Intro 16 Expo — 97 Devel — 152 Recap — 227

97 (.427)	55 (.243)	75 (.330)
GS^2 = 87	GS = 140	GS^2 = 87

Symph. 91

Intro 20 Expo — 119 Devel — 191 Recap — 268

119 (.444)	72 (.269)	77 (.287)
GS^2 = 102	GS = 166	GS^2 = 102

Without introduction, .001 from FN 2:3

Symph. 92

Intro 20 Expo — 82 Devel — 124 Recap — 190 Coda — 232

82 (.353)	42 (.181)	108 (.466)
GS^2 = 89	GS = 143	GS^2 = 89

Symph. 93

Intro 20 Expo — 108 Devel — 182 Recap — 263

108 (.411)	74 (.281)	81 (.308)
GS^2 = 100	GS = 163	GS^2 = 100

Symph. 94

Intro 17 Expo — 108 Devel — 156 Recap — 258

108 (.419)	48 (.186)	102 (.395)
GS^2 = 99	GS = 159	GS^2 = 99

Figure 5.6. Late Haydn symphonies (With introductions) (Cont)

Symph. 95

Expo	61 Devel	118.5 Recap	165
61 (.370)	57.5 (.348)	46.5 (.282)	
$GS^2 = 63$	GS = 102	$GS^2 = 63$	

Development longer than recapitulation

Symph. 96

Intro 17 Expo	83 Devel	153 Recap	203
83 (.409)	70 (.345)	50 (.246)	
$GS^2 = 78$	GS = 125	$GS^2 = 78$	

Devel longer than recap. Without intro, devel longer than both expo and recap

Symph. 97

Intro 13 Expo	107 Devel	166 Recap	293
107 (.365)	59 (.202)	127 (.433)	
$GS^2 = 112$	GS = 181	$GS^2 = 112$	

Symph. 98

Intro 15 Expo	131 Devel	208 Recap	320
131 (.409)	77 (.241)	112 (.350)	
$GS^2 = 122$	GS = 198	$GS^2 = 122$	

Symph. 99

Intro 18 Expo	89 Devel	137 Recap	202
89 (.441)	48 (.237)	65 (.322)	
$GS^2 = 77$	GS = 125	$GS^2 = 77$	

Symph. 100

Intro 23 Expo	124 Devel	201 Recap	289
124 (.429)	77 (.267)	88 (.304)	
$GS^2 = 110$	GS = 179	$GS^2 = 110$	

Symph. 101

Intro 23 Expo	122 Devel	218 Recap	347
122 (.359)	96 (.269)	129 (.372)	
$GS^2 = 133$	GS = 214	$GS^2 = 133$	

Figure 5.6. Late Haydn symphonies (With introductions) (Cont)

Symph. 102

Intro 22 Expo 110 Devel 226 Recap 311

110 (.354) | 116 (.373) | 85 (.273)

$GS^2 = 119$ | GS = 192 | $GS^2 = 119$

Development longer than both exposition and recapitulation

Symph. 103

Intro 39 Expo 94 Devel 159 Recap 229

94 (.410) | 65 (.284) | 70 (.306)

$GS^2 = 87$ | GS = 142 | $GS^2 = 87$

Without introduction, the development is longer than the exposition

Symph. 104

Intro 16 Expo 123 Devel 192 Recap 294

123 (.418) | 69 (.235) | 102 (.347)

$GS^2 = 112$ | GS = 182 | $GS^2 = 112$

Table 5.6a. Summary of the late Haydn symphonies (With introductions)

Symph No.	At GS or GS^2	1 meas.	Tri-sected	Too far	Sid's adj.	Comments
88		.387		.673		
89				.337/.640		
90				.427/.670	.47/.69	
91				.444/.713	.48/.73	
92				.353/.534	.41/.65	
93				.411/.692	.48/.73	
94				.419/.605	.48/.64	
95				.370/.718		Long devel
96				.409/.754	.46/.78	Long devel
97				.365/.567	.41/.60	
98				.409/.650	.43/.66	
99				.441/.678	.49/.71	
100				.429/.696	.49/.73	
101				.359/.628	.45/.68	
102				.354/.727	.41/.75	Long devel
103				.410/.694	.48/.70	
104				.418/.653	.47/.69	

1/17 = 5.9% within one measure. 0/17 = 0% exactly symmetrical.

Table 5.6b. Summary of the late Haydn symphonies (Without introductions)

Symph No.	At GS or GS^2	1 meas.	Tri-sected	Too far	Sid's adj.	Comments
88				.348/.652		Symmetrical
89				.640/.337		
90	.384			.645	.47/.69	
91				.399/.690	.48/.73	.001 from 2:3
92				.292/.491	.41/.65	
93			.667	.362	.48/.73	
94		.377		.577	.48/.64	
95				.370/.718		Long devel
96				.355/.731	.46/.78	Long devel
97			.336	.546	.41/.60	
98	.380			.633	.43/.66	
99		.386		.647	.49/.71	
100		.380	.669		.49/.73	
101				.306/.602	.45/.68	
102				.304/.706	.41/.75	Long devel
103				.289/.632	.48/.70	Long devel
104		.385		.633	.47/.69	

2/17 = 11.8% at GS^2
6/17 = 35.3% within one measure. This is not as high as Mozart
1/17 = 5.9% exactly symmetrical
4/17 = 23.5% long development
These pieces are better balanced without the introductions
There is more variation in the lengths of the developments than in Mozart
None of Sidorowicz's adjustments was closer to GS, as shown in the additional column

Figure 5.7. Haydn string quartets

Op. 9 no. 1, 2, 3, 4, 6. Not available for review. Sidorowicz's values are used
Op. 9 no. 5, op. 17 no. 3, op. 55 no. 2, op. 76 no. 5. Not in sonata form

Op. 9 no. 1

Expo	? Devel	? Recap	?
(.41) $GS^2 = .38$	(.24) $GS^3 = .24$	(.35) $GS^2 = .38$	

Figure 5.7. Haydn string quartets (Cont)

Op. 9 no. 2

Expo	? Devel	? Recap ?
(.38)	(.29)	(.33)
$GS^2 = .38$	$GS^3 = .24$	$GS^2 = .38$

GS^2

Op. 9 no. 3

Expo	? Devel	? Recap ?
(.42)	(.22)	(.36)
$GS^2 = .38$	$GS^3 = .24$	$GS^2 = .38$

Op. 9 no. 4

Expo	? Devel	? Recap ?
(.45)	(.28)	(.27)
$GS^2 = .38$	$GS^3 = .24$	$GS^2 = .38$

Op. 9 no. 6

Expo	? Devel	? Recap ?
(.38)	(.26)	(.36)
$GS^2 = .38$	$GS^3 = .24$	$GS^2 = .38$

GS^2

Op. 17 no. 1

Expo	43 Devel	62 Recap 110
43 (.391)	19 (.173)	48 (.436)
$GS^2 = 42$	$GS = 68$	$GS^2 = 42$

One from GS^2

Op. 17 no. 2

Expo	38 Devel	70 Recap 100
38 (.380)	32 (.320)	30 (.300)
$GS^2 = 38$	$GS = 62$	$GS^2 = 38$

GS^2. Development is longer than recapitulation

Op. 17 no. 4

Expo	53 Devel	92 Recap 120 Coda 130
53 (.408)	39 (.300)	38 (.292)
$GS^2 = 50$	$GS = 80$	$GS^2 = 50$

Development is longer than recapitulation

Figure 5.7. Haydn string quartets (Cont)

Op. 17 no. 5

Expo	33 Devel	68 Recap	90
33 (.367)	35 (.389)	22 (.244)	
$GS^2 = 34$	$GS = 56$	$GS^2 = 34$	

One from GS^2. Development is longer than both exposition and recapitulation

Op. 17 no. 6

Expo	73 Devel	116 Recap	200
73 (.365)	43 (.215)	84 (.420)	
$GS^2 = 76$	$GS = 124$	$GS^2 = 76$	

Op. 20 no. 1

Expo	42 Devel	71 Recap	106
42 (.396)	29 (.304)	35 (.300)	
$GS^2 = 40$	$GS = 66$	$GS^2 = 40$	

Op. 20 no. 2

Expo	47 Devel	80 Recap	106
47 (.443)	33 (.312)	26 (.245)	
$GS^2 = 40$	$GS = 66$	$GS^2 = 40$	

The development is longer than the recapitulation

Op. 20 no. 3

Expo	94 Devel	164 Recap	270
94 (.348)	70 (.259)	106 (.393)	
$GS^2 = 103$	$GS = 167$	$GS^2 = 103$	

Omitted by Sidorowicz and added.

Op. 20 no. 4

Expo	112 Devel	215 Recap	298
112 (.376)	103 (.345)	83 (.279)	
$GS^2 = 114$	$GS = 184$	$GS^2 = 114$	

.001 from FN 3:5

Op. 20 no. 5

Expo	48 Devel	84 Recap	136 Coda 159
48 (.302)	36 (.226)	75 (.472)	
$GS^2 = 61$	$GS = 98$	$GS^2 = 61$	

Figure 5.7. Haydn string quartets (Cont)

Op. 20 no. 6

Expo	64 Devel	107 Recap	164
64 (.390)	43 (.262)	57 (.348)	
$GS^2 = 63$	GS = 101	$GS^2 = 63$	

One from GS^2

Op. 33 no. 1

Expo	37 Devel	58 Recap	91
37 (.407)	21 (.230)	33 (.363)	
$GS^2 = 35$	GS = 56	$GS^2 = 35$	

Op. 33 no. 2

Expo	32 Devel	62 Recap	90
32 (.356)	30 (.333)	28 (.311)	
$GS^2 = 34$	GS = 56	$GS^2 = 34$	

Op. 33 no. 3

Expo	59 Devel	107 Recap	151 Coda 167
59 (.353)	48 (.288)	60 (.359)	
$GS^2 = 64$	GS = 103	$GS^2 = 64$	

Op. 33 no. 4

Expo	31 Devel	51 Recap	85 Coda 90
31 (.344)	20 (.223)	39 (.433)	
$GS^2 = 34$	GS = 56	$GS^2 = 34$	

Op. 33 no. 5

Expo	95 Devel	181 Recap	271 Coda 305
95 (.311)	86 (.282)	124 (.407)	
$GS^2 = 116$	GS = 189	$GS^2 = 116$	

Op. 33 no. 6

Expo	58 Devel	77 Recap	152 Coda 164
58 (.354)	19 (.116)	87 (.530)	
$GS^2 = 63$	GS = 101	$GS^2 = 63$	

A real recapitulation never occurs, and Sidorowicz placed it after measure 77, making the development disconcertingly short

Figure 5.7. Haydn string quartets (Cont)

Op. 42

Expo	39 Devel	65 Recap	105
39 (.371)	26 (.248)	40 (.381)	
$GS^2 = 40$	GS = 65	$GS^2 = 40$	

GS and one from GS^2

Op. 50 no. 1

Expo	60 Devel	109 Recap	164
60 (.366)	49 (.299)	55 (.335)	
$GS^2 = 63$	GS = 101	$GS^2 = 63$	

Op. 50 no. 2

Expo	106 Devel	175 Recap	290
106 (.366)	69 (.237)	115 (.397)	
$GS^2 = 111$	GS = 179	$GS^2 = 111$	

Op. 50 no. 3

Expo	44 Devel	81 Recap	113? 133
44 (.331)	37 (.278)	52 (.391)	
$GS^2 = 51$	GS = 82	$GS^2 = 51$	

One measure from GS. Omitted by Sidorowicz, presumably because of the false recapitulation after measure 113

Op. 50 no. 4

Expo	64 Devel	115 Recap	146 Coda 184
64 (.348)	51 (.277)	69 (.375)	
$GS^2 = 70$	GS = 114	$GS^2 = 70$	

One from GS and FN 5:3

Op. 50 no. 5

Expo	65 Devel	101 Recap	170
65 (.382)	36 (.212)	69 (.406)	
$GS^2 = 65$	GS = 105	$GS^2 = 65$	

GS^2

Op. 50 no. 6

Expo	54 Devel	116 Recap	164
54 (.329)	62 (.378)	48 (.293)	
$GS^2 = 63$	GS = 101	$GS^2 = 63$	

Devel longer than both expo and recap. Is there also a false recap?

Figure 5.7. Haydn string quartets (Cont)

Op. 54 no. 1

Expo	47 Devel	86 Recap	131
47 (.359)	39 (.297)	45 (.344)	
$GS^2 = 50$	GS = 81	$GS^2 = 80$	

Op. 54 no. 2

Expo	87 Devel	125 Recap	233
87 (.373)	38 (.163)	108 (.464)	
$GS^2 = 89$	GS = 144	$GS^2 = 89$	

Op. 54 no. 3

Expo	58 Devel	106 Recap	189
58 (.307)	48 (.254)	83 (.439)	
$GS^2 = 72$	GS = 117	$GS^2 = 72$	

Op. 55 no. 1

Expo	64 Devel	112 Recap	174
64 (.368)	48 (.276)	62 (.356)	
$GS^2 = 66$	GS = 108	$GS^2 = 66$	

Op. 55 no. 3

Expo	75 Devel	128 Recap	197
75 (.381)	53 (.269)	69 (.350)	
$GS^2 = 75$	GS = 122	$GS^2 = 75$	

GS^2

Op. 64 no. 1

Expo	60 Devel	97 Recap	174
60 (.345)	37 (.212)	77 (.443)	
$GS^2 = 66$	GS = 108	$GS^2 = 66$	

Op. 64 no. 2

Expo	40 Devel	67 Recap	108
40 (.370)	27 (.250)	41 (.380)	
$GS^2 = 41$	GS = 67	$GS^2 = 41$	

GS and one from GS^2

Figure 5.7. Haydn string quartets (Cont)

Op. 64 no. 3

Expo	69 Devel	125 Recap 170
69 (.406)	56 (.329)	45 (.265)
$GS^2 = 65$	GS = 105	$GS^2 = 65$

The development is longer than the recapitulation

Op. 64 no. 4

Expo	38 Devel	60 Recap 99
38 (.384)	22 (.222)	39 (.394)
$GS^2 = 38$	GS = 61	$GS^2 = 38$

GS^2 and one from GS. Sidorowicz's recap at measure 64 cannot be correct

Op. 64 no. 5

Expo	59 Devel	104 Recap 179
59 (.330)	45 (.251)	75 (.419)
$GS^2 = 68$	GS = 111	$GS^2 = 68$

Op. 64 no. 6

Expo	45 Devel	97 Recap 144
45 (.313)	52 (.361)	47 (.326)
$GS^2 = 55$	GS = 89	$GS^2 = 55$

Development is longer than both exposition and recapitulation

Op. 71 no. 1

Expo	69 Devel	112 Recap 155
69 (.445)	43 (.278)	43 (.277)
$GS^2 = 59$	GS = 96	$GS^2 = 59$

Op. 71 no. 2

Intro 4 Expo	52 Devel	70 Recap 125
52 (.416)	18 (.144)	55 (.440)
$GS^2 = 48$	GS = 77	$GS^2 = 48$

Without the introduction, it divides at .397. Of the three quartets with introductions, this is the only one whose adjustment makes a difference

Op. 71 no. 3

Expo	109 Devel	182 Recap 287 Coda 315
109 (.346)	73 (.232)	133 (.422)
$GS^2 = 120$	GS = 195	$GS^2 = 120$

Figure 5.7. Haydn string quartets (Cont)

Op. 74 no. 1

Expo	54 Devel	96 Recap	155
54 (.348)	42 (.271)	59 (.381)	
$GS^2 = 59$	GS = 96	$GS^2 = 59$	

GS

Op. 74 no. 2

Intro 8 Expo	100 Devel	174 Recap	260
100 (.385)	74 (.284)	86 (.331)	
$GS^2 = 99$	GS = 161	$GS^2 = 99$	

One from GS^2

Op. 74 no. 3

Expo	78 Devel	127 Recap	197
78 (.396)	49 (.249)	70 (.355)	
$GS^2 = 75$	GS = 122	$GS^2 = 75$	

Op. 76 no. 1

Expo	88 Devel	139 Recap	225
88 (.391)	51 (.227)	86 (.382)	
$GS^2 = 86$	GS = 139	$GS^2 = 86$	

GS

Op. 76 no. 2

Expo	56 Devel	98 Recap 138 Coda	154
56 (.364)	42 (.272)	56 (.364)	
$GS^2 = 59$	GS = 95	$GS^2 = 59$	

Exactly symmetrical

Op. 76 no. 3

Expo	44 Devel	78 Recap	121
44 (.364)	34 (.281)	43 (.355)	
$GS^2 = 46$	GS = 75	$GS^2 = 46$	

Op. 76 no. 4

Expo	68 Devel	107 Recap	187
68 (.364)	39 (.208)	80 (.428)	
$GS^2 = 71$	GS = 116	$GS^2 = 71$	

Figure 5.7. Haydn string quartets (Cont)

Op. 77 no. 1

Expo	62 Devel	127 Recap	189
62 (.328)	65 (.344)	62 (.328)	
$GS^2 = 72$	$GS = 117$	$GS^2 = 72$	

Symmetrical. Development is longer than both exposition and recapitulation

Op. 77 no. 2

Expo	58 Devel	114 Recap	172
58 (.337)	56 (.326)	58 (.337)	
$GS^2 = 66$	$GS = 106$	$GS^2 = 66$	

Exactly symmetrical

Table 5.7. Summary of the Haydn string quartets

Opus No.	At GS or GS^2	1 meas.	Bi-or tri-sected	Too far	Comments
9–1				.41/.65	
9–2	.38		.67		
9–3				.42/.64	
9–4				.45/.73	
9–6	.38			.64	
17–1		.391		.564	
17–2	.380			.700	Long development
17–4				.408/.708	Long development
17–5		.367		.756	Long development
17–6				.365/.580	
20–1				.396/.700	
20–2				.443/.755	Long development
20–3				.348/.607	
20–4				.376/.721	.001 from 3:5
20–5				.302/.528	
20–6		.390		.652	
33–1				.407/.637	
33–2				.356/.689	
33–3				.353/.641	
33–4				.344/.567	
33–5				.311/.593	
33–6				.354/.470	Short development
42	.619	.371			

Opus No.	At GS or GS^2	1 meas.	Bi-or tri-sected	Too far	Comments
50–1			.665	.366	
50–2				.366/.603	
50–3		.609	.331	.850	False recapitulation?
50–4		.625		.348	5:3
50–5	.382			.594	
50–6				.329/.707	Long development
54–1				.359/.656	
54–2				.373/.536	
54–3				.307/.561	
55–1				.368/.644	
55–3	.381			.650	
64–1				.345/.557	
64–2	.620	.370			
64–3				.406/.735	Long development
64–4	.384	.606			
64–5			.330	.581	
64–6				.313/.674	Long development
71–1				.445/.723	
71–2				.416/.560	
71–3				.346/.578	
74–1	.619			.348	
74–2		.385	.669		
74–3				.396/.645	
76–1	.618			.391	
76–2				.364/.636	Symmetrical
76–3				.364/.645	
76–4				.364/.572	
77–1				.328/.672	Long development
77–2				.337/.663	Symmetrical

10/52 = 19.2% at GS or GS^2
16/52 = 30.8% within one measure
2/52 = 3.8% exactly symmetrical
8/52 = 15.4% long development

Figure 5.8. Haydn piano sonatas

Numbers 1–17. Omitted because New Grove's found them to be spurious[9]
Numbers 39, 40, 42, 48. Not in sonata form

H. XVI no. 18

Expo	38 Devel	73 Recap	116
38 (.328)	35 (.301)	43 (.371)	
$GS^2 = 44$	$GS = 72$	$GS^2 = 44$	

One from GS

H. XVI no. 19

Expo	42 Devel	68 Recap	103
42 (.408)	26 (.252)	35 (.340)	
$GS^2 = 39$	$GS = 64$	$GS^2 = 39$	

H. XVI no. 20

Expo	37 Devel	68 Recap	100
37 (.370)	31 (.310)	32 (.320)	
$GS^2 = 38$	$GS = 62$	$GS^2 = 38$	

One from GS^2

H. XVI no. 21

Expo	56 Devel	94 Recap	148
56 (.378)	38 (.257)	54 (.365)	
$GS^2 = 57$	$GS = 91$	$GS^2 = 57$	

One from GS^2

H. XVI no. 22

Expo	24 Devel	41 Recap	75
24 (.320)	17 (.227)	34 (.453)	
$GS^2 = 29$	$GS = 46$	$GS^2 = 29$	

H. XVI no. 23

Expo	46 Devel	85 Recap	127
46 (.362)	39 (.307)	42 (.331)	
$GS^2 = 49$	$GS = 78$	$GS^2 = 49$	

H. XVI no. 24

Expo	51 Devel	98 Recap	155
51 (.329)	47 (.303)	57 (.368)	
$GS^2 = 59$	$GS = 96$	$GS^2 = 59$	

Figure 5.8. Haydn piano sonatas (Cont)

H. XVI no. 25

Expo	27 Devel	50 Recap	71
27 (.380)	23 (.324)	21 (.296)	
GS^2 = 27	GS = 44	GS^2 = 27	

GS^2. Development longer than recapitulation

H. XVI no. 26

Expo	29 Devel	58.5 Recap	82
29 (.354)	29.5 (.359)	23.5 (.287)	
GS^2 = 31	GS = 51	GS^2 = 31	

Development longer than both exposition and recapitulation

H. XVI no. 27

Expo	57 Devel	86 Recap	143
57 (.399)	29 (.202)	57 (.399)	
GS^2 = 55	GS = 88	GS^2 = 55	

Exactly symmetrical and .001 from both FN 2:3 and FN 3:2

H. XVI no. 28

Expo	58 Devel	97 Recap	155
58 (.374)	39 (.252)	58 (.374)	
GS^2 = 59	GS = 96	GS^2 = 59	

One from both GS^2 and GS, symmetrical, and .001 from FN 3:5 and FN 5:3

H. XVI no. 29

Expo	31 Devel	59 Recap	90
31 (.344)	28 (.312)	31 (.344)	
GS^2 = 34	GS = 56	GS^2 = 34	

Exactly symmetrical

H. XVI no. 30

Expo	60 Devel	107 Recap	161
60 (.373)	47 (.292)	54 (.335)	
GS^2 = 61	GS = 100	GS^2 = 61	

One from GS^2

H. XVI no. 31

Expo	24 Devel	41 Recap	65
24 (.369)	17 (.262)	24 (.369)	
GS^2 = 25	GS = 40	GS^2 = 25	

One from both GS^2 and GS and exactly symmetrical

Figure 5.8. Haydn piano sonatas (Cont)

H. XVI no. 32

Expo	28 Devel	47 Recap 70
28 (.400)	19 (.271)	23 (.329)
$GS^2 = 27$	GS = 43	$GS^2 = 27$

One from GS^2 and FN 2:3

H. XVI no. 33

Expo	68 Devel	116 Recap 193
68 (.352)	48 (.249)	77 (.399)
$GS^2 = 74$	GS = 119	$GS^2 = 74$

.001 from FN 3:2

H. XVI no. 34

Expo	45 Devel	76 Recap 125
45 (.360)	31 (.248)	49 (.392)
$GS^2 = 48$	GS = 77	$GS^2 = 48$

One from GS

H. XVI no. 35

Expo	67 Devel	103 Recap 170
67 (.394)	36 (.212)	67 (.394)
$GS^2 = 65$	GS = 105	$GS^2 = 65$

Exactly symmetrical

H. XVI no. 36

Expo	33 Devel	64 Recap 97
33 (.340)	31 (.320)	33 (.340)
$GS^2 = 37$	GS = 60	$GS^2 = 37$

Exactly symmetrical

H. XVI no. 37

Expo	40 Devel	60 Recap 103
40 (.388)	20 (.195)	43 (.417)
$GS^2 = 39$	GS = 64	$GS^2 = 39$

One from GS^2

H. XVI no. 38

Expo	28 Devel	48 Recap 76
28 (.368)	20 (.264)	28 (.368)
$GS^2 = 29$	GS = 47	$GS^2 = 29$

One from both GS^2 and GS and exactly symmetrical

Figure 5.8. Haydn piano sonatas (Cont)

H. XVI no. 41

Expo	55 Devel	96 Recap 151
55 (.364)	41 (.272)	55 (.364)
$GS^2 = 58$	GS = 93	$GS^2 = 58$

Exactly symmetrical

H. XVI no. 43

Expo	55 Devel	100 Recap 148
55 (.372)	45 (.305)	48 (.323)
$GS^2 = 57$	GS = 91	$GS^2 = 57$

H. XVI no. 44

Expo	30 Devel	51 Recap 77
30 (.390)	21 (.272)	26 (.338)
$GS^2 = 29$	GS = 48	$GS^2 = 29$

One from GS^2

H. XVI no. 45

Expo	44 Devel	61 Recap 96
44 (.458)	17 (.177)	35 (.365)
$GS^2 = 37$	GS = 59	$GS^2 = 37$

With one repeat, the development ends at one from GS

H. XVI no. 46

Expo	38 Devel	79 Recap 114
38 (.333)	41 (.360)	35 (.307)
$GS^2 = 44$	GS = 70	$GS^2 = 44$

With one repeat, the piece is bisected; with two repeats, trisected. Development is longer than both exposition and recapitulation

H. XVI no. 47

Expo	32 Devel	54 Recap 93
32 (.344)	22 (.237)	39 (.419)
$GS^2 = 36$	GS = 57	$GS^2 = 36$

H. XVI no. 49

Expo	64 Devel	131 Recap 195 Coda 218
64 (.294)	67 (.307)	87 (.399)
$GS^2 = 83$	GS = 135	$GS^2 = 83$

.001 from FN 3:2. Devel is longer than exposition

Figure 5.8. Haydn piano sonatas (Cont)

H. XVI no. 50

Expo — 53 Devel — 101 Recap — 150

53 (.353) | 48 (.320) | 49 (.327)

$GS^2 = 57$ | $GS = 93$ | $GS^2 = 57$

H. XVI no. 51

Expo — 43 Devel — 79 Recap — 111

43 (.387) | 36 (.325) | 32 (.288)

$GS^2 = 42$ | $GS = 69$ | $GS^2 = 42$

One from GS^2. Development is longer than recapitulation

H. XVI no. 52

Expo — 43 Devel — 78 Recap — 116

43 (.371) | 35 (.301) | 38 (.328)

$GS^2 = 44$ | $GS = 72$ | $GS^2 = 44$

One from GS^2

Table 5.8. Summary of the Haydn piano sonatas

H. No. XVI:	At GS or GS^2	1 meas.	Bi- or trisected	Too far	Comments
18		.629		.328	
19				.408/.660	
20		.370		.680	
21		.378		.635	
22				.320/.547	
23			.669	.362	
24				.329/.632	
25	.380			.704	Long development
26				.354/.713	Long development
27				.399/.601	Sym., .001 from 3:2
28		.374/.626			Sym., .001 from 3:5
29				.344/.656	Symmetrical
30		.373	.665		
31		.369/.631			Symmetrical
32		.400		.671	2:3
33				.352/.601	.001 from 3:2
34		.608		.360	
35				.394/.606	Symmetrical

H. No. XVI:	At GS or GS^2	1 meas.	Bi- or trisected	Too far	Comments
36				.340/.660	Symmetrical
37		.388		.583	
38		.368/.632			Symmetrical
41				.364/.636	Symmetrical
43				.372/.677	
44		.390		.662	
45				.458/.635	
46			.333	.693	Long development
47				.344/.581	
49				.294/.601	.001 from 3:2, Long
50				.353/.673	
51		.387		.712	Long development
52		.371		.672	

1/31= 3.2% at GS^2

14/31 = 45.2% within one measure. This seems very high

8/31 = 25.6% exactly symmetrical. This is also high

5/31 = 16.1% long development

Numbers 27, 28, 29, 31, 35, 36, 38, 41. With two repeats, developments end at .500 and are exactly symmetrical. This does not seem to be accidental

Beethoven

Figure 5.9a. Beethoven symphonies (Without repeats)

Symph. 1

Intro 12 Expo 109 Devel 177 Recap 298

109 (.366) | 68 (.228) | 121 (.406)

GS^2 = 114 | GS = 184 | GS^2 = 114

Symph. 2

Intro 33 Expo 134 Devel 218 Recap 300 Coda 363

134 (.369) | 84 (.232) | 145 (.399)

GS^2 = 139 | GS = 224 | GS^2 = 139

.001 from FN 3:2

Symph. 3

Intro 4 Expo 152 Devel 398 Recap 556 Coda 691

152 (.220) | 246 (.356) | 293 (.424)

GS^2 = 264 | GS = 427 | GS^2 = 264

Development is longer than exposition

Figure 5.9a. Beethoven symphonies (Without repeats) (Cont)

Symph. 4

Intro 44 Expo 184 Devel 336 Recap 498

184 (.369) | 152 (.306) | 162 (.325)

$GS^2 = 190$ | GS = 308 | $GS^2 = 190$

Without the introduction, the development is longer than the exposition

Symph. 5

Expo 124 Devel 247 Recap 482 Coda 502

124 (.247) | 123 (.245) | 255 (.508)

$GS^2 = 192$ | GS = 310 | $GS^2 = 192$

Symph. 6

Expo 138 Devel 311 Recap 413 Coda 512

138 (.270) | 173 (.337) | 201 (.393)

$GS^2 = 196$ | GS = 316 | $GS^2 = 196$

Devel is longer than expo and recap without coda. Sidorowicz put the recap at measure 277, but it probably should be at measure 311

Symph. 7

Intro 62 Expo 176 Devel 299 Recap 390 Coda 450

176 (.391) | 123 (.273) | 151 (.336)

$GS^2 = 172$ | GS = 278 | $GS^2 = 172$

Without the intro, the recap is at .611 and the devel is longer than the expo

Symph. 8

Expo 103 Devel 189 Recap 230 Coda? 373

103 (.276) | 127 (.341) | 143 (.383)

230 (.617)

GS = 231 | $GS^2 = 142$

One from GS at a major structural break even though it does not seem to be the recap. With the recap at 189, it is two and a half measures from bisected

Symph. 9

Intro 80 Expo 163 Devel 300 Recap 426 Coda 547

163 (.298) | 137 (.250) | 247 (.452)

$GS^2 = 209$ | GS = 338 | $GS^2 = 209$

Without intro, devel longer than expo. Proportion should be .55 instead of .58

Table 5.9a. Summary of the Beethoven symphonies (Without repeats)

Symph No.	At GS or GS^2	1 meas.	Tri-sected	Too far	Sid's adj.	Comments on Sid's adj.
1				.366/.594	.45/.65	Far from GS
2				.369/.601	.48/.67	Far from GS
3				.220/.576		
4				.369/.675	.51/.75	Far from GS
5				.247/.492		
6				.270/.607		
7			.664	.391	.52/.70	Far from GS
8		.617		.276/.507		Not recap
9				.298/.548		

1/9 = 11.1% within one measure
0/9 = 0% exactly symmetrical
These pieces are far different from either Mozart or Haydn

Figure 5.9b. Beethoven symphonies (With repeats)

Symph. 1

Intro 12 Expo 109 Rpt 206 Devel 274 Recap 395

206 (.522) | 68 (.172) | 121 (.306)

$GS^2 = 151$ | GS = 244 | $GS^2 = 151$

Symph. 2

Intro 33 Expo 134 Rpt 231 Devel 315 Recap 397 Coda 460

231 (.502) | 84 (.183) | 145 (.315)

$GS^2 = 176$ | GS = 284 | $GS^2 = 176$

Symph. 3

Intro 4 Expo Rpt 303 Devel 549 Recap 707 Coda 842

303 (.361) | 246 (.291) | 293 (.348)

$GS^2 = 322$ | GS = 520 | $GS^2 = 322$

Symph. 4

Intro 44 Expo 198 Rpt 338 Devel 490 Recap 652

338 (.518) | 152 (.234) | 162 (.248)

$GS^2 = 249$ | GS = 403 | $GS^2 = 249$

Figure 5.9b. Beethoven symphonies (With repeats) (Cont)

Symph. 5

Expo 124 Rpt 248 Devel 371 Recap 601 Coda 626

248 (.396) | 123 (.197) | 255 (.407)

GS^2 = 239 | GS = 387 | GS^2 = 239

Symph. 6

Expo 138 Rpt 276 Devel 449 Recap 551 Coda 650

276 (.425) | 173 (.266) | 201 (.309)

GS^2 = 248 | GS = 402 | GS^2 = 248

Symph. 7

Intro 62 Expo 176 Rpt 290 Devel 413 Recap 504 Coda 564

290 (.514) | 123 (.218) | 151 (.268)

GS^2 = 215 | GS = 349 | GS^2 = 215

Symph. 8

Expo Rpt 207 Devel 293 Recap 476

207 (.435) | 86 (.181) | 183 (.384)

GS^2 = 182 | GS = 294 | GS^2 = 182

One from GS

Table 5.9b. Summary of the Beethoven symphonies (With repeats)

Symph No.	At GS or GS^2	1 meas.	Tri-sected	Too far	Sid's adj.	Comments on Sid's adj.
1				.522/.694	.57/.72	Far from GS
2				.502/.685	.58/73	Far from GS
3				.361/.652		
4				.518/.752	.61/.80	
5				.396/.593		
6				.425/.691		
7				.514/.732	.60/.75	
8		.616		.435		
9				.298/.548		

1/9 = 11.1% within one measure
0/9 = 0% exactly symmetrical
0/9 = 0% long developments

Pieces adjusted for slow introductions using Sidorowicz's formula are shown in the right-hand column. The adjustment seems excessive

The repeats seem to be needed in *3, 5, 6,* and *8* and not needed in *1, 2, 4, 7*

The symphonies are structurally more complex than the quartets and sonatas mainly because of the introductions. When the repeats are taken, the developments are always shorter than the expositions or the recapitulations. This is further evidence that the repeats should be taken

Figure 5.9c. Beethoven symphonies (With repeats, without introductions)

Symph. 1

Expo 97 Rpt 194 Devel 262 Recap 383

194 (.507) | 68 (.177) | 121 (.316)

$GS^2 = 146$ | GS = 237 | $GS^2 = 146$

Symph. 2

Expo 101 Rpt 198 Devel 282 Recap 364 Coda 427

198 (.464) | 84 (.197) | 145 (.339)

$GS^2 = 163$ | GS = 264 | $GS^2 = 163$

Symph. 4

Expo 154 Rpt 294 Devel 446 Recap 608

294 (.484) | 152 (.250) | 162 (.266)

$GS^2 = 232$ | GS = 376 | $GS^2 = 232$

Symph. 7

Expo 114 Rpt 228 Devel 351 Recap 442 Coda 502

228 (.454) | 123 (.245) | 151 (.301)

$GS^2 = 192$ | GS = 310 | $GS^2 = 192$

Table 5.9c. Summary of the Beethoven symphonies (With repeats, w/o intros)

Symph No.	At GS or GS^2	1 meas.	Tri-sected	Too far	Comments
1				.507/.684	
2				.464/.661	
3				.484/.714	
7				.454/.699	

Proportions of *Symphonies 1, 2, 3, 7* without intros are not improved much

Figure 5.10. Beethoven string quartets

Qtet. no. 1 op. 18 no. 1

Expo	114 Devel	177 Recap	280 Coda 312
114 (.365)	63 (.202)	135 (.433)	
$GS^2 = 119$	$GS = 193$	$GS^2 = 119$	

Qtet. no. 2 op. 18 no. 2

Expo	81 Devel	148 Recap	248
81 (.327)	67 (.270)	100 (.403)	
$GS^2 = 95$	$GS = 153$	$GS^2 = 95$	

Qtet. no. 3 op. 18 no. 3

Intro 2 Expo	109 Devel	159 Recap	240 Coda 269
109 (.405)	50 (.186)	110 (.409)	
$GS^2 = 103$	$GS = 166$	$GS^2 = 103$	

Qtet. no. 4 op. 18 no. 4

Expo	77 Devel	135 Recap	219
77 (.352)	58 (.264)	84 (.384)	
$GS^2 = 84$	$GS = 135$	$GS^2 = 84$	

GS. Recap is at GS (.616), but if the repeat is taken it is at .716

Qtet. no. 5 op. 18 no. 5

Expo	80 Devel	135 Recap	219 Coda 225
80 (.356)	55 (.244)	90 (.400)	
$GS^2 = 86$	$GS = 139$	$GS^2 = 86$	

FN 3:2. Double repeat should probably be taken

Qtet. no. 6 op. 18 no. 6

Expo	91 Devel	174 Recap	264
91 (.345)	83 (.314)	90 (.341)	
$GS^2 = 101$	$GS = 163$	$GS^2 = 101$	

Double repeat should probably be taken

Qtet. no. 7 op. 59 no. 1

Expo	102 Devel	253 Recap	400
102 (.255)	151 (.378)	147 (.367)	
$GS^2 = 153$	$GS = 247$	$GS^2 = 153$	

The piece is completely written out, with no repeats. Development is longer than both exposition and recapitulation

Figure 5.10. Beethoven string quartets (Cont)

Qtet. no. 8 op. 59 no. 2

Expo 71 Devel 142 Recap 209 Coda 255

71 (.278)	71 (.279)	113 (.443)
$GS^2 = 97$	GS = 158	$GS^2 = 97$

Devel is unusually long. With double repeat, proportions are .301|.395|.396

Qtet. no. 9 op. 59 no. 3

Intro 29 Expo 106 Devel 190 Recap 265

106 (.400)	84 (.317)	75 (.283)
$GS^2 = 101$	GS = 164	$GS^2 = 101$

FN 2:3. Without introduction, proportions are .326|.356|.318

Qtet. no. 10 op. 74

Intro 24 Expo 77 Devel 138 Recap 203 Coda 262

77 (.294)	61 (.233)	124 (.473)
$GS^2 = 100$	GS = 162	$GS^2 = 100$

Without introduction, proportions are .413|.193|.394

Qtet. no. 11 op. 95

59 82 128 Coda 151

59 (.391)	23 (.152)	69 (.457)
$GS^2 = 58$	GS = 93	$GS^2 = 58$

One from GS^2 although not in sonata form

Table 5.10. Summary of the Beethoven string quartets (No repeats)

Opus No.	At GS or GS^2	1 meas.	Tri-sected	Too far	Sid's value	Comments
18–1				.365/.567		
18–2				.327/.597		
18–3				.405/.591		
18–4	.616			.352		
18–5				.356/.600		FN 3:2
18–6				.345/.659		
59–1				.255/.633		Long devel
59–2				.278/.557		Long devel
59–3				.400/.717		FN 2:3
74				.294/.527		
95		.391		.543	.39/.54	Not sonata

1/11 = 9.1% at GS
2/11 = 18.2% within one measure
0/11 = 1% exactly symmetrical
2/11 = 18.2% long development
Op. 127–135. Not in sonata form

Most of the quartets do not have second repeats. Taking the first repeat makes a large difference in the proportions, but the results are still far from phi. Except for *Opus 74,* taking single repeats worsens the proportions

Beethoven's proportions are far from phi in all but two of these analyses

The quartets seem better organized than the symphonies

Figure 5.11. Beethoven piano sonatas

Numbers 12, 13, 14, 22, 26, 30, 31. Not in sonata form.

Sonata no. 1 op. 2 no. 1

Expo 48 Devel 100 Recap 152

48 (.316)	52 (.342)	52 (.342)
$GS^2 = 58$	$GS = 94$	$GS^2 = 58$

Development is unusually long

Sonata no. 2 op. 2 no. 2

Expo 121 Devel 224 Recap 336

121 (.360)	103 (.307)	112 (.333)
$GS^2 = 128$	$GS = 208$	$GS^2 = 128$

The exposition was cut short by Sidorowicz, but the recapitulation is correct

Sonata no. 3 op. 2 no. 3

Expo 90 Devel 138 Recap 232 Coda 257

90 (.350)	48 (.187)	119 (.463)
$GS^2 = 98$	$GS = 159$	$GS^2 = 98$

Sonata no. 4 op. 7

Expo 136 Devel 188 Recap 350 Coda 362

136 (.376)	52 (.143)	174 (.481)
$GS^2 = 138$	$GS = 224$	$GS^2 = 138$

.001 from 3:5

Sonata no. 5 op. 10 no. 1

Expo 105 Devel 167 Recap 270 Coda 284

105 (.370)	62 (.218)	117 (.412)
$GS^2 = 108$	$GS = 176$	$GS^2 = 108$

Figure 5.11. Beethoven piano sonatas (Cont)

Sonata no. 6 op. 10 no. 2

Expo — 66 Devel — 136 Recap — 202

Expo	Devel	Recap
66 (.327)	70 (.346)	66 (.327)
$GS^2 = 77$	GS = 125	$GS^2 = 77$

Symmetrical. Development longer than both exposition and recapitulation

Sonata no. 7 op. 10 no. 3

Expo — 124 Devel — 183 Recap — 286 Coda 344

Expo	Devel	Recap
124 (.360)	59 (.172)	161 (.468)
$GS^2 = 131$	GS = 213	$GS^2 = 131$

Sonata no. 8 op. 13

Intro 10 Expo — 136 Devel — 194 Recap — 294 Coda 310

Expo	Devel	Recap
136 (.439)	58 (.190)	116 (.371)
$GS^2 = 118$	GS = 192	$GS^2 = 118$

Recap is at .613 (one from GS) without introduction. Adjustment is too distant

Sonata no. 9 op. 14 no. 1

Expo — 61 Devel — 90 Recap — 147 Coda 162

Expo	Devel	Recap
61 (.377)	29 (.179)	72 (.444)
$GS^2 = 62$	GS = 100	$GS^2 = 62$

One from GS^2

Sonata no. 10 op. 14 no. 2

Expo — 63 Devel — 124 Recap — 200

Expo	Devel	Recap
63 (.315)	61 (.305)	76 (.380)
$GS^2 = 76$	GS = 124	$GS^2 = 76$

GS

Sonata no. 11 op. 22

Expo — 68 Devel — 127 Recap — 161 Coda 199

Expo	Devel	Recap
68 (.342)	59 (.296)	72 (.362)
$GS^2 = 76$	GS = 123	$GS^2 = 76$

Sonata no. 15 op. 28

Expo — 163 Devel — 267 Recap — 437 Coda 460

Expo	Devel	Recap
163 (.354)	104 (.226)	193 (.420)
$GS^2 = 176$	GS = 284	$GS^2 = 176$

Figure 5.11. Beethoven piano sonatas (Cont)

Sonata no. 16 op. 31 no. 1

Expo	113 Devel	196 Recap	295 Coda 325
113 (.348)	83 (.255)	129 (.397)	
$GS^2 = 124$	GS = 201	$GS^2 = 124$	

Sonata no. 17 op. 31 no. 2

Expo	92 Devel	142 Recap	228
92 (.404)	50 (.219)	86 (.377)	
$GS^2 = 87$	GS = 141	$GS^2 = 87$	

One from GS

Sonata no. 18 op. 31 no. 3

Expo	88 Devel	136 Recap	219 Coda 253
88 (.348)	48 (.190)	117 (.462)	
$GS^2 = 97$	GS = 156	$GS^2 = 97$	

Sonata no. 19 op. 49 no. 1

Expo	33 Devel	63 Recap	96 Coda 110
33 (.300)	30 (.273)	47 (.427)	
$GS^2 = 42$	GS = 68	$GS^2 = 42$	

Sonata no. 20 op. 49 no. 2

Expo	52 Devel	66 Recap	115 Coda 122
52 (.426)	14 (.115)	56 (.459)	
$GS^2 = 47$	GS = 75	$GS^2 = 47$	

Sonata no. 21 op. 53

Expo	89 Devel	155 Recap	248 Coda 302
89 (.295)	66 (.218)	147 (.487)	
$GS^2 = 115$	GS = 187	$GS^2 = 115$	

With the repeat, the recapitulation is at 240 making the proportion .620 (one from GS). It should probably be taken

Sonata no. 23 op. 57

Expo	78 Devel	135 Recap	203 Coda 261
78 (.299)	57 (.218)	126 (.483)	
$GS^2 = 100$	GS = 161	$GS^2 = 100$	

Miscounted by Sidorowicz, it should be .30/.22 instead of .25/.27

Figure 5.11. Beethoven piano sonatas (Cont)

Sonata no. 24 op. 78

Intro 4 Expo 38 Devel 56 Recap 105

38 (.362) 18 (.171) 49 (.467)

$GS^2 = 40$ GS = 65 $GS^2 = 40$

Sonata no. 25 op. 79

Expo 50 Devel 122 Recap 171 Coda 201

50 (.249) 72 (.358) 79 (.393)

$GS^2 = 77$ GS = 124 $GS^2 = 77$

Development is longer than exposition and recapitulation without coda

Sonata no. 27 op. 90

Expo 81 Devel 143 Recap 222 Coda 245

81 (.331) 62 (.253) 102 (.416)

$GS^2 = 94$ GS = 151 $GS^2 = 94$

Sonata no. 28 op. 101

Expo 34 Devel 57 Recap 96 Coda 102

34 (.333) 23 (.226) 45 (.441)

$GS^2 = 39$ GS = 63 $GS^2 = 39$

Sidorowicz's recapitulation should be at measure 57 instead of 52

Sonata no. 29 op. 106

Expo 123 Devel 226 Recap 361 Coda 405

123 (.304) 103 (.254) 179 (.442)

$GS^2 = 155$ GS = 250 $GS^2 = 155$

Sonata no. 32 op. 111

Intro 18 Expo 69 Devel 91 Recap 134 Coda 158

69 (.437) 22 (.139) 67 (.424)

$GS^2 = 60$ GS = 98 $GS^2 = 60$

Without the introduction, the proportions are .364|.157|.479

Table 5.11. Summary of the Beethoven piano sonatas

Opus No.	At GS or GS^2	1 meas.	Bi- or trisected	Too far	Comments
2–1				.316/.658	Long development
2–2			.667	.360	
2–3				.350/.537	

Opus No.	At GS or GS²	1 meas.	Bi- or trisected	Too far	Comments
7				.376/.519	.001 from 3:5
10–1				.370/.588	
10–2				.327/.673	Sym., long devel.
10–3				.360/.532	
13				.439/.629	
14–1		.377		.556	
14–2	.620			.315	
22				.342/.638	
28				.354/.580	
31–1				.348/.603	
31–2		.623		.404	
31–3				.348/.538	
49–1				.300/.573	
49–2				.426/.541	
53				.295/.513	
57				.299/.517	
78				.362/.533	
79				.249/.607	Long development
90			.331	.584	
101			.333	.559	
106				.304/.558	
111				.437/.576	

1/25 = 4.0% at GS
3/25 = 12.0% within one measure
3/25 = 12.0% long development
These pieces are much more balanced than the symphonies

Table 5.12 summarizes these three composers.

Table 5.12. Summary of the composers' percentages

Composer	At GS or GS²	1 meas.	Total	Bi/trisected	Exactly sym.	Long devel.
Mozart	14.5%	17.4%	31.9%	2.9%	7.2%	2.9%
Haydn	13.0%	23.0%	36.0%	11.0%*	10.0%	17.0%
Beethoven	4.4%	8.9%	13.3%	6.7%†	2.2%	11.1%†

*Without introductions †With repeats

These composers wrote very few long developments. Inverting the figures from the right-hand column, we see that their developments were shorter than the expositions and/or recapitulations in 97.1%, 83.0%, and 88.9% of the pieces. This overwhelmingly confirms our claim on the fourth page of this chapter that development length controls the structure.

Adkins

Donald Glenn Adkins' dissertation "Golden Mean Durational Analysis as a Guide to Orchestral Repeats and Tempos" attempted to strengthen statements made by Classic period authors that minuet and trio repeats should be taken, by showing that the proportions would be at or near phi.[10] This work is shown in tables 5.13 to 5.15.

Table 5.13. Minuet and trio movements in the Mozart symphonies

	Without da capo repeats			With da capo repeats		
K. No.	At GS	Near GS (1 meas.)	Too far	At GS or GS^2	Near GS (1 meas.)	Too far
17		.378	.811			.318/.682
43			.364/.818			.308/.692
45			.508/.746			.405/.595
48			.353/.824			.300/.700
73			.429/.786			.348/.638
110						.500/.750
112						.333/.667
114			.413/.794			.342/.658
124			.424/.788			.350/.650
130			.348/.826			.296/.704
132			.455/.773			.370/.630
133		.389	.806			.326/.674
134			.411/.795			.341/.659
183			.414/.763	.383/.617		
200			.506/.747			.404/.596
201			.457/.771			.372/.628
202			.500/.750			.400/.600
319			.500/.750			.400/.600
385		.375	.813			.316/.684
425			.444/.777			.363/.636
543			.489/.756			.393/.607
550			.400/.800			.333/.667

K. 48. Without the da capo repeats, the middle of the trio is at .500

K. 110. The da capo is written out. The trio begins at .500 and the middle of the trio is three measures from .600

K. 112. The notation clearly shows that the repeats are required. It is bisected and trisected, with the minuet dividing at .333, the middle of the trio at .500, and the da capo at .667

K. 202. FN 2:3 and 3:2 with repeats

K. 319. FN 2:3 and 3:2 with repeats. The middle of the trio is at one measure from GS without repeats

K. 385. The end of the minuet is at one measure from GS^2 without repeats

K. 550. Identical proportions to those of *K. 112* when the repeats are taken

Table 5.14. Minuet and trio movements in the late Haydn symphonies

	Without da capo repeats			With da capo repeats		
No.	At GS	Near GS (1 meas.)	Too far	At GS or GS^2	Near GS (1 meas.)	Too far
88			.478/.761		.386/.614	
89			.437/.782			.358/.642
90			.504/.748			.403/.597
91			.483/.759		.611	.388
92		.388	.806			.325/.675
93			.426/.787			.351/.649
94			.517/.742			.411/.589
95			.500/.750			.400/.600
96			.473/.764	.382/.618		
97						.370/.630
98			.458/.771			.373/.627
99			.493/.754			.395/.605
100			.480/.720			.400/.600
101			.438/.783			.357/.643
102			.475/.763		.384/.616	
103			.511/.745			.407/.593
104			.361/.787			.314/.686

No. 97. Repeats are written out

Table 5.15. Minuet and trio movements in the Beethoven symphonies

	Without da capo repeats			With da capo repeats		
No.	At GS	Near GS (1 meas.)	Too far	At GS or GS^2	Near GS (1 meas.)	Too far
1			.453/.774			.369/.631
2			.488/.756			.393/.607
3						.472/.706
4						.370/.735
5						.356/.649
6	-	-	-	-	-	-
7			.227/.625			.334/.676
8			.440/.780			.361/.639
9			.511/.678			.396/.525

Symph 3, 4, 5. The da capos are written out

Symph 6. There is no minuet form

Symph 7. "Extremely significant" golden proportion values, according to Adkins. The closest we can get is proportion .625 (four from GS)

Symph 8. Without the repeat, the middle of the trio is at .610 (two from GS)

The difference with respect to phi is slight, but one might ask why the values in the right-hand columns are usually complements of each other while those in the left-hand columns are not. The answer reveals the probable solution to this question: with the literal da capo repeats, the last section is identical to the first, giving an exact symmetry AABBAA.

Thus, the need for balance, rather than a need to approximate phi, is the most likely reason for the need to take both repeats.

Camp

Elinor Jane Perry Camp's dissertation "Temporal Proportion: A Study of Sonata Forms in the Piano Sonatas of Mozart" was thorough, but the use of ratios instead of proportions created faulty results. Tables 5.16 to 5.17 correct Camp's Tables A and B.[11]

Notice that some of the corrections are close to Camp's original numbers. This will always be the case when the numbers are sufficiently close to phi. Even so, her numbers do not give the relative durations of the sections, but rather, their complements. Other numbers, such as those for *K. 281* in table B (minuet and trio movements), are as much as 100% away from their correct values.

Table 5.16. Summary of the Mozart piano sonatas from Camp's Table A, with corrections

Piece K. No.	Mvt.	Ratio	Camp's re-sult	Correction 1st section	Correction 2nd section
279	I	152:248	.6129	.380	.620
	II	84:138	.6087	.378	.622
	III	112:204	.5490	.354	.646
280	I	168:264	.6364	.389	.611
	II	144:216	.6667	.400	.600
	III	231:339	.6814	.405	.595
281	I	80:138	.5797	.367	.633
	II	138:180	.7667	.434	.566
282	I	60:84	.7143	.417	.523
	III	78:126	.6190	.382	.618
283	I	159:201	.7910	.442	.558
	II	56:100	.5600	.359	.641
	III	306:525	.5829	.368	.632
284	I	204:304	.6711	.402	.598
309	I	232:388	.5979	.374	.626
310	I	196:336	.5833	.368	.632
	II	93:165	.5636	.360	.640
311	I	156:292	.5342	.348	.652
330	I	116:184	.6304	.387	.613
	III	136:206	.6602	.398	.602
332	I	279:408	.6838	.406	.594
	III	180:310	.5806	.367	.633
333	I	252:408	.6176	.382	.618
	II	93:153	.6078	.378	.622
457	I	296:444	.6667	.400	.600
533	I	408:548	.7445	.427	.573
	II	138:228	.6053	.377	.623
545	I	112:180	.6222	.384	.616
547a	I	234:354	.6610	.398	.602
570	I	237:390	.6077	.378	.622
576	I	116:204	.5686	.362	.638

Table 5.17. Summary of the Mozart piano sonatas from Camp's Table B

Piece K. No.	Mvt.	Ratio	Camp's result	Correction 1st section	Correction 2nd section
281	III	32:32	1.0000	.500	.500
282	II–1	36:60	.6000	.375 (FN)	.625 (FN)
	II–2	48:72	.6667	.400 (FN)	.600 (FN)
284	III	32:36	.8889	.471	.529
310	III	32:32	1.0000	.500	.500
330	II	120:264	.4545	.313	.687
331	I	8:10	.8000	.444	.556
	II	144:300	.4800	.324	.676
	III	16:32	.5000	.333	.667
332	II	80:80	1.0000	.500	.500
494	III	32:56	.5714	.364	.636
545	II	96:126	.7619	.432	.568
547a	III	16:16	1.0000	.500	.500
570	II*	16:32	.5000	.333	.667
	III*	32:48	.6667	.400 (FN)	.600 (FN)

*Partial movement

Do not be confused by the column entitled "Correction 2nd section." This differs from the second number in the other tables of this chapter in that it is the duration of the second section, as contrasted with the location of the recapitulation as in the other tables. This is the complement of the number in the column entitled "Correction 1st section."

Demaree

Robert William Demaree Jr.'s dissertation "The Structural Proportions of the Haydn Quartets" was ultimately disappointing for purposes of this book because it did not really give proportions. Rather, it gave nice illustrations that made it easy to construct proportions oneself, making up for the deficiency of Sidorowicz's doubling up of proportions in pieces with double repeats.

There was much discussion of background writings regarding the quartets, much of which was at odds with our findings. Lind, for example, was said to have written about Haydn's "false recapitulations" that were in the wrong key, or the right key with the wrong theme. There were indeed some of these (for example *Op. 50 no. 3*), but they were few and did not inhibit our ability to find the recapitulations. Usually they were clear, and at worst, it was only necessary to count backwards from the end of the recapitulation the number of measures in the exposition. The recapitulation was usually near this point, perhaps buried by an elision.[12]

Curiously, Mozart's *K. 133* and *K. 311*, written several years before the Haydn examples, also display false recapitulations. It would be interesting to know if there are other examples and who has precedence.[13]

During Haydn's long career, his structures did get more complex and the pieces longer. Even so, developments were still in approximately the same proportions to their expositions and recapitulations. The only major structural change had to do with longer and slower introductions in the symphonies that made finding the proportions difficult because of the tempo changes. This did not happen in the quartets or the piano sonatas and, when we removed the introductions from the symphonies, phi often appeared. See table 5.6.[14]

Oddly, the Beethoven symphonies are so different from the Mozart and Haydn works that they almost do not belong together. Beethoven's "minuet" forms became exceedingly long (1,698 measures with all the repeats in the *Ninth Symphony*) and complex (additional partial repeats of both the minuet and the trio in the *Fourth Symphony*).

In Demaree's defense, his values for Haydn's *Op. 20 no. 1* pointed out an error in our table 5.7. His data are wholly compatible, and will not be presented here.

Dorfman

Allen Arthur Dorfman's dissertation "A Theory of Form and Proportion in Music" presented a discussion of Gustav Ernest's paper "Some Aspects of Beethoven's Instrumental Forms" read to the Royal Musical Association in 1903. Ernest had studied fifty-five movements from Beethoven's piano and solo sonatas and found that phi could be applied to forty-two of them. Examining Dorfman's quote from Ernest of Beethoven's *Piano Sonata op. 10 no. 1*, we found the structure shown in figure 5.12.[15] This is nearly identical to the value in table 5.11 above (Ernest may have miscounted by one measure). Three measures from GS is only 1.8% and would have been close enough in his day, but not in ours.

Figure 5.12. Beethoven's *Piano Sonata Op. 10 no. 1* Mvt I (Ernest)

105	166	283
105 (.371)	61 (.216)	117 (.413)
	178 (.629)	
$GS^2 = 108$	GS = 175	

Three measures from GS and GS^2

Dorfman also presented an example from Webster's article "Golden-Mean Form in Music" of Beethoven's *Quartet op. 18 no. 4.*[16] His numbers, without the repeat, are the same as given here in table 5.10, that is, the recapitulation falls at .616 (GS). With the repeat, this proportion becomes .716, which, of course, is far from phi.

Dorfman criticized Pascoe's and Rogers' dissertations but repeating his remarks would add nothing to this chapter.[17] Otherwise, these authors' work covers the same ground and needs no comment. Rothwell is omitted for the same reason.

Kokkonen

Joonas Kokkonen's article "Les Sonates pour Piano de Mozart et la Section d'Or" contained a table of calculated versus actual values for sonata movements in most of Mozart's piano sonatas.[18] Eight of seventeen first movements are close enough (one measure from phi). This is in approximate agreement with table 5.5, but Kokkonen's tolerances were too loose, in that he accepted values as far as eleven measures from phi.

Conclusion

Phi remains elusive in this music. Possibly these composers were working with other principles than assumed by phi researchers. We proposed that they knew they were using phi, but it appears that Mozart and Haydn did not consider it important. However, it is perfectly clear that Beethoven hardly used phi at all.

Figures 5.3 to 5.5 indicate that Mozart's forms were tightly bound, with expositions and recapitulations of nearly equal lengths separated by short developments often confined to the distance between GS^2 and GS which we might call "GS^3 phi-space". Thus, they seem highly symmetrical. Perhaps phi had a good chance of occurring as a natural result of working in such a tight form. But, how much of this is merely wishful thinking?

As seen in figures 5.6 to 5.8, Haydn's forms, too, were structured symmetrically around the GS^3 phi-space, but with more variability than with Mozart. Especially in his quartets, Haydn frequently wrote expositions and/or recapitulations that lay outside the bounds of GS^2 and GS. How, then can we account for Haydn having a higher quantity of values within one measure of phi than Mozart?

Complicating this is the slow introductions in Haydn's symphonies that make it difficult to calculate the proportions. Sidorowicz's complicated formula for converting them to a length comparable to the rest of the movement may be correct in terms of perception, but it greatly complicates comprehension of the structures. Little pattern can be seen until the introductions are removed.

At the beginning of this study, we adamantly opposed omitting any part of the music to determine structure. However, in light of Haydn's symphonies, we now incline toward accepting the omission of long and/or slow introductions. There seems to be little evidence that Haydn considered the introductions to be integral to the structure. (But we still oppose removing codas!)

Beethoven's introductions are problematic, as well. Here, however, there is an additional problem: the great variability of recapitulation lengths. In fact, the extended developments and recapitulations cause the expositions to appear too short if the repeat is not taken. Repeating the exposition makes the movement more symmetrical, with the development usually shorter than either the exposition or the recapitulation.

Thus, there is evidence that the repeat should be taken in Beethoven, but not in order to achieve phi proportion, but rather, to achieve symmetry. This is supported by Adkins' study of the minuet and trio forms in the symphonies. As seen in tables 5.13 to 5.15, there is little evidence of phi proportions when the da capo repeats are taken, but symmetry is pervasive. This concept may easily be applied to the question of taking the repeat(s) in sonata forms.

In conclusion, it is not clear whether these composers used GS knowingly. Whenever phi proportions turned up, they seemed to be byproducts of symmetry. But, Beethoven certainly did not use a Fibonacci sequence in the first movement of his *Fifth Symphony*!

Chapter Notes

[1] [Hof p. 46–48] [Hayl] [Garl87 p. 38] [Garl95 p. 116] Haylock mentioned a course given at Open University in 1977 by John Rutter, Course A421 "Elements of Music," Unit 15 (Milton Keynes, 1977). Please advise if you can get the text.

[2] GS/GS^2 should be 600 × .618 = 370.8 (371) and 600 × .382 = 229.2 (229). The fact that the recapitulation begins at Haylock's GS (measure 372) may have convinced him that it was valid to remove the coda, and if he had not used the mottos as frames, he might have gotten away with it. Nonetheless, in the context of 626 measures it is not phi.

[3] It would have been better to append 233 directly to 377 and drop the final motto statement.

[4] [BeethSym]

[5] [Sid p. 86]

[6] [Sid p. 76–7]

[7] This is not to say that the repeats should not be taken in performance; other considerations than phi proportion may take precedence.

[8] Then are we wrong to prohibit removing parts of the music?

[9] [HaydSon Publ. note] See also article in New Grove's Dictionary.

[10] [Adkins p. 57–67]

[11] [Camp p. 84–105] See section on method in chapter 1.

[12] [Dem p. 41] I have used the return to tonic as the point of beginning of the recapitulation.

[13] Lind (PhD dissertation "The Evolution of the Fausse Reprise in Haydn String Quartets and Symphonies," p. 32–33, quoted in Demaree p. 41) says that Haydn started this practice before 1771. This would apparently give Haydn precedence because Mozart would have been only fifteen years old.

[14] I speculate that the reason Haydn had introductions in the symphonies and not in the quartets or sonatas is that he had to deal with different types of audiences. For the symphonies, the audiences would have been large and perhaps needed time to settle down. For the quartets and sonatas the audiences would have been much smaller and probably did not need such special treatment.

[15] [Dor p. 102]

[16] [Dor p. 106]

[17] [Dor p. 114–131]

[18] [Kok p. 114]

Chapter 6
Romantic

Introduction

The Romantic period saw many new forms of music. Song forms were widely used by composers such as Brahms, Chopin, Mendelssohn, Schubert, Schumann, and Wolf in response to the aphorisms and adages called fragments that were prevalent in literature;[1] the symphony was gradually transformed into the symphonic poem in response to epic poetry;[2] and the opera became truly grandiose in the hands of Wagner.

Yet, similarities remained. Schubert, Schumann, Mendelssohn, and Brahms wrote symphonies that harked back to Classic forms. Sonata forms continued to have expositions, developments, and recapitulations, although composers struggled with them: recapitulations often entered with the "wrong" melodies in the "wrong" keys and "withheld" resolutions until the end. This made symmetry, let alone golden proportions, elusive.

Several of the authors referred to in earlier chapters analyzed Romantic works. Adkins, Dorfman, Pascoe, Rogers, Rothwell, and Webster all appear in this chapter. In addition, Kirk's dissertation on Chopin's *Preludes op. 28* will appear here.

Unfortunately, a massive set of analyses like Sidorowicz provided for the Classic period is not available. Thus, the previous method of reviewing analyses and approving or refuting them will have to do. In cases of disagreement or particular interest, the whole set will be analyzed.

Chopin caught the attention of six authors, which is curious, since he wrote only three piano sonatas and did not use phi in them. Perhaps the attraction was the regularity of tempo within each piece. It will be interesting to see how well these analyses stand up to scrutiny.

Other composers' use of four-movement symphonic forms seems an attractive avenue of research. Consequently, works of Brahms, Schubert, Schumann, and Tchaikowsky will be carefully examined. Other less orthodox composers such as Berlioz, Rimsky-Korsakov, Saint Saëns, and Wagner were considered by Adkins and will be discussed.

The Romantic period has a much larger body of composers to draw from than previously. In addition to not having inherited a clear tradition of phi proportions from the Classic period, their freedom of expression would seem to work against any regimentation into Fibonacci or phi proportions knowingly.

However, the noisy work of Gustav Fechner with regard to phi in visual relationships and Adolph Zeising's 1854 book on phi in the harmonic sequence could have come to the attention of composers in the second half of the period, but this would not explain how Chopin and Mendelssohn could have known about it. Still, Brahms and Tchaikovsky were the only composers in the second half of the period to come within one measure of GS; Brahms appears to have used it quite a lot in his piano music. The chapter is organized by frequency of mention of the various composers.

Chopin

Frédéric Chopin did sometimes come very close to phi in his *Preludes* and *Nocturnes*.

Preludes

Allen Arthur Dorfman and Kenneth Patrick Kirk analyzed all the *Preludes op. 28.*[3] Both used forms of tension and release to describe the structures of these pieces. Dorfman's diagrams described the shape of accumulated tension, whereas Kirk used the term "turning point" (TP) to indicate the point at which tension began to be released.[4]

Analyses of all 24 *Preludes* follow in figures 6.1 to 6.24. Appropriate comments accompany any figures that are not self-evident.

Figure 6.1. Chopin's *Prelude op. 28 no. 1*

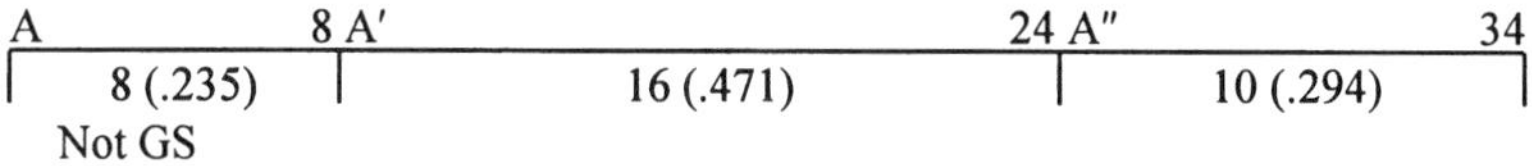

Figure 6.2. Chopin's *Prelude op. 28 no. 2*

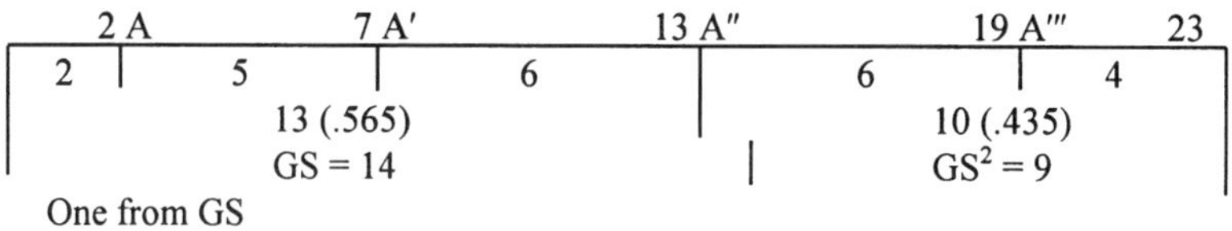

Figure 6.3. Chopin's *Prelude op. 28 no. 3*

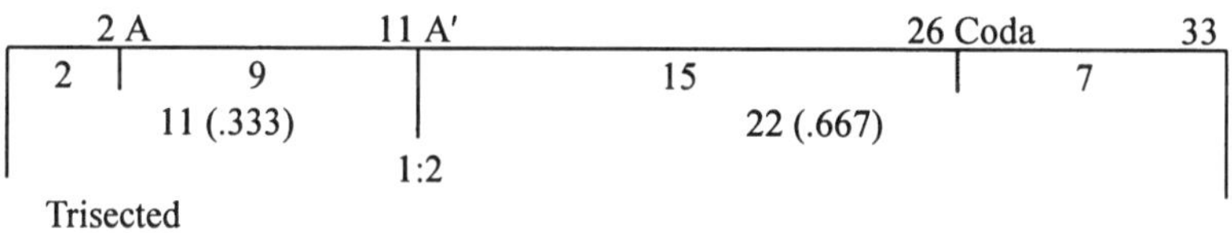

Rogers said that the dramatic peak of *Prelude no. 4* occurred at measure 16 of 25.[5] This is correct, for the leap from $a\sharp^1$ to g^2 is after beat 62, which is GS. He also correctly stated that the piece was *structurally* not phi, as shown in figure 6.4.

Figure 6.4. Chopin's *Prelude op. 28 no. 4* (Rogers)

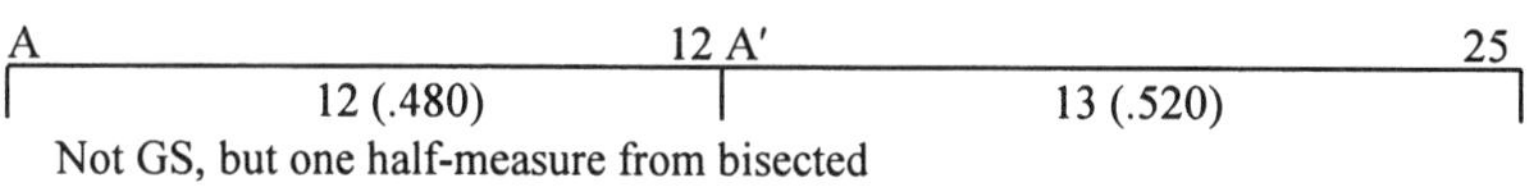

Figure 6.5. Chopin's *Prelude op. 28 no. 5*

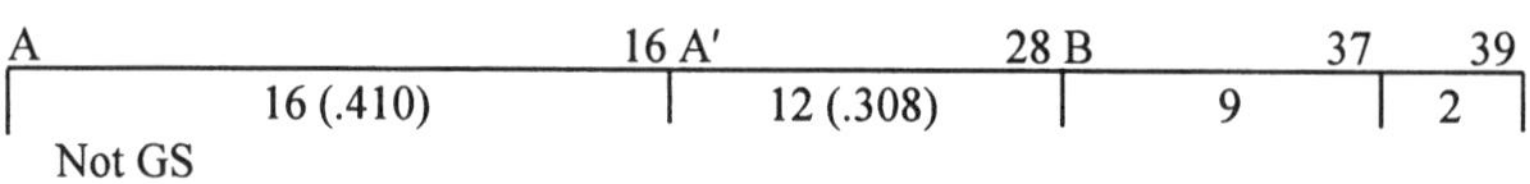

The analysis by Clive Pascoe of *Prelude no. 6* was weak, dividing at measures 8 and 22 out of 26, which has proportions .308 and .846.[6]

Figure 6.6. Chopin's *Prelude op. 28 no. 6* (Pascoe)

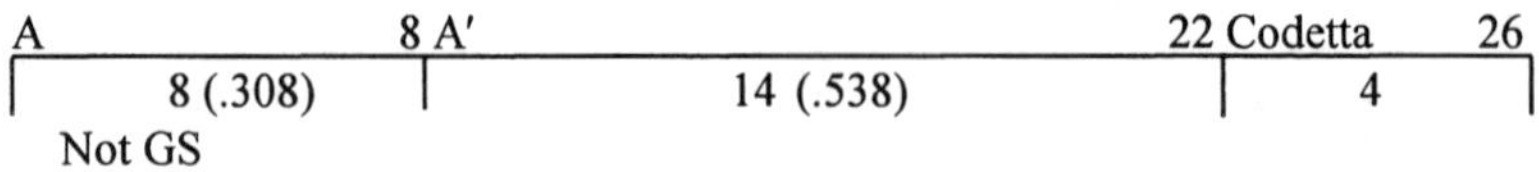

Figure 6.7. Chopin's *Prelude op. 28 no. 7*

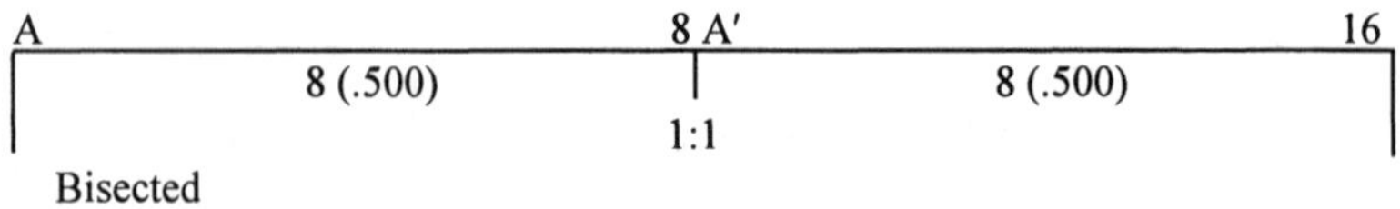

Figure 6.8. Chopin's *Prelude op. 28 no. 8*

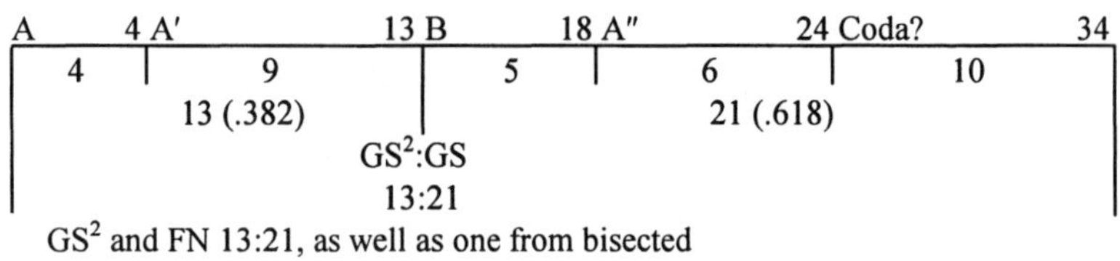

Rothwell and Dorfman said that *no. 9* was divided at phi, with a *dynamic climax* at beat 29.[7] Structurally, however, it is trisected.

Figure 6.9. Chopin's *Prelude op. 28 no. 9* (Madden)

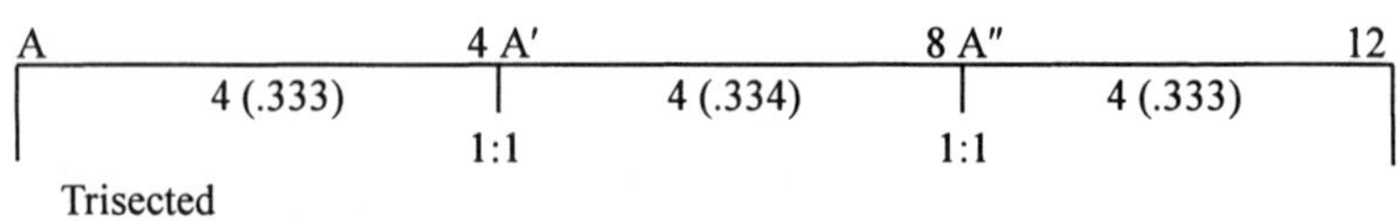

No. 10 is trisected as well.

Figure 6.10. Chopin's *Prelude op. 28 no. 10*

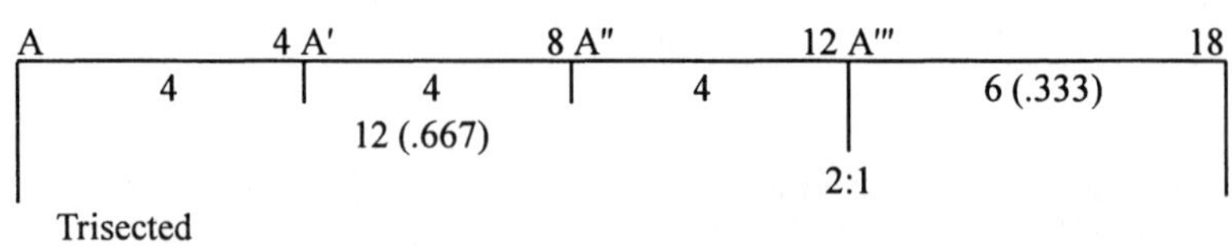

As is *No. 11*.

Figure 6.11. Chopin's *Prelude op. 28 no. 11*

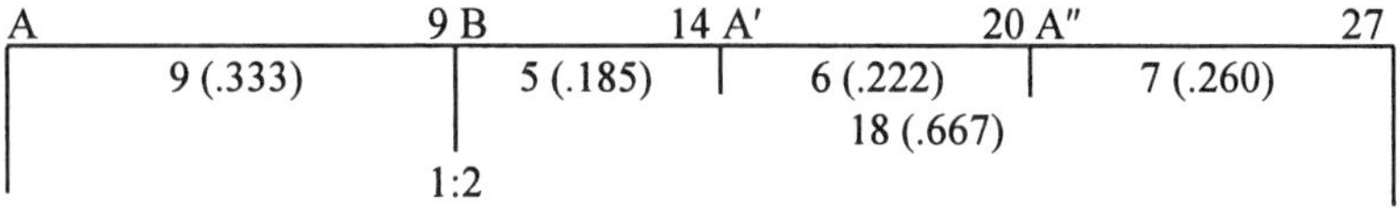

Trisected and one half-measure from bisected

Prelude no. 12 is either divided after measure 28, which is not phi, or after measure 40, which is nearly bisected. See figures 6.12.

Figure 6.12. Chopin's *Prelude op. 28 no. 12*

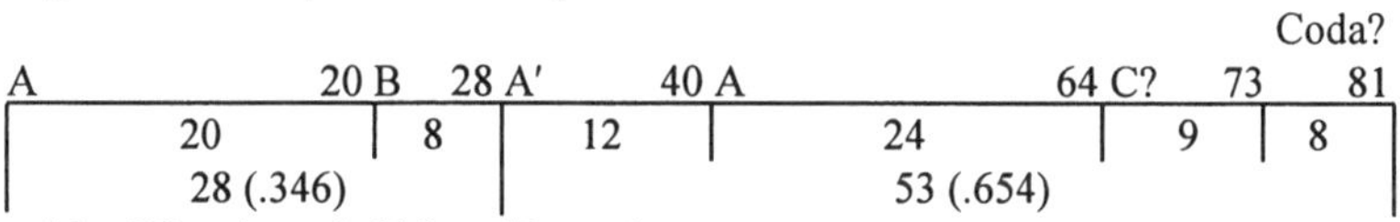

Not GS and one-half from bisected

Figure 6.13. Chopin's *Prelude op. 28 no. 13*

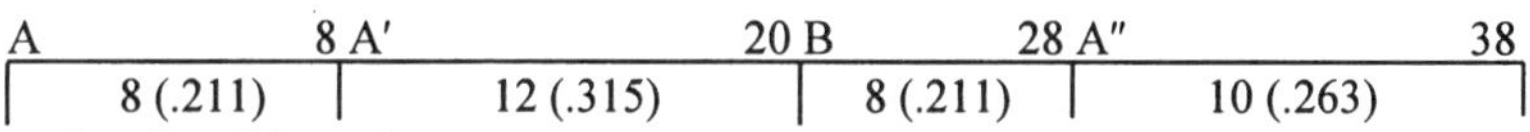

One from bisected

Figure 6.14. Chopin's *Prelude op. 28 no. 14*

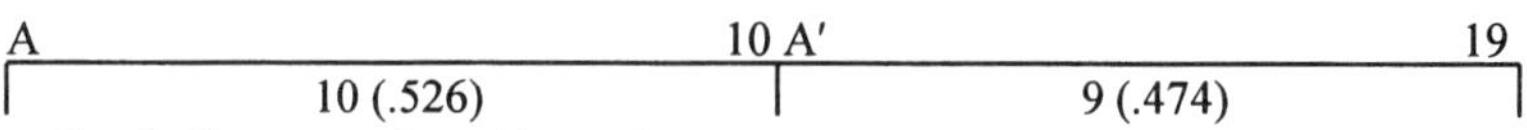

One half-measure from bisected

Figure 6.15. Chopin's *Prelude op. 28 no. 15*

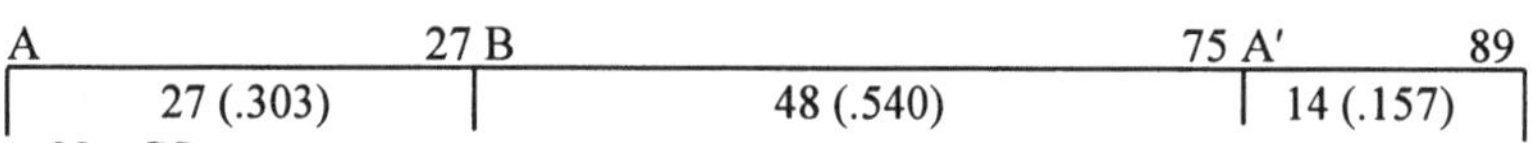

Not GS

Prelude no. 16 is arguable, as shown in figure 6.16. After measure 17, Part A returns with constant modifications and no return to a strong dominant or tonic until the very end. Nonetheless, it divides at one from both GS and GS^2.

Figure 6.16. Chopin's *Prelude op. 28 no. 16*

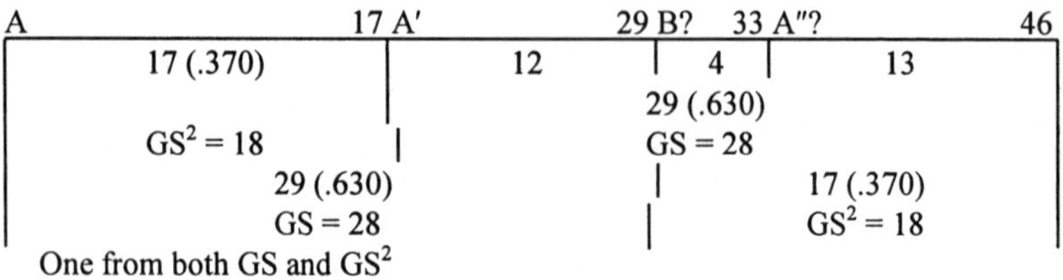

No. 17 clearly divides at GS^2. See figure 6.17.

Figure 6.17. Chopin's *Prelude op. 28 no. 17*

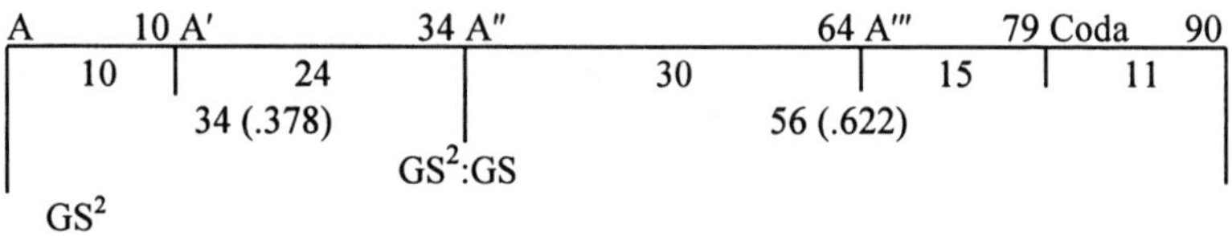

Prelude no. 18 has similar problems to *No. 16*. Part B apparently begins in the middle of measure 9. Thus, the piece is counted in half-measures.[8] This is shown in figure 6.18.

Figure 6.18. Chopin's *Prelude op. 28 no. 18* (Half-measures)

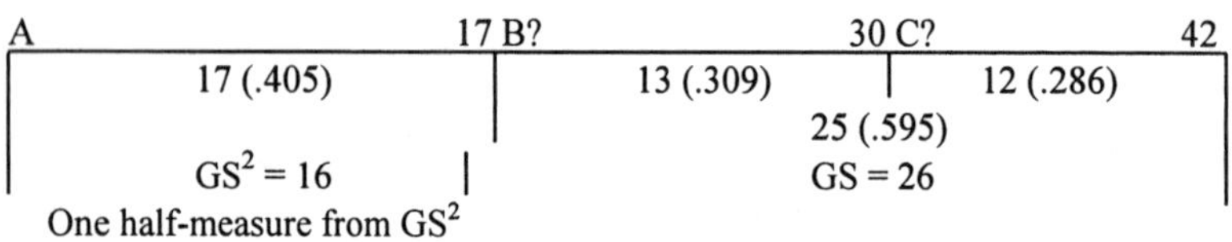

Figure 6.19. Chopin's *Prelude op. 28 no. 19*

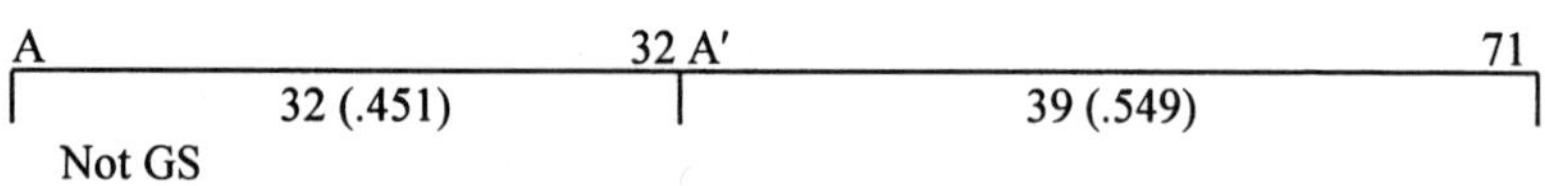

Prelude no. 20 is clearly divided at GS and FN, as shown in figure 6.20.

Figure 6.20. Chopin's *Prelude op. 28 no. 20*

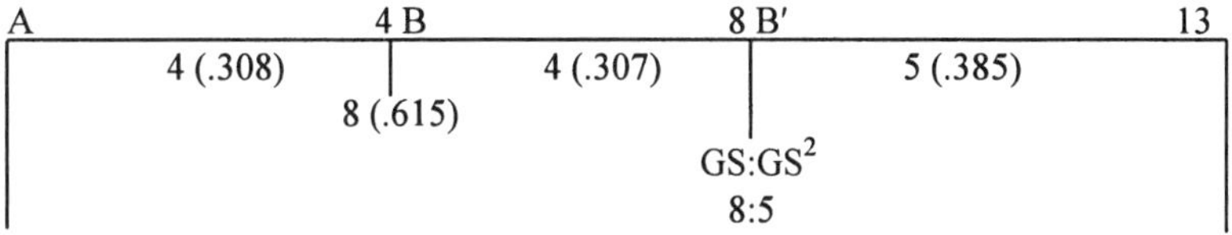

GS and FN 8:5

Figure 6.21. Chopin's *Prelude op. 28 no. 21*

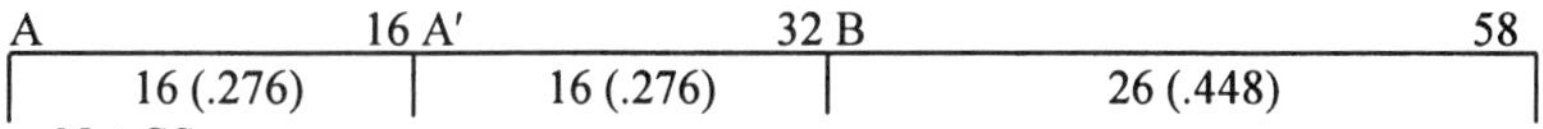

Not GS

Figure 6.22. Chopin's *Prelude op. 28 no. 22* (Half-measures)

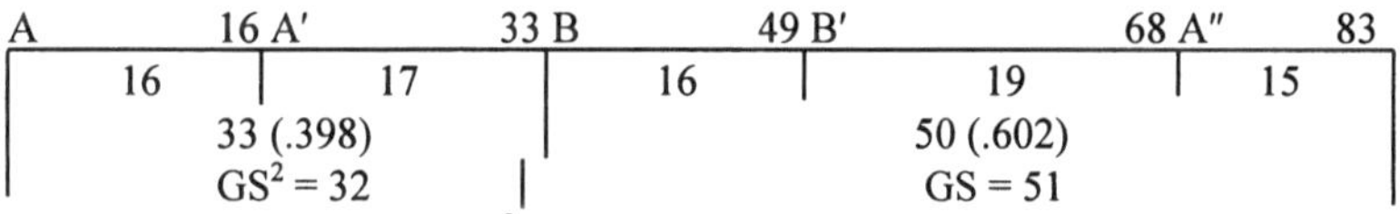

One half-measure from GS^2

Figure 6.23. Chopin's *Prelude op. 28 no. 23*

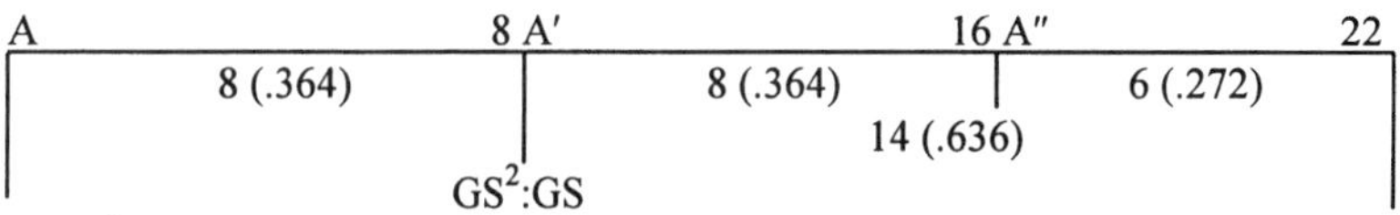

GS^2 although the percentages are very poor

Prelude no. 24 may have structural points after measures 15, 18, 20, 33, 42, and 67, and definitely does after 50 and 64, but they are not phi.

Figure 6.24. Chopin's *Prelude op. 28 no. 24*

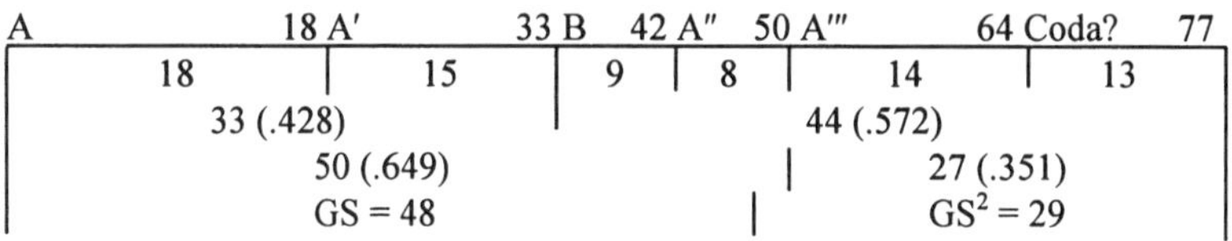

Not GS

A summary of the Chopin Preludes appears in table 6.1. Although there is some use of phi, it is not heavy.

Table 6.1. Summary of the Chopin *Preludes op. 28*

No.	At GS or GS^2	1 meas.	Bi- or tri-sected	Too far	FN	Notes
1				.235/.706		
2				.304/.565		
3			.333	.788		
4				.480		
5				.410/.718		
6				.308/.846		
7			.500			AA′
8	.382			.529	13:21	
9			.333/.667			
10			.667	.444		
11			.333	.518		
12				.346/.494		
13				.526/.737		
14				.526		AA′
15				.303/.843		
16		.370/.630				
17	.378			.711		Poor
18		.405		.714		
19				.451		AA′
20	.615			.308	8:5	
21				.276/.552		
22		.398		.590		
23	.364			.728		Very poor
24				.428/.649		

4/24 = 16.7% at GS or GS^2
7/24 = 29.2% within one measure
5/24 = 30.8% bisected or trisected

Nocturnes

Nocturne op. 9 no. 1 is shown in figure 6.25. It is not fib or phi.

Figure 6.25. Chopin's *Nocturne op. 9 no. 1*

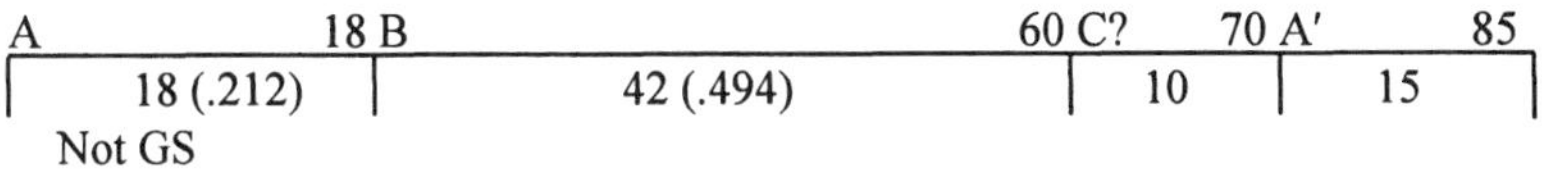

Webster and Pascoe said that *Nocturne op. 9 no. 2* was divided in phi proportions.[9] It is shown in figure 6.26. Although it is one measure from both GS and GS^2, the percentages are very poor.

Figure 6.26. Chopin's *Nocturne op. 9 no. 2* (Webster, Pascoe)

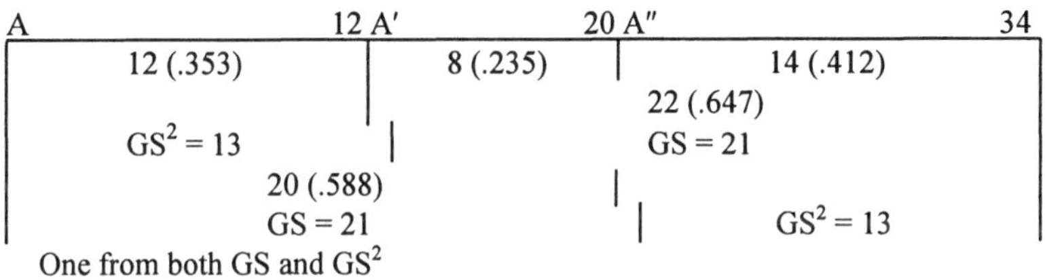

Nocturnes op. 9 no. 3 and op. 15 no. 1 have not been analyzed because of the difficulties introduced by tempo changes.

Op. 15 no. 2 divides at GS^2 if the tempo change (Doppio movimento) is ignored. However, if the tempo change is taken, the proportions are .480 and .520, as shown in figures 6.27a and b.

Figure 6.27a. Chopin's *Nocturne op. 15 no. 2* (without tempo change)

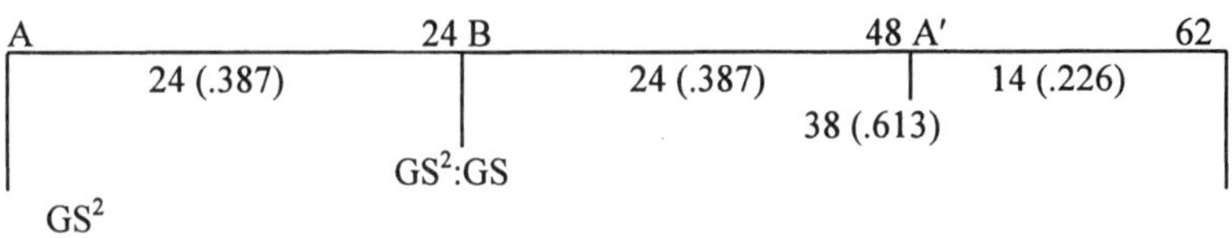

Figure 6.27b. Chopin's *Nocturne op. 15 no. 2* (with tempo change)

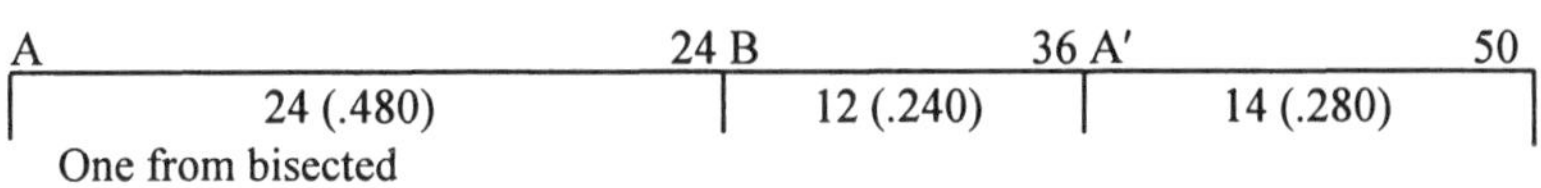

Figure 6.28. Chopin's *Nocturne op. 15 no. 3*

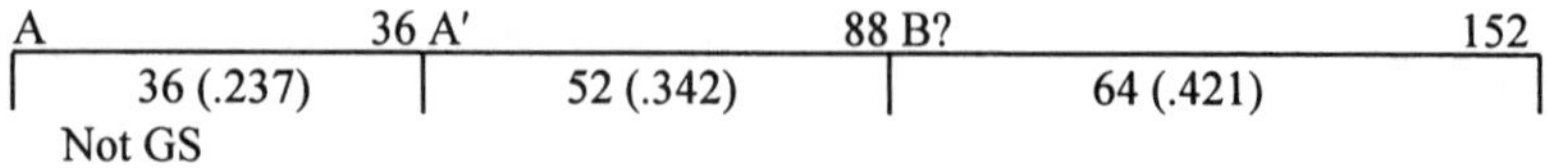

Nocturne op. 27 no. 1 has not been analyzed because of the difficulties introduced by tempo changes.

Figure 6.29. Chopin's *Nocturne op. 27 no. 2*

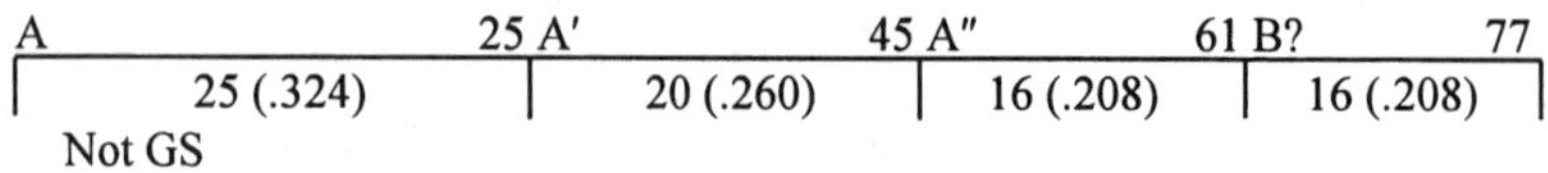

Figure 6.30. Chopin's *Nocturne op. 32 no. 1*

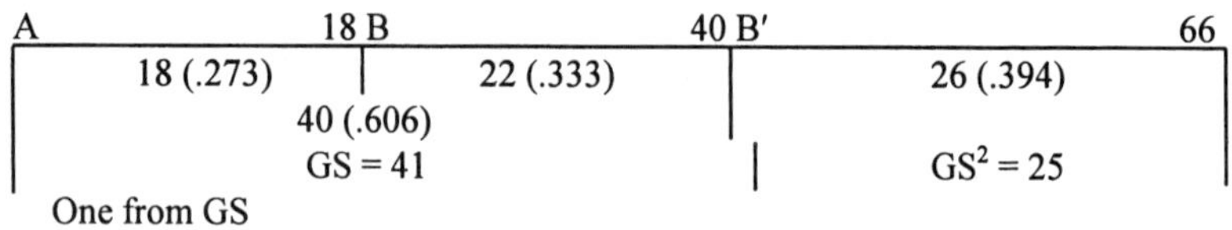

Nocturne op. 32 no. 2 is shown in figure 6.31, ignoring tempo changes.

Figure 6.31. Chopin's *Nocturne op. 32 no. 2*

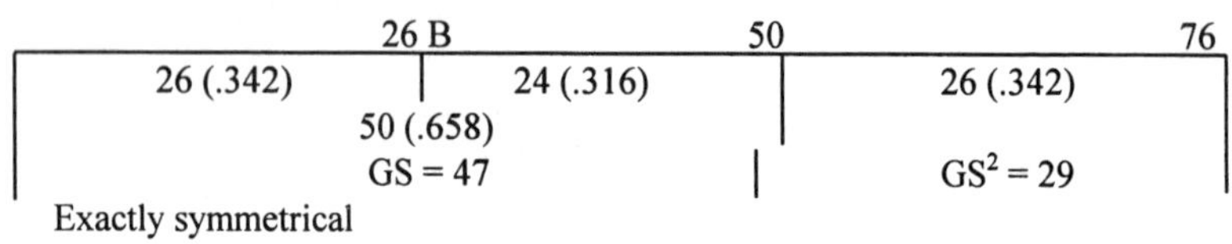

Figure 6.32. Chopin's *Nocturne op. 37 no. 1*

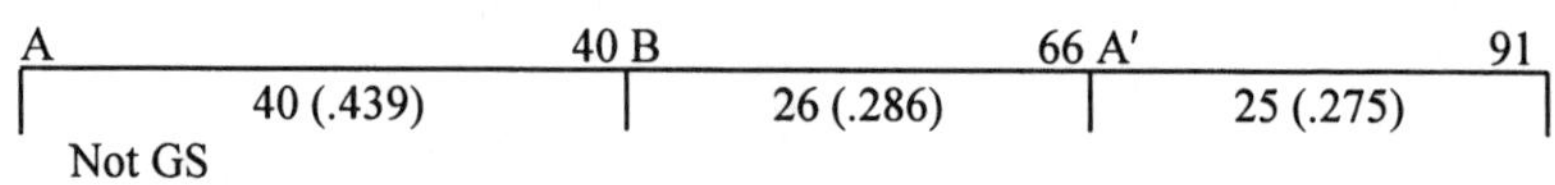

Nocturne op. 37 no. 2, shown in figure 6.33, is not at all clear. The final tonic does not appear until measure 125, which is proportion .893. Nonetheless, an important structural point seems to appear at FN 3:2.

Figure 6.33. Chopin's *Nocturne op. 37 no. 2*

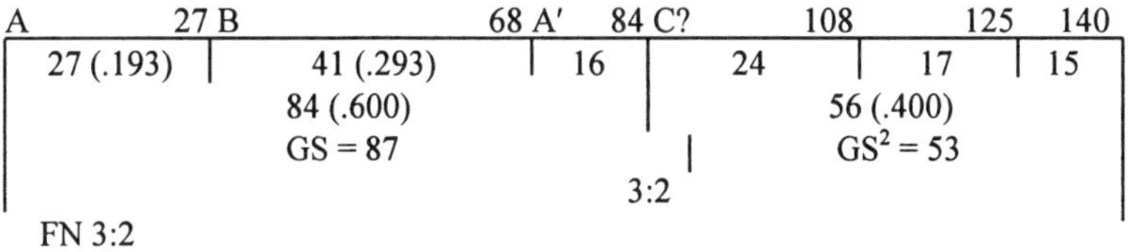

Webster said that *op. 48 no. 1* is in phi proportion, and it is.[10]

Figure 6.34. Chopin's *Nocturne op. 48 no. 1* (Webster)

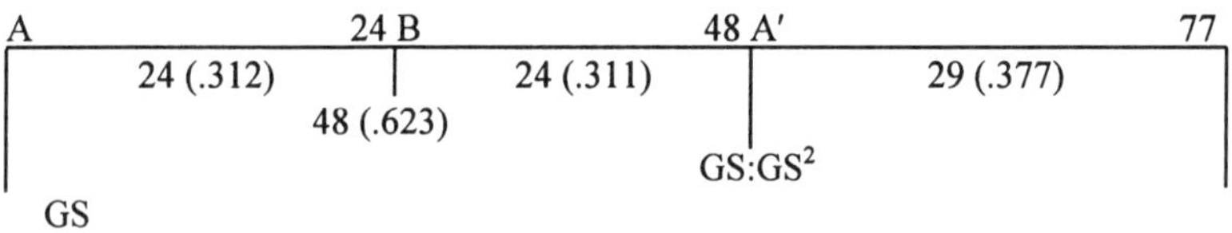

Nocturne op. 48 no. 2 has not been analyzed because of the difficulties introduced by tempo changes.

Figure 6.35. Chopin's *Nocturne op. 55 no. 1*

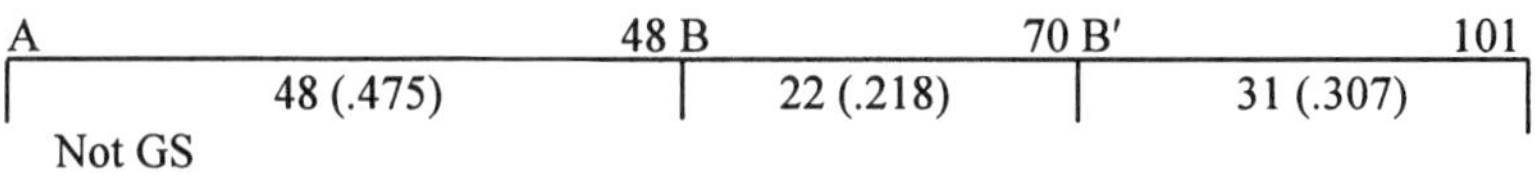

Figure 6.36. Chopin's *Nocturne op. 55 no. 2*

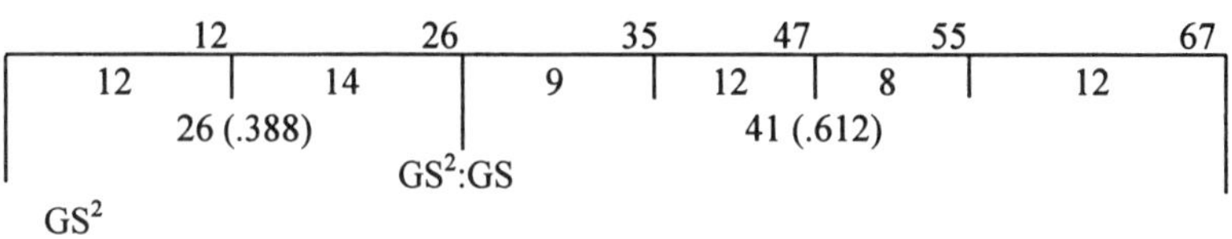

Figure 6.37. Chopin's *Nocturne op. 62 no. 1*

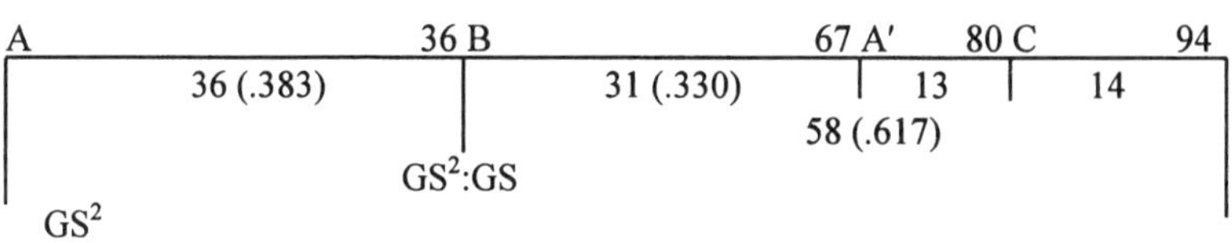

Figure 6.38. Chopin's *Nocturne op. 62 no. 2*

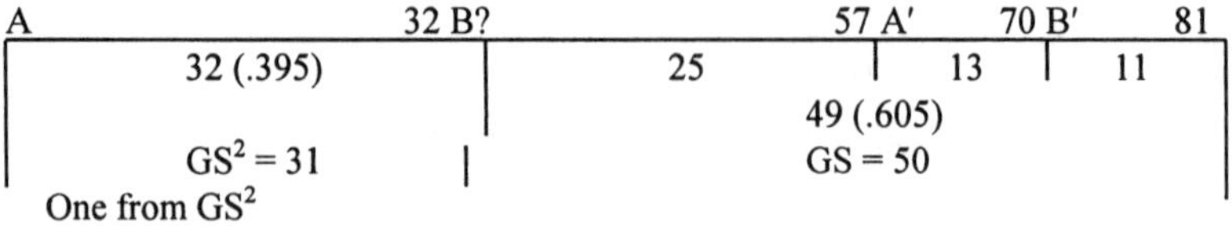

Figure 6.39. Chopin's *Nocturne op. 72 no. 1*

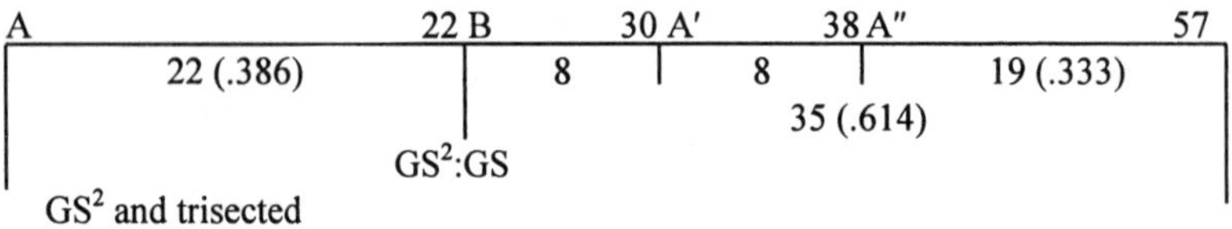

Table 6.2 summarizes the *Nocturnes*.

Table 6.2. Summary of the Chopin *Nocturnes*

Opus/ No.	At GS or GS^2	1 meas.	Bi/tri-sected	Too far	FN	Notes
9/1				.212/706		
9/2		.353/.588				Poor
9/3						Not analyzed
15/1						Not analyzed
15/2	.387			.774		Unadjusted
				.480/.720		Adjusted
15/3				.237/.579		
27/1						Not analyzed
27/2				.324/.584		
32/1		.606		.273		
32/2				.342/.658		Symmetrical
37/1				.439/.725		
37/2				.486/.600	3:2	
48/1	.623			.312		
48/2						Not analyzed
55/1				.475/.693		
55/2	.388			.522/.701		
62/1	.383			.713		
62/2		.395		.704		
72/1	.386		.667	.526		

5/15 = 33.3% at GS or GS^2
8/15 = 53.3% within one measure
This is very high

Brahms

Johannes Brahms was discussed by five of the authors in this book: Adkins, Pascoe, Rogers, Rothwell, and Webster. Many of the analyses were unpersuasive, so only the first movements of the four symphonies will be discussed here and even they do not have phi proportions.

Adkins analyzed the four symphonies of Brahms.[11] He felt that knowing where the phi proportions are would aid in conducting the music so as to be more satisfying than not knowing. Given the wide variation in tolerances in the placement of phi in Brahms' symphonies, the usefulness of this concept is tenuous at best.

Webster also discussed Brahms' symphonies, but all his values were out of tolerance for this book. In addition, he did not take the repeats and sometimes left off the codas. Nevertheless, they are shown here.

The proportions in figures 6.40a–c are calculated on the basis of measures without making adjustments for tempo changes.

Figure 6.40a. Brahms' *Symphony no. 1* Mvt I (Webster) (Without repeat)

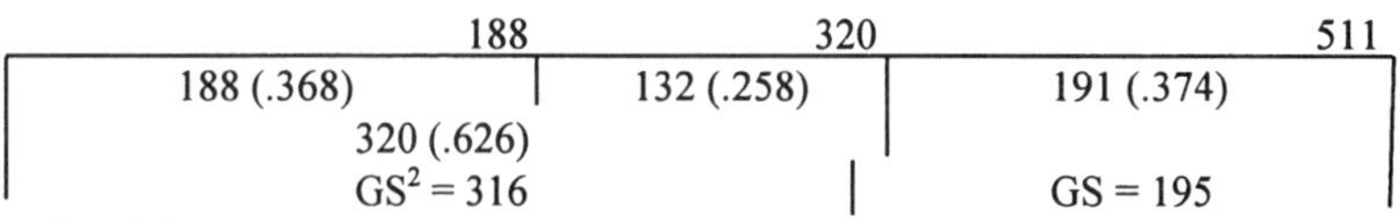

Not GS

Figure 6.40b. Brahms' *Symphony no. 1* Mvt I (Madden) (Without repeat)

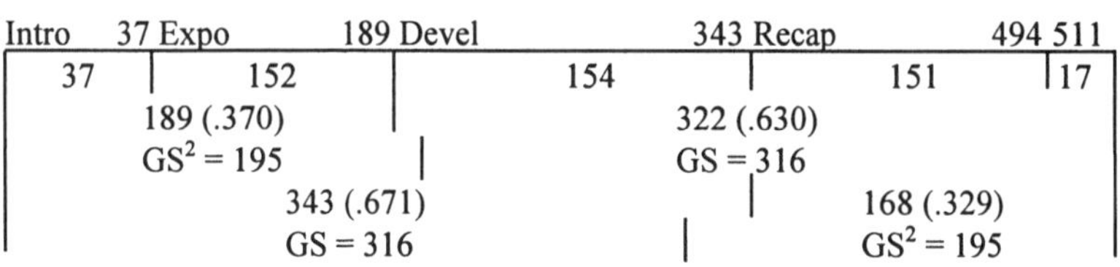

Not GS

Figure 6.40c. Brahms' *Symphony no. 1* Mvt I (Madden) (With repeat)

Intro 37 Expo 188 Rpt 340 Devel 494 Recap 645 662

37 | 151 | 152 | 154 | 151 | 17

340 (.514) | 322 (.486)

Not GS

Symphony no. 2 is shown in figure 6.41a–c. Curiously, the proportions of the main sections are the same in both 6.40c and 6.41c. Is this just coincidence?

Figure 6.41a. Brahms' *Symphony no. 2* Mvt I (Webster) (Without repeat)

186 302 523

186 (.356) | 116 (.222) | 221 (.422)

Not GS

Figure 6.41b. Brahms' *Symphony no. 2* Mvt I (Madden) (Without repeat)

Expo 179 Devel 301 Recap 523

179 (.342) | 122(.233) | 222 (.425)

301 (.575)

GS = 323 | $GS^2 = 200$

Not GS

Figure 6.41c. Brahms' *Symphony no. 2* Mvt I (Madden) (With repeat)

Expo 185 Rpt 364 Devel 486 Recap 708

185 (.261) | 179 (.253) | 122 (.172) | 222 (.314)

364 (.514) | 344 (.486)

Not GS

For *Symphony no. 3*, Adkins admitted that GS has no bearing on the structure. However, it is exactly bisected when counted by measures and ignoring meter changes.

Figure 6.42a. Brahms' *Symphony no. 3* Mvt I (Adkins) (Without repeat)

72 120 224

72 (.322) | 48 (.214) | 104 (.464)

Not GS

Figure 6.42b. Brahms' *Symphony no. 3* Mvt I (Madden) (Without repeat)

Expo — 76 Devel — 124 Recap — 224

76 (.339) | 48 (.214) | 100 (.447)

Not GS

Figure 6.42c. Brahms' *Symphony no. 3* Mvt I (Madden) (With repeat)

Expo — 72 Rpt — 148 Devel — 196 Recap — 296

72 (.243) | 76 (.257) | 48 (.162) | 100 (.338)

148 (.500) | 148 (.500)

1:1

Bisected and nearly trisected at 196

Webster removed the coda in *Symphony no. 4*. With it added back, the movement is still not in phi proportion.

Figure 6.43a. Brahms' *Symphony no. 4* Mvt I (Webster) (Without coda)

145 — 246 — 393

145 (.369) | 101 (.257) | 147 (.374)

$GS^2 = 150$

Not GS

Figure 6.43b. Brahms' *Symphony no. 4* Mvt I (Madden)

Expo — 142 Devel — 258 Recap — 393 Coda — 440

142 (.323) | 116 (.263) | 135 (.307) | 47 (.107)

258 (.586) | 182 (.414)

Not GS

Rothwell found FN internally between portions of the *Intermezzo op. 10 no. 3.*[12] Although 80:50 (8:5), 50:20 (5:2), and 20:12 (5:3) do appear between sections, it is not phi overall, as shown in figure 6.44.

Figure 6.44. Brahms' *Intermezzo op. 10 no. 3* (Madden)

2 — 40 Rpt — 80 — 130 — 150 — 162

2 | 38 | 40 | 50 | 20 | 12

80 (.494) | 82 (.506)

One from bisected

Rothwell felt that *Walz op. 39 no. 5* was divided in 5/8 ratio.[13] We found it to be one from GS^2, as shown in figure 6.45.

Figure 6.45. Brahms' *Walz op. 39 no. 5* (Madden)

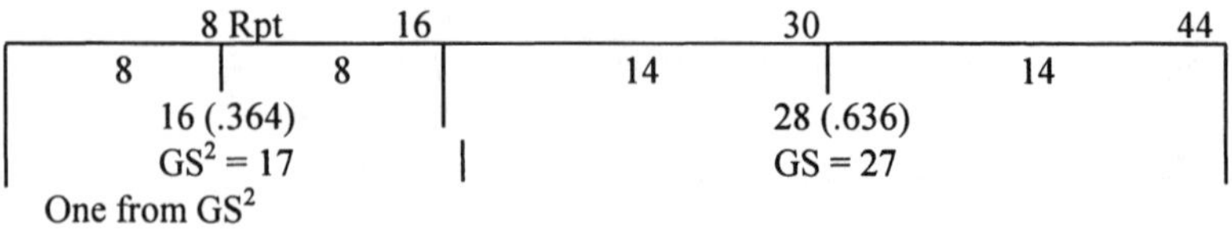

Rothwell miscounted *Walz op. 39 no. 6*, finding it to be 68 measures in length. As such, it would be divided at GS, but it is 69 measures long and divided at one from GS, as shown in figure 6.46a and b.

Figure 6.46a. Brahms' *Walz op. 39 no. 6* (Rothwell)

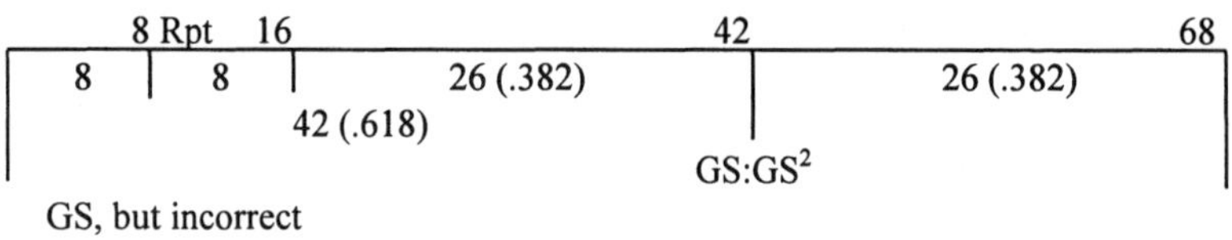

Figure 6.46b. Brahms' *Walz op. 39 no. 6* (Madden)

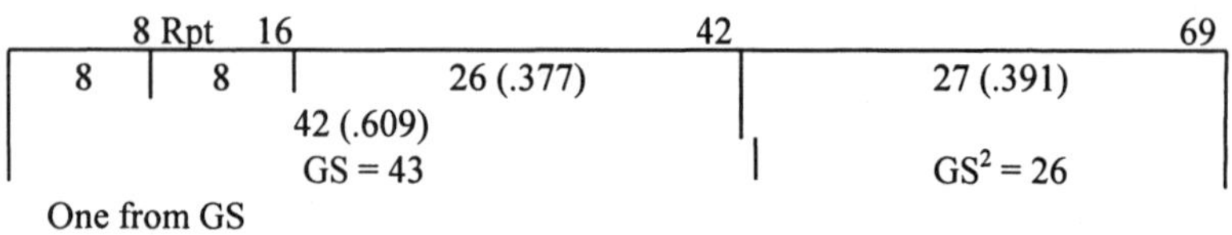

Rothwell was mostly right about *Walz op. 39 no. 11* except that it is divided at 2/5, rather than 3/5, as shown in figure 6.47.

Figure 6.47. Brahms' *Walz op. 39 no. 11* (Madden)

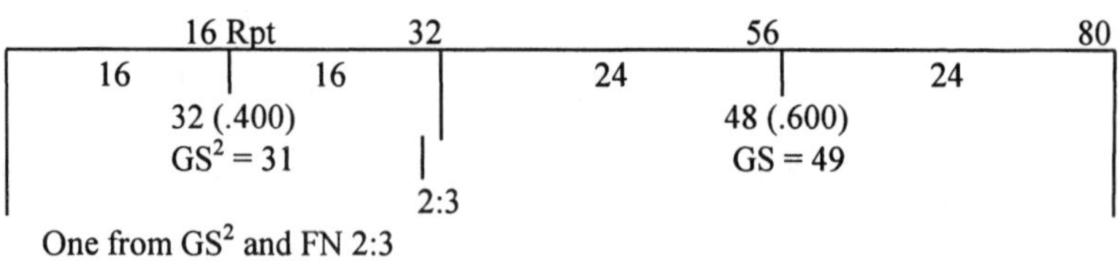

Walz op. 39 no. 12 is divided at both GS^2 and FN 3:5. See figure 6.48.

Figure 6.48. Brahms' *Walz op. 39 no. 12* (Madden)

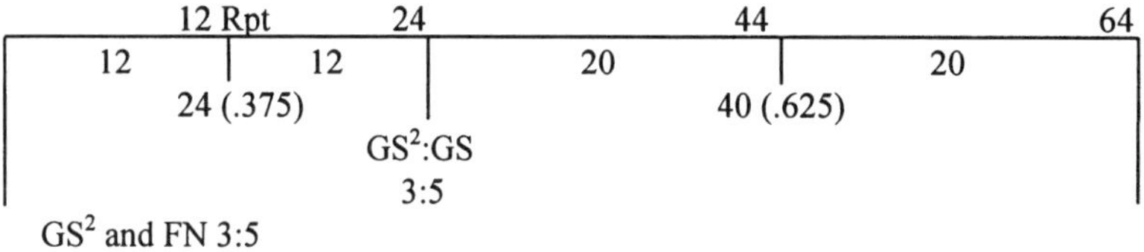

Figure 6.49. Brahms' *Walz op. 39 no. 15* (Madden)

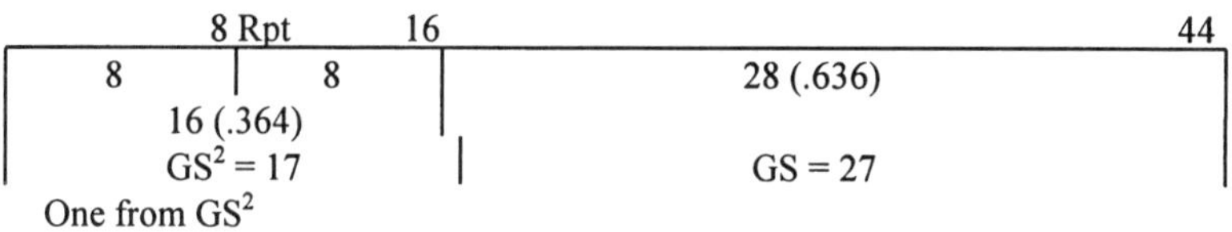

Brahms' *Capriccio op. 76 no. 1* is not GS, but it is FN 3:2.[14]

Figure 6.50. Brahms' *Capriccio op. 76 no. 1* (Rothwell)

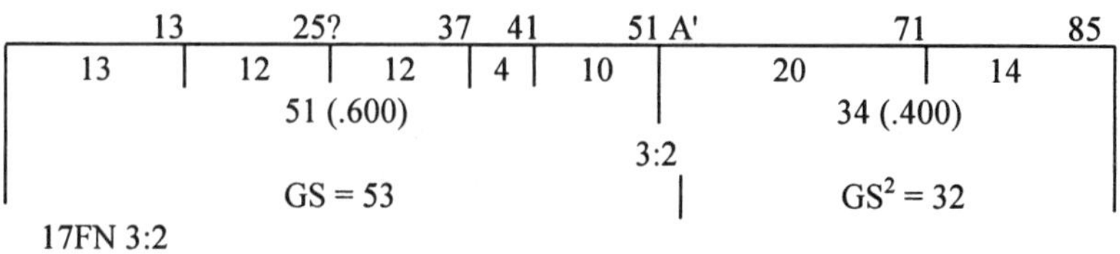

Table 6.4 summarizes Brahms. It shows some use of GS and FN.

Table 6.4. Summary of Brahms

Piece *Symph.*	At GS or GS^2	One meas.	Bi/tri-sected	Too far	FN	Other
No. 1				.368/.626		Web. w/o rpt
				.370/.671		Mad. w/o rpt
				.514/.746		Mad. w/rpt
No. 2				.356/.578		Web. w/o rpt
				.342/.575		Mad. w/o rpt

Piece *Symph.*	At GS or GS^2	One meas.	Bi/tri- sected	Too far	FN	Other
				.514/.686		Mad. w/rpt
No. 3				.322/.536		Web. w/o rpt
				.339/.553		Mad. w/o rpt
			.500	.662		Mad. w/rpt
No. 4				.369/.626		Web. w/o coda
				.323/.586		Mad. w/coda
Op. 10–3				.494		
Op. 39–5		.364		.682		
Op. 39–6	.618					Roth. Incorrect
		.609				Madden
Op. 39–11		.400		.700	2:3	
Op. 39–12	.375			.688	3:5	
Op. 39–15		.364				
Op. 76–1				.600	3:2	2 meas.

5/11 = 45.5% within one measure

Schubert

Franz Schubert was discussed by four of the authors in this book: Adkins, Pascoe, Rothwell, and Webster. We were not satisfied with most of the analyses and will analyze only the first movements of three of Schubert's symphonies.

Symphony no. 4 is calculated on the basis of measures, without adjusting for tempo changes.[15] Although it is three measures from GS, it is quite close to phi in terms of points (7 points ≈ 1.1%). The analysis with the repeat is shown in figure 6.51a. Adkins generally agreed with this assessment, but it probably does not make sense to take the repeat because the recapitulation is so short.[16] It is more like a coda. An analysis with the repeat is shown in 6.51b. Incidentally, the recapitulation is in the key of the dominant.

Figure 6.51a. Schubert's *Symphony no. 4* Mvt I (Without repeat)

Intro 29 Expo 134 Devel 243 Recap 286

29 | 105 | 109 | 43

134 (.469) | 152 (.531)

GS = 177 | GS^2 = 109

Not GS

Figure 6.51b. Schubert's *Symphony no. 4* Mvt I (With repeat)

Intro 29 Expo 134 Rpt 239 Devel 348 Recap 391

29 | 105 | 105 | 109 | 43

239 (.611) | 152 (.389)

GS = 242 | GS^2 = 149

Not GS

Symphony no. 5 is straightforward, but it needs to be shown in two versions, without the repeat of the exposition, and with it, as shown in figures 6.52a and b. In 6.52a the proportions are fairly close to GS and in 6.52b the piece is trisected. Also, note that the development section is very short, reminiscent of the Classic period. Here again it does not seem to make sense to take the repeat.

Figure 6.52a. Schubert's *Symphony no. 5* Mvt I (without repeat)

Expo 117 Devel 160 Recap 299

117 (.391) | 43 (.144) | 139 (.465)

182 (.609)

GS^2 = 114 | GS = 185

Not GS

Figure 6.52b. Schubert's *Symphony no. 5* Mvt I (with repeat)

Expo 117 Rpt 234 Devel 277 Recap 416

117 | 117 | 43 | 139 (.334)

277 (.666)

Trisected

In *Symphony no. 8*, shown in figure 6.53, the sections are approximately equal, with the exception of the coda. Webster claimed that this piece is close to phi, but 8 measures is too far for this book and Adkins stated that phi is insignificant in this piece.[17] We agree.

Figure 6.53a. Schubert's *Symphony no. 8* Mvt I (without repeat)

Expo 110 Devel 217 Recap 327 Coda 368

110 (.299) | 107 (.291) | 110 (.299) | 41

217 (.590) | 151 (.410)

Not GS

Figure 6.53b. Schubert's *Symphony no. 8* Mvt I (with repeat)

Expo 110 Rpt 219 Devel 326 Recap 436 Coda 477
110 | 109 | 107 | 110 | 41
219 (.459) | 258 (.541)
Not GS

Symphony no. 9 was discussed by Adkins, with the suggestion that GS occurs at measure 348, just before the recapitulation.[18] This is outside our tolerance of one measure. Because of the several tempo changes, this piece has not been analyzed here.

Lawrence Moss analyzed Schubert's *Moments Musicaux.*[19] It is structured with Fibonacci numbers in profusion and divided at one from both GS and GS^2, as shown in figure 6.54.

Figure 6.54. Schubert's *Moments Musicaux* (Moss) (without repeats)

A 8 29 B 37 58 66 A 74 95
8 | 21 | 8 | 21 | 8 | 8 | 21
58 (.611) | 37 (.389)
GS = 59 | GS^2 = 36
37 (.389) | 58 (.611)
GS^2 = 36 | GS = 59
One from both GS and GS^2 and contains numerous Fibonacci numbers

Pascoe placed the recapitulation of the first movement of the *Piano Sonata op. 147* at measure 89, which is only close to phi. This is shown in figure 6.55a. With the repeat, it also is not phi, as shown in figure 6.55b.[20]

Figure 6.55a. Schubert's *Piano Sonata op. 147* Mvt I (Madden) (Without repeat)

14 24 40 59 Devel 89 Recap 148
14 | 10 | 16 | 19 | 3 | 11 | 8 | 8 | 14 | 12 | 14 | 19
59 (.399) | 30 (.202) | 59 (.399)
89 (.601)
GS^2 = 57 | GS = 91
89 (.601) | 59 (.399)
GS = 91 | GS^2 = 57
.001 from both FN 2:3 and FN 3:2

Figure 6.55b. Schubert's *Piano Sonata op. 147* Mvt I (Madden) (With repeat)

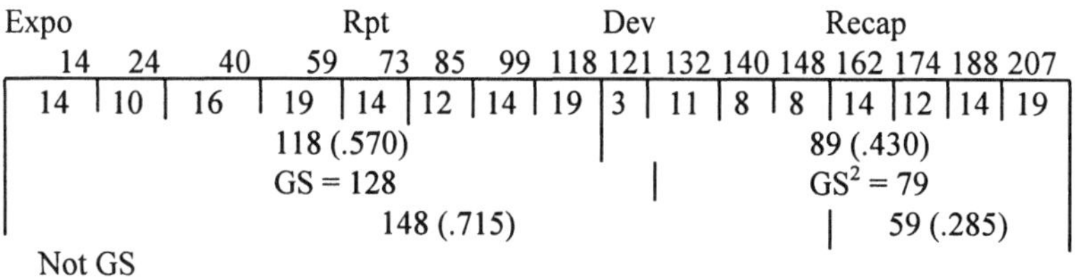

However, the second movement is startlingly symmetrical and divides at both GS and GS^2 as shown in figure 6.56.

Figure 6.56. Schubert's *Piano Sonata op. 147* Mvt II (Madden)

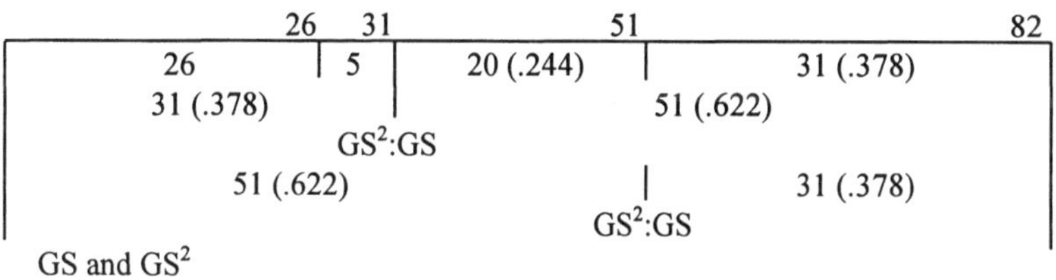

Pascoe showed the fourth movement to be very close to phi as shown in figure 6.57.[21] This is true if the repeat is not taken, which seems like evidence that it should not.

Figure 6.57a. Schubert's *Piano Sonata op. 147* Mvt IV (Pascoe) (W/o repeat)

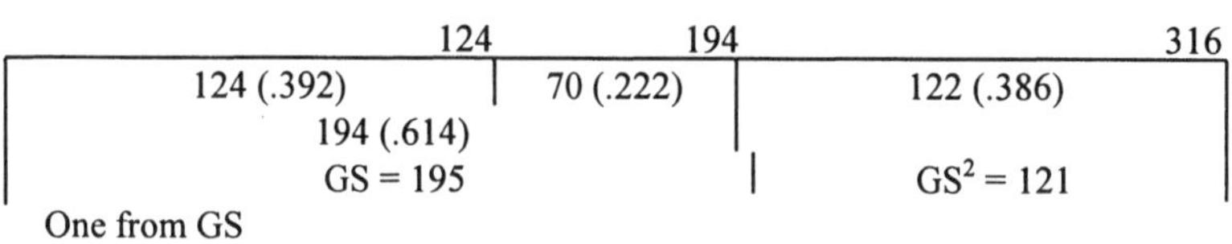

Figure 6.57b. Schubert's *Piano Sonata op. 147* Mvt IV (Madden) (With repeat)

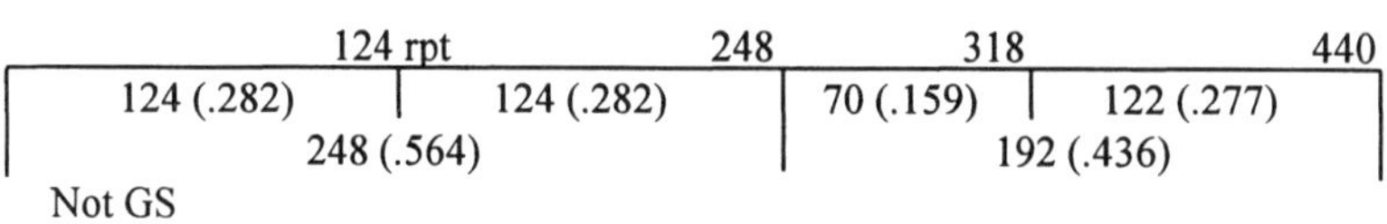

Table 6.5 summarizes the music of Schubert studied here. Although in *Symphonies 4, 5,* and *8* Schubert seems to have not been concerned to use phi proportions, he did use them in some of his piano music.

Table 6.5. Summary of Schubert

Piece *Symphony*	At GS or GS^2	One meas.	Bi/tri-sected	Too far	Other
No. 4				.469	W/o rpt
				.389/.611	W/rpt
No. 5				.391/.535	W/o rpt
			.666	.563	W/rpt
No. 8				.299/.590	W/o rpt
				.459/.683	W/rpt
Mom. Mus.		.389/.611			W/o rpt
Op. 147 I				.399/.601	.001 from 2:3, 3:2
				.570/.715	W/rpt
Op. 147 II	.378/.622				
Op. 147 IV		.614		.392	W/o rpt
				.282/.564	W/rpt

1/7 = 14.3% at GS or GS^2
3/7 = 42.9% within one measure

Schumann

Robert Schumann was discussed by authors Webster and Adkins. Webster mentioned only the *Piano Quintet op. 44*, saying that it measured 338 measures divided at 116 (where GS^2 = 130) and 217 (where GS = 210).[22] Thus, it could not be divided at GS.

Adkins mentioned the *Piano Concerto op. 54*, shown in figure 6.58a and b.[23] It, too, is too far from phi.

Figure 6.58a. Schumann's *Piano Concerto op. 54* Mvt I (Adkins)

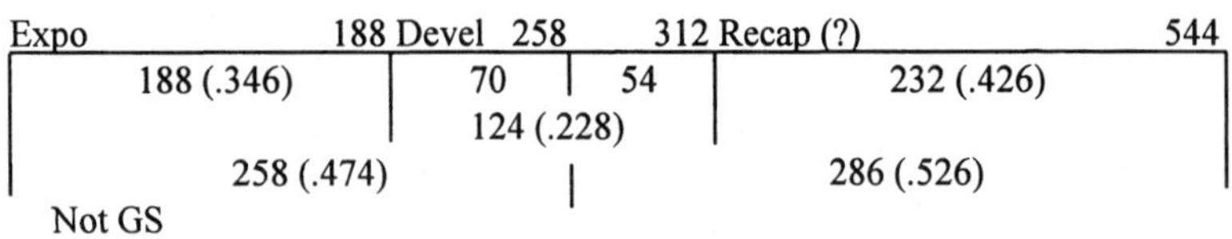

Figure 6.58b. Schumann's *Piano Concerto op. 54* Mvt I (Madden)

Expo	184 Devel	258	311 Recap (?)	544
184 (.338)	74	53	233 (.428)	
	127 (.234)			
258 (.474)		286 (.526)		

Not GS

Adkins also mentioned *Symphonies no. 2* and *no. 3* and found them to not have significant phi proportions.[24] We studied the four symphonies and present them in figures 6.59 to 6.62.

Analyzing *Symphonies no. 1* and *no. 2* is perilous due to changes of tempo. In neither case does adjusting for the introduction or leaving it off result in phi proportions. However, disregarding the tempo change of *no. 1* and counting it by measures brings it to one measure from GS with good percentages, as shown in figure 6.59.

Figure 6.59. Schumann's *Symphony no. 1* Mvt I (disregarding tempos)

Intro	38 Expo	133 Rpt	232 Devel	411 Recap	610
38	95	99	179	199	
232 (.380)			378 (.620)		
$GS^2 = 233$			GS = 377		

One measure from GS^2 and one from FN 233:377

For *Symphony no. 2*, disregarding the tempos does not help much.

Figure 6.60. Schumann's *Symphony no. 2* Mvt I (disregarding tempos) (W/rpt)

Intro	49 Expo	106 Rpt	162 Devel	267 Recap (?)	448
49	57	56	105 (.234)	181 (.404)	
162 (.362)			286 (.638)		
$GS^2 = 171$			GS = 277		

Not GS

Symphony no. 3 is a little easier because there is no introduction. Still, the location of the recapitulation is uncertain. The tonic key appears at both measures 366 and 472, without any clear statement of the theme. We prefer it to be the earlier point.

Figure 6.61. Schumann's *Symphony no. 3* Mvt I

Expo 184 Devel 366 Rec (?) 472 Rec (?) 585

184 (.315) | 182 (.311) | 106 (.181) | 113 (.193)

366 (.626) | 219 (.374)

GS = 362 | GS^2 = 223

Not GS but .001 from 5:3

Symphony no. 4 again disregards the tempo and meter changes. Still, that does not bring it close to GS, as shown in figure 6.62.

Figure 6.62. Schumann's *Symphony no. 4* Mvt I (disregarding tempos) (W/rpt)

Intro 28 Exp 86 Rpt 144 Devel 212 Recap 370 Coda 416

28 | 58 | 58 | 68 | 158 (.380) | 46

144 (.346) | 272 (.654)

GS^2 = 159 | GS = 257

Not GS

1/5 = 20% within one measure
1/5 = 20% very close to FN

Although one of the five pieces is near GS and another one is near FN, it does not appear that Schumann was seriously interested in phi.

Tchaikovsky

Adkins and Webster studied some of the symphonies of Peter Ilyitch Tchaikovsky with mixed results. We studied *Symphonies no. 5* and *no. 6* and found disagreements, so all six symphonies are presented.

Figure 6.63. Tchaikovsky's *Symphony no. 1* Mvt I

Expo 189 Devel 385 449 Recap? 521 663 723

189 | 196 | 64 | 72 | 142 | 60

449 (.621) | 274 (.379)

GS = 447 | GS^2 = 276

Not GS

Figure 6.64. Tchaikovsky's *Symphony no. 2* Mvt I

53 | 138? Devel? | 219? Recap? | 318? Coda? | 368

53 | 85 | 81 | 99 | 50

138 (.375) | 230 (.625)

$GS^2 = 141$ | GS = 227

3:5

FN 3:5

Figure 6.65. Tchaikovsky's *Symphony no. 3* Mvt I

33 56 79 | 173 | 323 | 472

173 (.367) | 150 (.318) | 149 (.315)

$GS^2 = 180$ | GS = 292

Not GS

Symphony no. 4 is ambiguous and many questions remain. Figure 6.66 is a possible solution. Something happens after measures 168 and 272 (one from GS), but they do not seem to be major sections.

Figure 6.66. Tchaikovsky's *Symphony no. 4* Mvt I

Rec? | Coda

Intro 26 Expo | 133 B major | 205 Devel | 294 | 334 354 380 | 438

133 | 72 | 89 | 40 | 20 | 26 | 58

294 (.671) | 144 (.329)

GS = 271 | $GS^2 = 167$

Not GS

Adkins and Webster found no golden proportions in the first movement of Tchaikovsky's *Symphony no. 5* but we did if there are no adjustments for tempo changes.[25] This is shown in figure 6.67.

Figure 6.67. Tchaikovsky's *Symphony no. 5* Mvt I (disregarding tempos)

Intro 37 Expo | 193 Devel | 336 Recap | 450 Coda | 542

37 | 156 | 143 | 114 | 92

336 (.620) | 206 (.380)

GS = 335 | $GS^2 = 207$

One measure from GS and percentage is good

Adkins also studied *Symphony no. 6* and found highly significant phi proportions but we did not.[26]

Figure 6.68. Tchaikovsky's *Symphony no. 6* Mvt I (disregarding tempos)

Coda
Intro 18 Expo 88 Rpt 129 Devel 248 Recap 334 354
18 | 70 | 41 | 119 | 86 | 20
129 (.364) 225 (.636)
GS^2 = 135 GS = 219
Not GS

1/6 = 16.7% within one of GS or GS^2
1/6 = 16.7% at FN

Thus, some of Tchaikovsky's first movements are fairly close to GS, but there seems to be nothing to suggest conscious use of phi.

Mendelssohn

Webster mentioned Felix Mendelssohn's *Songs Without Words no. 20* and *no. 35*, showing that they were close to GS. Figures 6.69 and 6.70 show that this is true. Table 6.6 shows all of the collection.[27] Some have repeats, while others do not, as indicated in the "Other" column.

Figure 6.69. Mendelssohn's *Songs Without Words no. 20* (Webster)

32 48 77
32 (.416) | 16 (.207) | 29 (.377)
48 (.623)
GS:GS^2
GS

Figure 6.70. Mendelssohn's *Songs Without Words no. 35* (Webster)

5 13 23 33 37
5 | 8 | 10 | 10 | 4
23 (.622) 14 (.378)
GS:GS^2
GS

Some of these pieces are not close to phi and percentages are often poor, and many of them are so short that phi might be approached by accident. Nonetheless, it is hard to refrain from thinking that he must have had some awareness of phi.

Table 6.6. Summary of Mendelssohn's *Songs Without Words* (Repeats as marked)

No.	At GS or GS²	1 meas.	Bi/tri-sected	Too far	FN	Other
1				.452/.661		W/rpt
2	.615			.319		
3		.629		.419		W/rpt
4				.417/.683		
5	.384			.721		
				.555/.798		W/rpt
6				.543		2-part form
7		.603		.338		Two rpts
8				.288/.704		W/rpt
9	.625			.411		FN 5:3
10	.376			.584		.001 from 3:5
11	.614	.404				W/rpt
12		.400		.782	2:3	
				.286/.571		W/rpt
13		.635		.302		
14		.397		.705		W/rpt
15	.616			.452/.726		
16	.367/.633					% is poor
17				.315/.639		
18	.618			.411		
19		.367	.664			
20	.623			.416		
21				.544/.696		
22				.300/.567		
23		.373		.699		
24				.524/.700		
25				.537/.695		
26		.632		.325		
27			.667	.417		
28				.526/.895		
29		.593		.356/.763		
30	.378			.544/.778		
31		.595		.270		
32			.664	.310/.487		
33				.286/.531		
34			.333	.450/.680		
35	.622	.351				Poor

No.	At GS or GS^2	1 meas.	Bi/tri-sected	Too far	FN	Other
36				.522/.673		
37				.340/.780		Also .580
38				.484/.736		
39				.475/.656		W/rpt
40				.514/.784		
41			.333	.583		
42				.338/.595		
43		.600		.275	3:2	
44		.393		.656		
45				.453/.698		W/rpt
46				.288/.763		Also .538
47				.247/.521		
48				.424/.727		Also .545
49		.387/.626				
50	.386	.596				

12/50 = 24% at GS or GS^2
13/50 = 26% at one measure
25/50 = 50% within one measure
This seems like a high percentage

Adkins mentioned Mendelssohn's *Symphony no. 3*.[28] Using what we thought was the same edition, our results were far from his; all his rehearsal letters seem to be in the wrong places. Perhaps he adjusted for tempos. Our adjusted counts are shown here. *Symphonies no. 4* and *no. 5* are analyzed here as well.

Figure 6.71a. Mendelssohn's *Symphony no. 3* Mvt I (Madden) (Measures) (without repeat)

Coda
Intro 63 Expo 212 Devel 333 Recap 512 522
63 | 149 (.285) | 121 (.232) | 179 (.343) | 10
212 (.406) | 310 (.594)
Not GS

Figure 6.71b. Mendelssohn's *Symphony no. 3* Mvt I (Madden) (with repeat)

Coda
Intro 63 Expo 211 Rpt 360 Devel 481 Recap 660 670
63 | 148 (.221) | 149 (.222) | 121 (.181) | 179 (.267) | 10
360 (.537) | 310 (.463)
Not GS

Figure 6.72. Mendelssohn's *Symphony no. 3* Mvt II (Madden)

104 175 273
104 (.381) | 71 (.260) | 98 (.359)
169 (.619)
GS^2:GS
GS^2

Figure 6.73. Mendelssohn's *Symphony no. 3* Mvt III (Madden)

37 52 74 93 150
37 (.247) | 15 | 22 (.147) | 19 (.127) | 57 (.380)
93 (.620)
GS:GS^2
GS

Figure 6.74. Mendelssohn's *Symphony no. 3* Mvt IV (Madden)

146 181 303 490
146 (.298) | 35 (.071) | 122 (.249) | 187 (.382)
303 (.618)
GS:GS^2
GS

Figure 6.75a. Mendelssohn's *Symphony no. 4* Mvt I (Madden) (without repeat)

Expo 187 Devel 345 Recap 563
187 (.332) | 158 (.281) | 218 (.387)
345 (.613)
GS = 348 | GS^2 = 215
Not GS but nearly trisected at 187

Figure 6.75b. Mendelssohn's *Symphony no. 4* Mvt I (Madden) (with repeat)

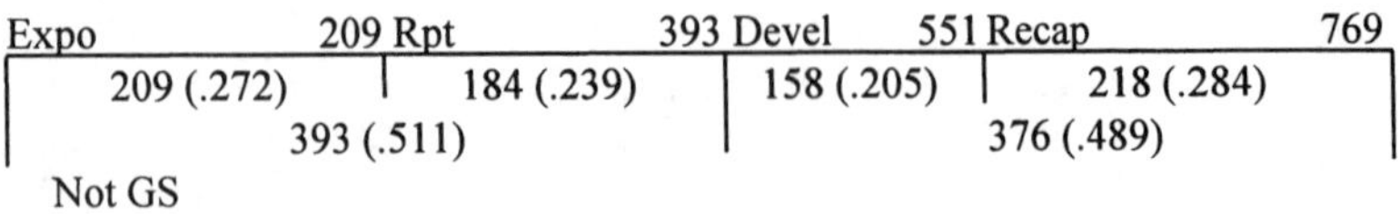

Figure 6.76. Mendelssohn's *Symphony no. 4* Mvt II (Madden)

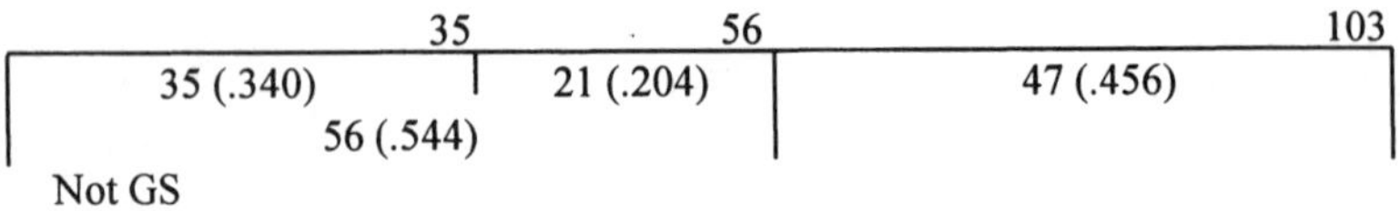

Figure 6.77a. Mendelssohn's *Symphony no. 4* Mvt III (Madden) (w/o repeats)

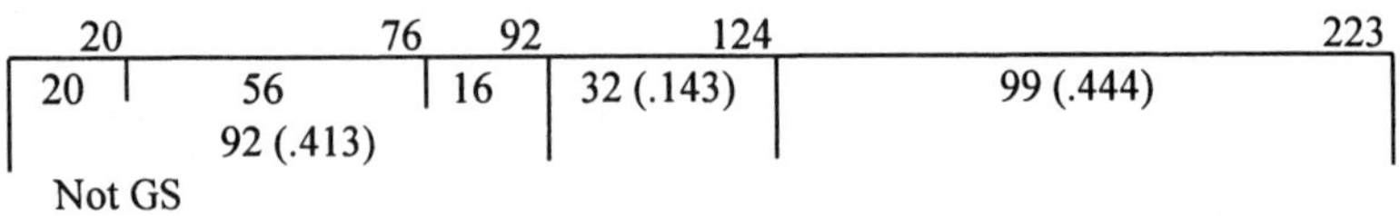

Figure 6.77b. Mendelssohn's *Symphony no. 4* Mvt III (Madden) (with repeats)

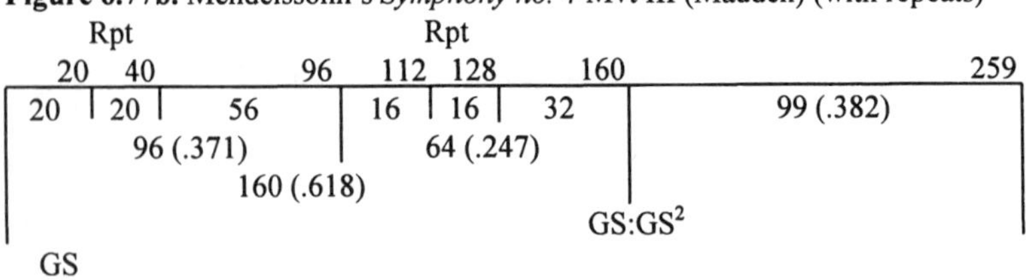

Movement IV has a cadence at measure 159 extended to measure 163, making it GS and GS^2, as shown in figure 6.78.

Figure 6.78. Mendelssohn's *Symphony no. 4* Mvt IV (Madden)

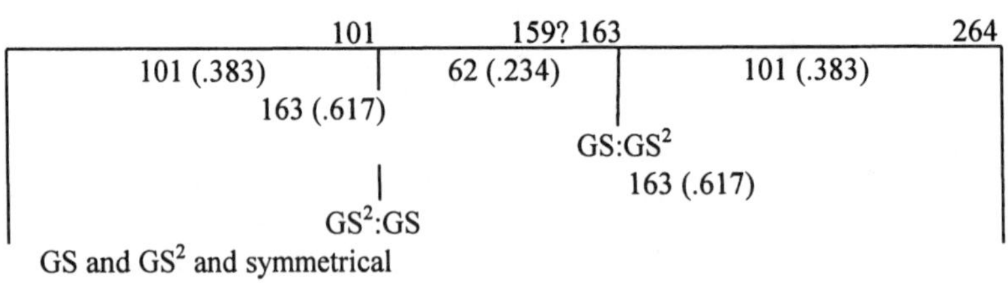

Figure 6.79. Mendelssohn's *Symphony no. 5* Mvt I (Madden)

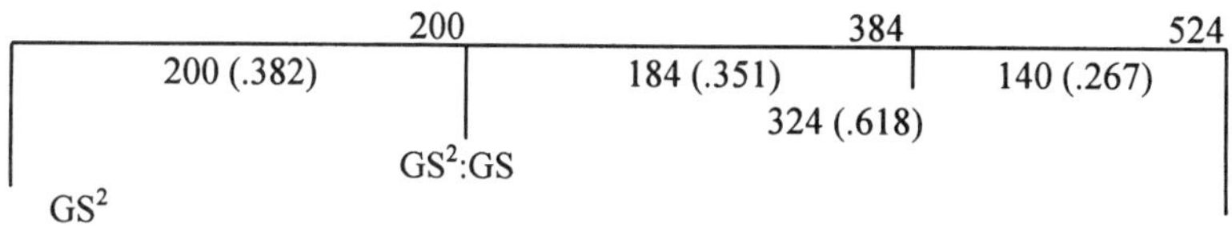

Symphony no. V Mvt II is one measure from GS^2 both with and without repeats, as shown in figure 6.80a and b.

Figure 6.80a. Mendelssohn's *Symphony no. 5* Mvt II (Madden) (Without repeat)

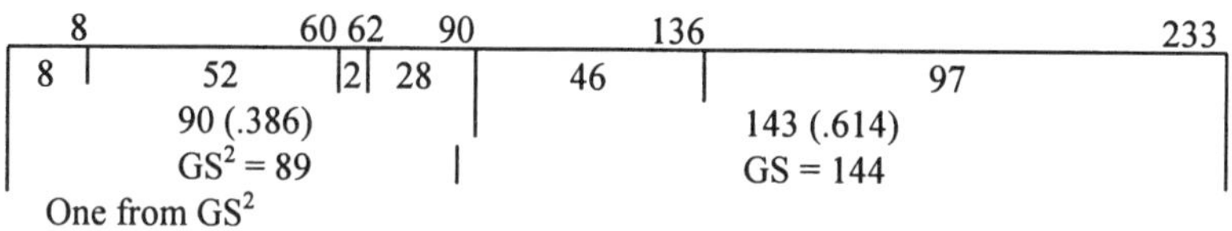

Figure 6.80b. Mendelssohn's *Symphony no. 5* Mvt II (Madden) (With repeats)

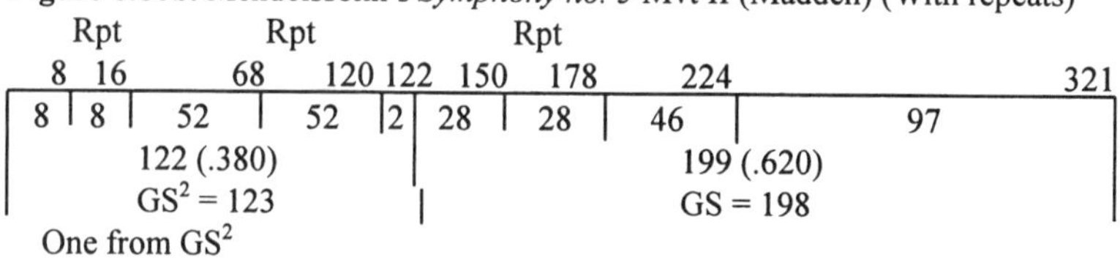

Figure 6.81. Mendelssohn's *Symphony no. 5* Mvt III (Madden)

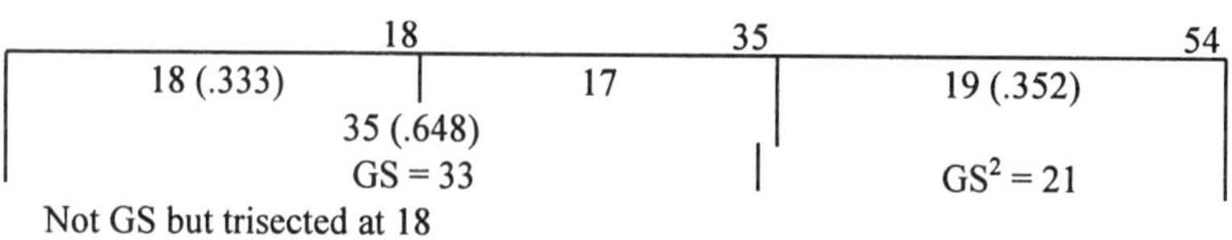

Figure 6.82. Mendelssohn's *Symphony no. 5* Mvt IV (Madden)

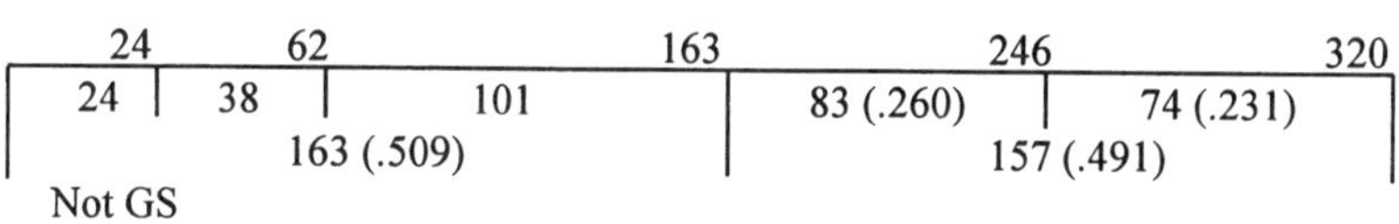

Table 6.7 summarizes Mendelssohn's music studied here.

Table 6.7. Summary of Mendelssohn's symphonies

Piece	At GS or GS^2	One meas.	Bi/tri-sected	Too far	Other
Symph. no. 3 Mvt I				.406/.638	
				.537/.718	W/rpt
Symph. no. 3 Mvt II	.381			.641	
Symph. no. 3 Mvt III	.620				
Symph. no. 3 Mvt IV	.618			.369	
Symph. no. 4 Mvt I			.332	.613	
				.511/.716	W/rpt
Symph. no. 4 Mvt II				.340/.544	
Symph. no. 4 Mvt III				.413/.556	
	.618			.371	W/rpts
Symph. no. 4 Mvt IV	.383/.617				Sym.
Symph. no. 5 Mvt I	.382			.733	
Symph. no. 5 Mvt II		.386		.584	
		.380		.555/.698	W/rpt
Symph. no. 5 Mvt III			.333	.648	
Symph. no. 5 Mvt IV				.509/.769	

6/12 = 50.0% at GS or GS^2
8/12 = 66.7% within one measure

From tables 6.6 and 6.7, it seems that Mendelssohn must have known about phi.

Wagner

Pascoe mentioned the "Prelude" to Richard Wagner's *Mastersingers of Nuremberg*.[29] He timed it, but there are no metronome markings in the score with which to check the analysis. An analysis using measures follows in figure 6.83. From this one piece it might be assumed that Wagner did not use phi, but it is not enough on which to base an absolute judgment.

Figure 6.83. Wagner's "Prelude" from *The Mastersingers of Nuremberg*

40 96 Devel 108 121 150 Recap 220

40 | 56 | 12 | 13 | 29 | 70 (.318)

96 (.436) 54 (.246)

Not GS

Berlioz

Adkins analyzed Hector Berlioz' *Fantastic Symphony* as shown in figures 6.84–88 and Berlioz' *Harold in Italy*, as shown in figures 6.89–91.[30] His analyses correctly showed that Berlioz was usually far from phi.

Figure 6.84a. Berlioz' *Fantastic Symphony* Mvt I (without repeat) (Adkins)

Intro 71 Expo 166 Devel 322 409 Recap 525

71 | 95 | 156 | 87 | 116 (.221)

166 (.316) 243 (.463)

Not GS

Figure 6.84b. Berlioz' *Fantastic Symphony* Mvt I (with repeat) (Adkins)

Intro 71 Expo 166 Rpt 262 Devel 418 505 Recap 621

71 | 95 | 96 | 156 | 87 | 116 (.187)

262 (.422) 243 (.391)

Not GS

Figure 6.85. Berlioz' *Fantastic Symphony* Mvt II (Adkins)

35–38 91 120 175 227–232 319 368

35 | 56 | 29 | 55 | 52 | 92 | 49

227 (.617) 141 (.383)

GS:GS^2

GS

Figure 6.86. Berlioz' *Fantastic Symphony* Mvt III (Adkins)

19 48 64 86–87 116–7 130–1 142 150 199

19 | 29 | 16 | 22 | 30 | 14 | 12 | 8 | 49

64 (.322) 52 (.261) 83 (.417)

Not GS

Figure 6.87a. Berlioz' *Fantastic Symphony* Mvt IV (without repeat) (Adkins)

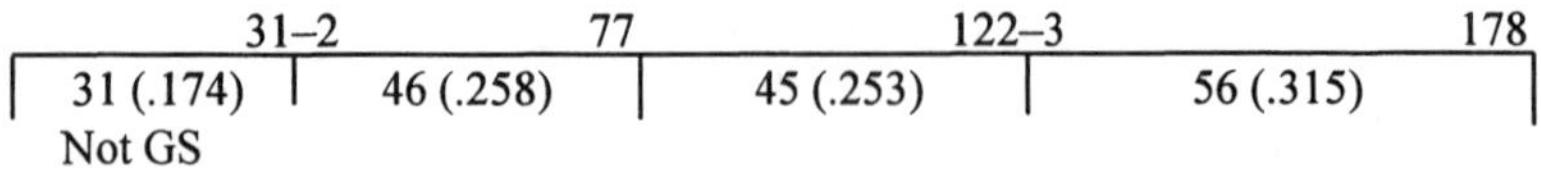

Figure 6.87b. Berlioz' *Fantastic Symphony* Mvt IV (with repeat) (Adkins)

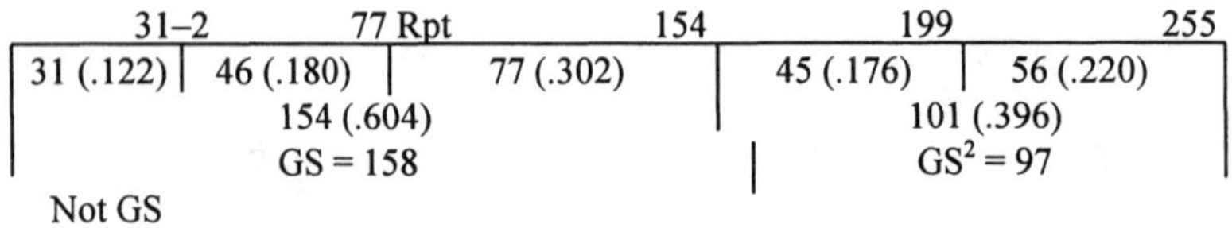

Figure 6.88. Berlioz' *Fantastic Symphony* Mvt V (Adkins)

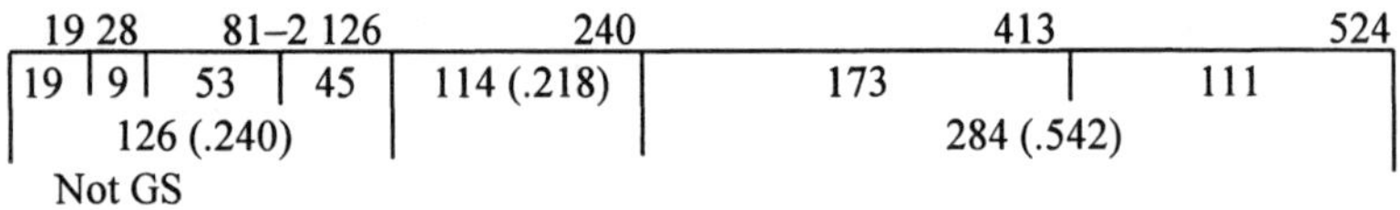

Figure 6.89a. Berlioz' *Harold in Italy* Mvt I (without repeat) (Adkins)

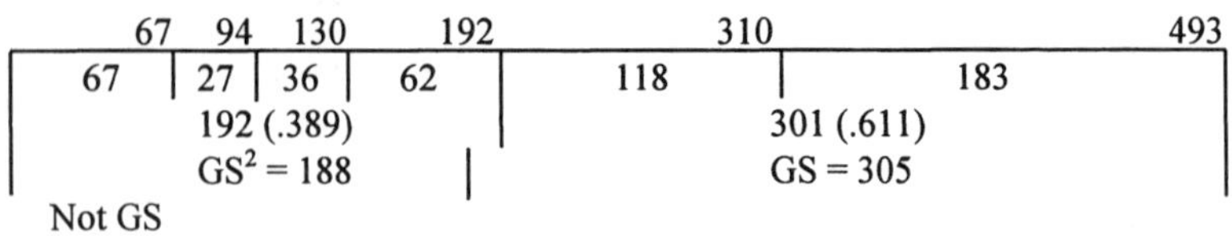

Figure 6.89b. Berlioz' *Harold in Italy* Mvt I (with repeat) (Adkins)

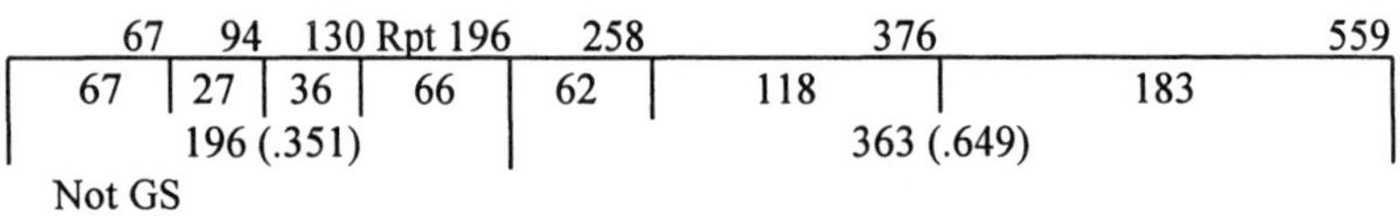

Figure 6.90. Berlioz' *Harold in Italy* Mvt II (Adkins)

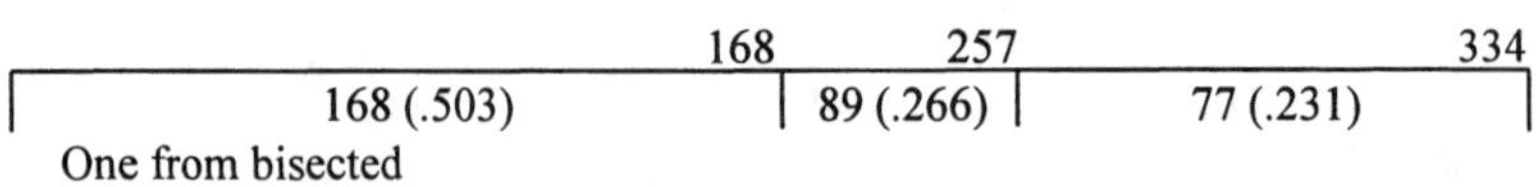

Figure 6.91. Berlioz' *Harold in Italy* Mvt III (Adkins)

31 | 135 | 165 | 208

31 | 104 | 30 | 43

135 (.649) | 73 (.351)

GS = 129 | GS^2 = 79

Not GS

The analysis for Movement IV was inconclusive, but nothing happens at measures GS^2 = 223 or GS = 360 to suggest structural points.

Thus, Berlioz used phi only once in these eight pieces and is seen to not be a frequent user of phi.

Rimsky-Korsakov

Adkins studied Nikolai Rimsky-Korsakov's *Capriccio Espagnol.*[31] We show only Movements I and II. Analyses follow in figures 6.92–3. There are many questions about the analysis of the second movement. Nothing happens at GS.

Figure 6.92. Rimsky-Korsakov's *Capriccio Espagnol* Mvt I (Adkins)

26 | 40 | 72

26 (.361) | 14 (.194) | 32 (.445)

GS^2 = 27.5 | 46 (.639)

GS = 44.5

Marginally GS with poor percentages

Figure 6.93. Rimsky-Korsakov's *Capriccio Espagnol* Mvt II (Adkins)

40? | 88? | 128

40 (.312) | 48 (.375) | 40 (.313)

Not GS but symmetrical

Movement III has the same structure as Movement I and will not be shown. Movements IV and V are very free and Adkins was noncommittal with regard to GS. It seems clear that Rimsky-Korsakov was not an important user of phi proportions.

Saint-Saëns

Adkins also analyzed Camille Saint-Saëns' *Symphony no. 3.*[32] These are shown in figures 7.94–7. In Movement I, there seem to be structural points near GS^2 and GS but they are not the development and recapitulation. Saint-Saëns appears to not have been an important user of GS.

Figure 6.94. Saint-Saëns' *Symphony no. 3* Mvt I (Adkins)

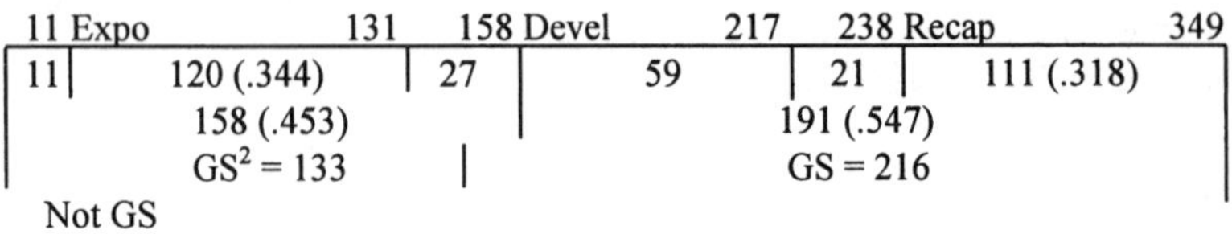

Figure 6.95. Saint-Saëns' *Symphony no. 3* Mvt II (Adkins)

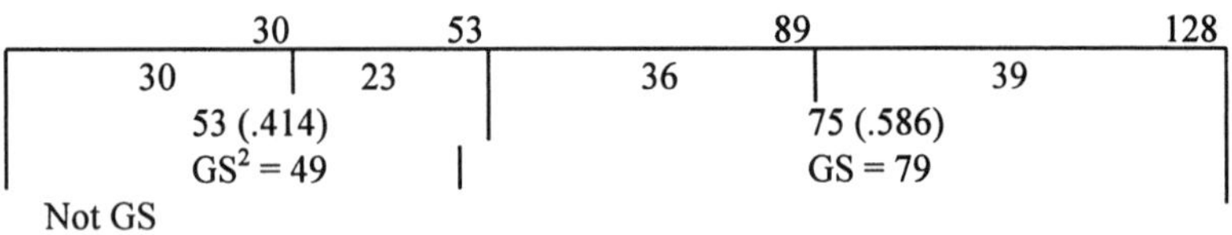

Figure 6.96. Saint-Saëns' *Symphony no. 3* Mvt III (Adkins)

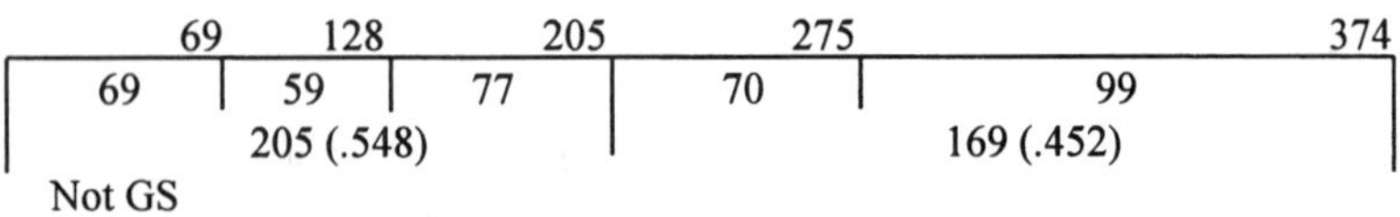

Figure 6.97. Saint-Saëns' *Symphony no. 3* Mvt IV

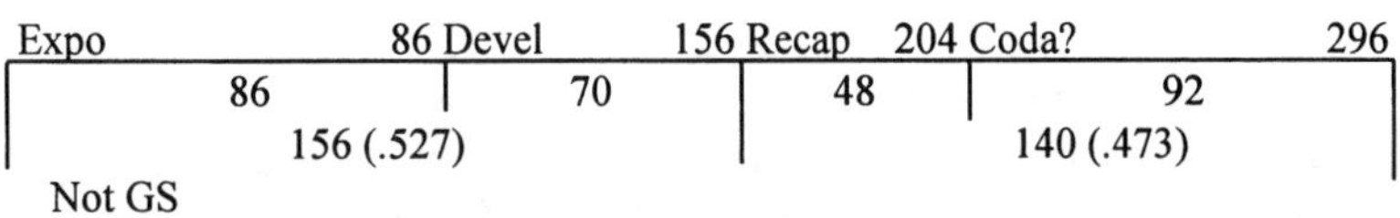

Conclusions

Phi proportions are found fairly often in the Romantic period. In the four movement symphony the golden proportion was found in several movements of Mendelssohn and one movement each of Berlioz, Schubert, and Tchaikovsky. Yet, even those pieces that seemed most Classical differed greatly from Classical structures. Often the developments were very long and the recapitulations were so short as to seem like codas. Symmetry was therefore not a concern for most of these composers.

Brahms and Schubert used phi in much of their piano music as did Chopin and Mendelssohn. Other composers, such as Rimsky-Korsakov, Saint-Saëns, Schumann, and Wagner were not important users of golden proportions.

Chapter Notes

[1] [Ros95 p. 48]
[2] Coleridge et al.
[3] [Dor] [Kirk]
[4] Tension/release and turning points belong to the topic of attractors, which will be discussed in Book 3, *Attractors in Music*. Here, we will use the familiar cadences and melodic returns to define structure.
[5] [Rog p. 60]
[6] [Pas p. 160]
[7] [Rothwell p. 125] [Dor p. 128]
[8] Where the parts begin and end is probably open to dispute.
[9] [Webster p. 244] [Pas p. 159]
[10] [Webster p. 244]
[11] [Adkins p. 105]
[12] [Rothwell p. 123]
[13] [Rothwell p. 113]
[14] [Rothwell p. 121]
[15] Some readers will no doubt dispute the authenticity of this approach, but cutting off the introduction does not improve the values, and adjusting for tempo changes is very complicated and probably not worth the aggravation.
[16] [Adkins p. 111]
[17] [Webster p. 243] [Adkins p. 112]
[18] [Adkins p. 112]
[19] [Moss p. 141, 165]
[20] [Pas p. 201]
[21] [Pas p. 201]
[22] [Webster p. 244]
[23] [Adkins p. 113]
[24] [Adkins p. 47]
[25] [Adkins p. 55]
[26] [Adkins p. 56]
[27] [Webster p. 244] [MenSong]
[28] [Adkins p. 40] [MenSym]
[29] [Pas p. 208]
[30] [Adkins p. 33, 35]
[31] [Adkins p. 44]
[32] [Adkins p. 45]

Chapter 7

20th Century

Introduction

Often called the Modern period, twentieth century music is said to have begun with Debussy in the late nineteenth century. It contains a noticeable increase in the use of Fibonacci numbers and the golden proportion. This is discussed in several dissertations (Adkins, Camp, Hoffman, Pascoe, Rogers, Rothwell, and Wardell) and books (Cogan, Escot, Gillies, Griffiths, Howat, Kramer, Lendvai, Parks, Schillinger, and others).

The number of composers discussed is vast, including Barber, Bartók, Berg, Cage, Dallapiccola, Debussy, Hindemith, Ligeti, Nono, Penderecki, Prokofieff, Rachmaninoff, Ravel, Satie, Schillinger, Schoenberg, Sibelius, Stockhausen, Varèse, Webern, Xenakis, and others. Only a few of these have given any indication of such use in their writings or interviews, and even these statements have usually been tenuous.

Yet, very strong circumstantial evidence of fib and phi is found in the work of several of these composers. They will be discussed in two sections, each in approximately alphabetical order, but with Bartók and Debussy reversed because of the relative importance of their chronology.

Major Users

Debussy

Considerable mystery surrounds Claude Debussy, the answers to which could help explain how he came to write his music as he did. Words such as *impressionism* and *symbolism* are attached to him and there is well-researched speculation that he was a mystic.

Impressionism is the best-known term used to describe his music. This term has become well established in art history to describe a means of conveying "an 'impression' of the scene . . . by means of blurred outlines," and in music history to convey "moods and emotions aroused by the subject."[1] This is a widely accepted designation for Debussy and will not be discussed further here.

Symbolism is another term associated with Debussy. A symbol has been described as a means of indicating thought and feeling. Symbols, then, would be those musical gestures that Debussy used to convey his impressions of his subjects.

The Symbolism movement was formed by Charles Henry, who, among other things, emphasized the need for proportion and balance by means of small-number proportions. That he also proposed the use of the golden proportion has been claimed by José Argüelles, but this has been disputed by Roger Herz-Fischler. [2]

Henry's symbolism was related to Gustav Fechner's ideas about the golden rectangle and some researchers claimed that Georges Seurat and other painters used phi extensively in their work. If so, Debussy would have known about it, but the use by Seurat is in doubt.[3]

It is known that Debussy associated with painters, poets, literary figures, and other highly intellectual people in Paris even more than with musicians. He attended soirees at people's homes and at bookshops where the latest mystical and intellectual schemes were discussed. Whatever was known to them would have been known to him.[4]

It is also well documented that Debussy met mystics as a child and may have been raised by them.[5] He may also have been a Rosicrucian and was thought for a time to have been the grand master of the Priory of Sion.[6] Since he is known to have consorted with Rosicrucian grand master Joséphin Péladan, his use of numbers from the Rosicrucian lore, such as Fibonacci numbers 5, 8, and 13, as well as Lucas numbers 4, 7, and 11, suggests a direct influence.

On the other hand, there may be a simpler explanation than mysticism. Simply paying attention to balance, as proposed by Charles Henry, might have had an equivalent result. A simple example will suffice.

A section of music that is divided into measures of 2 + 3 + 2 consists of Fibonacci numbers, but adds to 7, which is a Lucas number from the additive sequence 1, 3, 4, 7, 11, . . . On the other hand the Fibonacci sequence 3 + 2 + 3 adds to Fibonacci number 8, as does the non-Fibonacci sequence 2 + 4 + 2.[7]

Then there are the so-called Generalized Fibonacci sequences 1, 4, 5, 9, . . . and 1, 5, 6, 11, . . . and the Evangelist sequence 2, 5, 7, 12, 19, . . . The sums using Fibonacci numbers 2 + 5 + 2 and 5 + 2 + 5 form Generalized Fibonacci number 9 and Evangelist number 12 respectively. It is easy to see that numerous combinations of thus-balanced sequences add to one or another of these numbers.

Recall from chapter 1 that the ratios of adjacent Fibonacci, Lucas, Evangelist, or Generalized Fibonacci numbers approach the golden proportion as the sequences are followed to the right. Thus, with a tripartite balance scheme like that described above, the golden proportion is likely to appear wherever we look.

If use of the golden proportion in the Symbolist movement is uncertain, it may be presumed that the part of symbolism that affected Debussy was attention to balance and proportion, with or without further attention to the golden proportion.

In addition to Fibonacci-like numbers, exact bisections—often found in Debussy's music—are sometimes juxtaposed to additive sequences for contrast. In some cases, Fibonacci-like numbers are themselves bisected.

Thus, if we see these numbers in Debussy, can we conclude that he used the golden proportion? Not necessarily. The real determination may have to be based more on whether the piece as a whole is divided in golden proportions than on whether Fibonacci-like numbers are found internally. The above conjectures will be tested in the following analyses by Roy Howat and Richard S. Parks.

Beginning with Debussy's song "Spleen" from the *Ariettes oubliées* (composed in 1888), shown in figure 7.1, we see that the structure balances Fibonacci numbers with a bisection. The Fibonacci numbers mark the structural sections, while the bisection marks a change of key. The exactitude shows that Debussy knew about the Fibonacci numbers as early as 1888 and that his use was intentional.[8]

Figure 7.1. Debussy's "Spleen" from *Ariettes oubliées* (Howat, Madden)

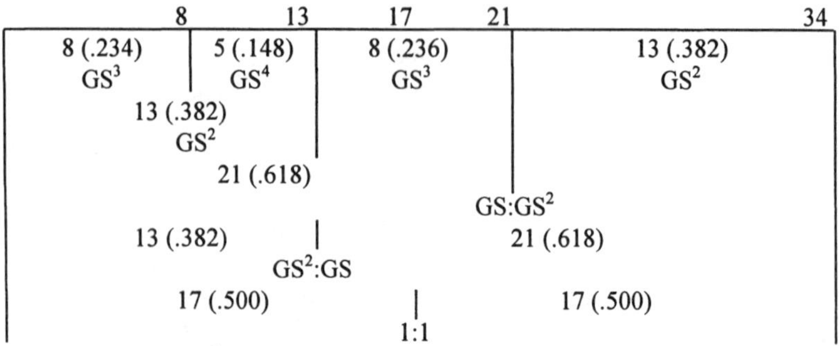

Bisected, GS, GS^2, and FN 13:8:13

Debussy's "Mouvement" from the piano collection *Images* (1905) has a similar form to "Spleen." Here, the proportions are Fibonacci numbers multiplied by 22. The result is one measure from GS^2. See figure 7.2a. However, this is incorrect. Howat dropped measure 177, and measure 110 is dynamic, not structural. Our analysis is shown in figure 7.2b.[9]

Figure 7.2a. Debussy's "Mouvement" from *Images* ("ternary") (Howat)

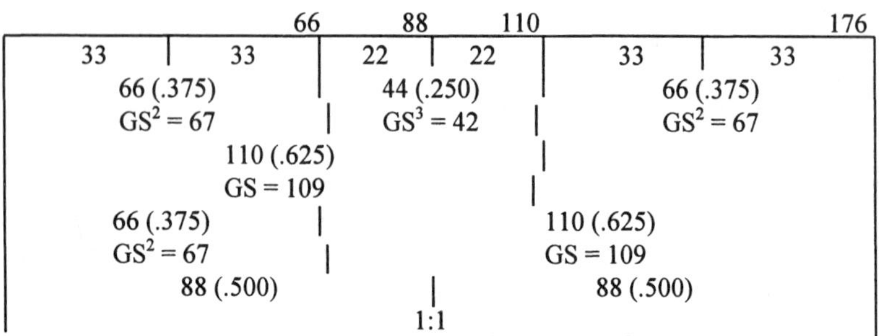

Bisected, 22FN 3:2:3 and one measure from GS and GS^2 but disputed

Figure 7.2b. Debussy's "Mouvement" from *Images* (formal) (Madden)

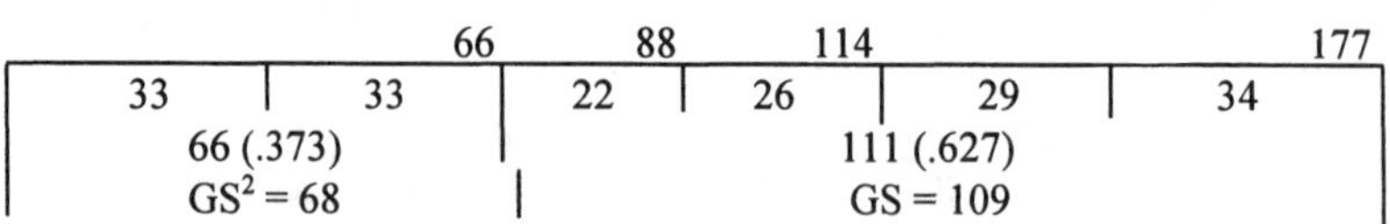

Not GS but one-half from bisected

Howat hedged on the analysis of Debussy's "Reflets dans l'eau" from *Images* (1905), with good reason.[10] At the dynamic level only, the piece divides at GS, as shown in figure 7.3a.

Figure 7.3a. Debussy's "Reflets dans l'eau" from *Images* (dynamics) (Howat)

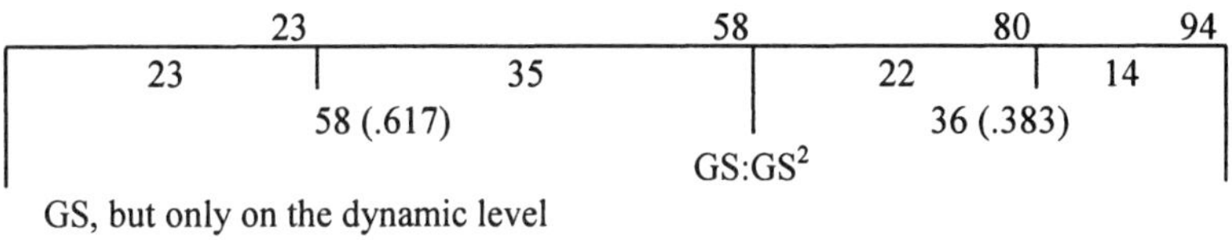

On the tonal level (obvious key changes), the structure is best described as bisected, as shown in figure 7.3b. Despite the presence of Fibonacci numbers 5, 8, 13, 2×13, 2×21, and 2×34, it is arguable whether this piece is really built on the Fibonacci sequence.

Figure 7.3b. Debussy's "Reflets dans l'eau" from Images (tonalities) (Howat)

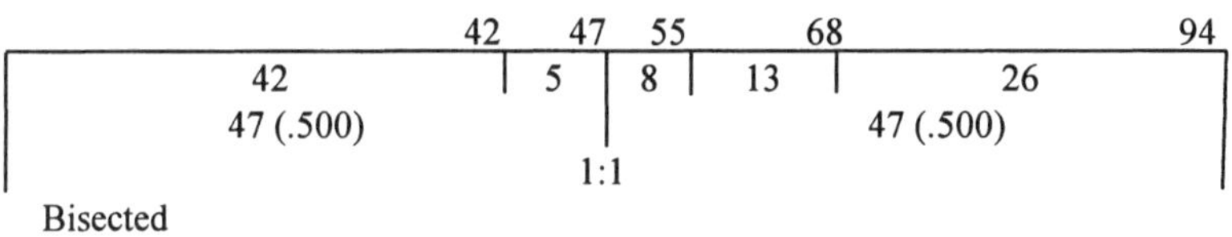

On the formal level, the structure is very clearly not based on fib or phi, despite the incidental presence of Fibonacci numbers 34 and 21, as shown in figure 7.3c. Thus, phi does not govern the form.

Figure 7.3c. Debussy's "Reflets dans l'eau" from Images (formal) (Madden)

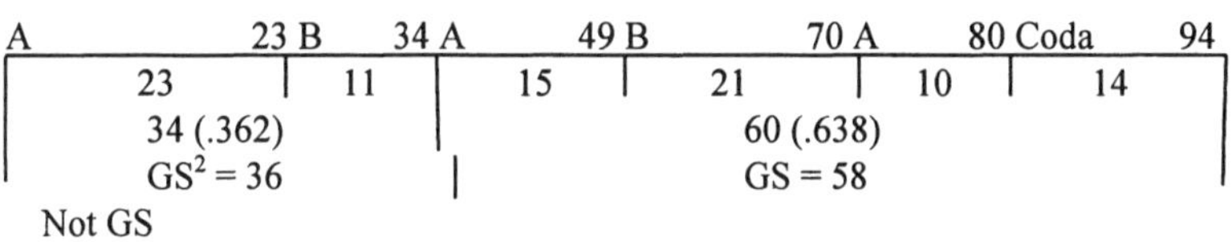

Howat stated that Debussy's "Clair de lune" from the *Suite bergamesque* (1890?) "shows an enormous advance in subtlety."[11] It is so subtle, in fact, that he had to split it into four parallel streams that mix LN and FN, as shown in figure 7.4a.

Figure 7.4a. Debussy's "Clair de lune" from *Suite bergamesque* (Howat)

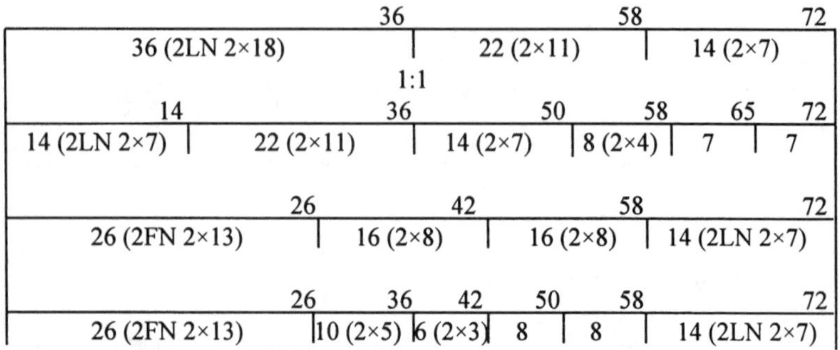

Bisected, 2LN, and 2FN, but not GS

This is not persuasive. A sectional analysis appears in figure 7.4b. Although it is one measure from GS^2, the percentages are poor.

Figure 7.4b. Debussy's "Clair de lune" from *Suite bergamesque* (Madden)

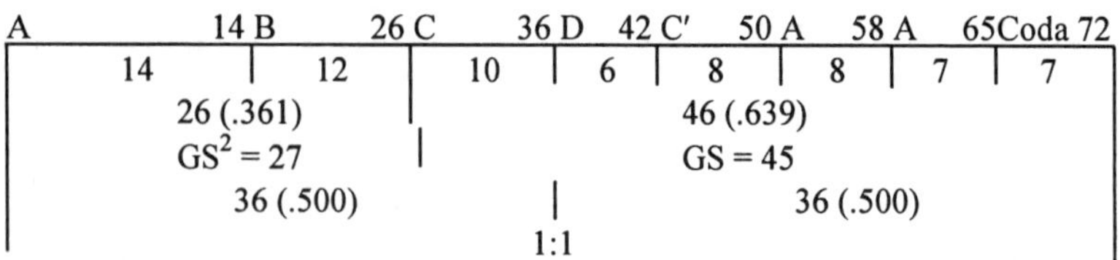

Bisected, one from GS^2

Howat's analysis of Debussy's *L'isle joyeuse* (1904), contained a blizzard of illustrations on several levels.[12] Picking out his figures (in units of 3/8 meter) on the formal level reveals that the piece is divided at GS^2, as shown in figure 7.5.

Figure 7.5. Debussy's *L'isle joyeuse* (Howat, Madden)

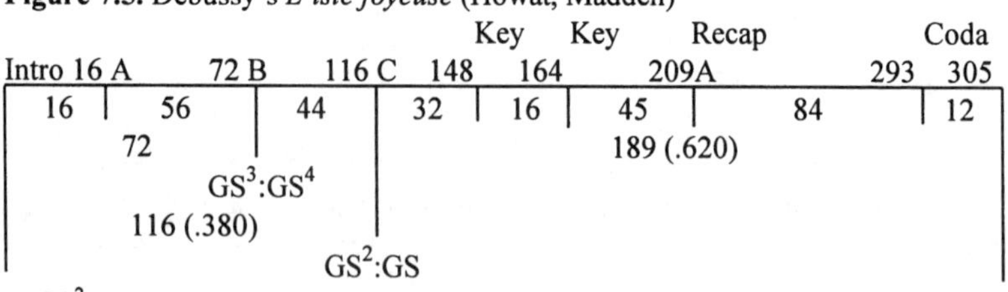

GS^2

Richard S. Parks extended and clarified Howat's work. His illustrations were particularly clear with few points to dispute.

Nuits d'étoiles (ca. 1880) is a five-part ABABA form with a high degree of symmetricality.[13] It is very close to the fraction 5/8, and also one measure from GS, as shown in figure 7.6a.

Figure 7.6a. Debussy's *Nuits d'étoiles* (Parks)

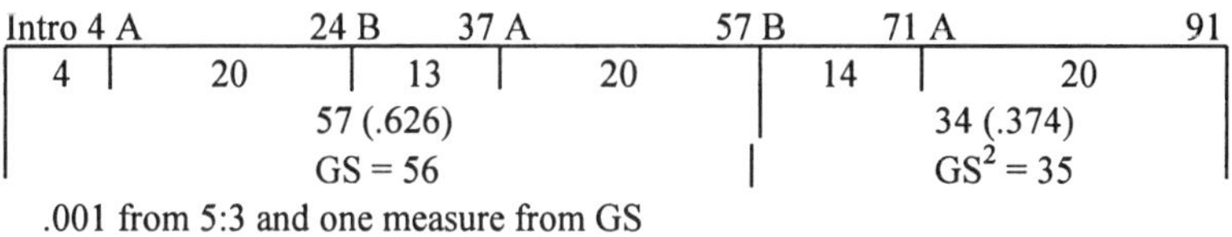

An alternative view is shown in figure 7.6b, which is divided at GS.

Figure 7.6b. Debussy's *Nuits d'étoiles* (Madden)

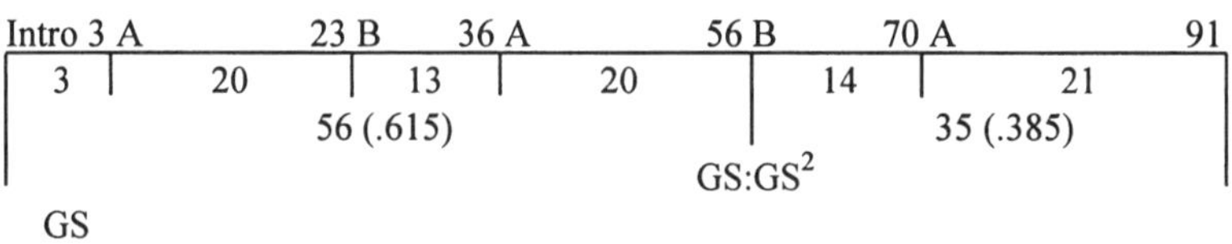

Mandoline (1882) is an ABA form with many Lucas numbers in Parks' analysis.[14] It clearly divides at GS^2 as shown in figure 7.7.

Figure 7.7. Debussy's *Mandoline* (Parks, Madden) (Simplified)

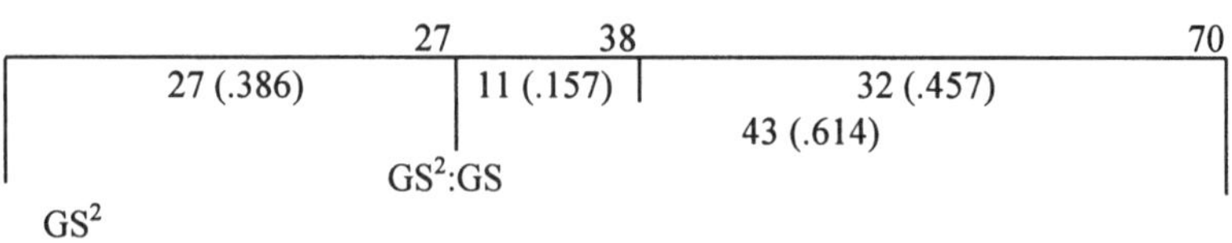

Parks discussed the first and second *Arabesques*, but they do not show convincing use of GS-like numbers, nor do they divide at GS.[15] They will be omitted here.

The String Quartet (1893), on the other hand, divides neatly at GS^2, as shown in figure 7.8.[16]

Figure 7.8. Debussy's *String Quartet* Mvt I (Parks, Madden)

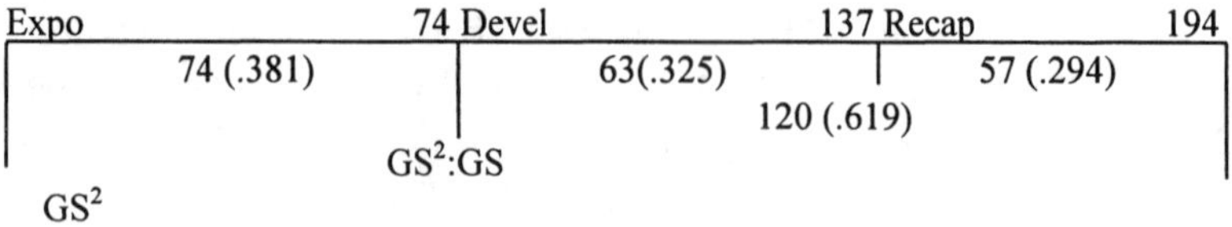

Debussy's *Prelude to the Afternoon of a Faun* (1892-94) does not show golden proportions, but it is nearly bisected.[17] See figure 7.9.

Figure 7.9. Debussy's *Prelude to the Afternoon of a Faun* (Parks, Madden)

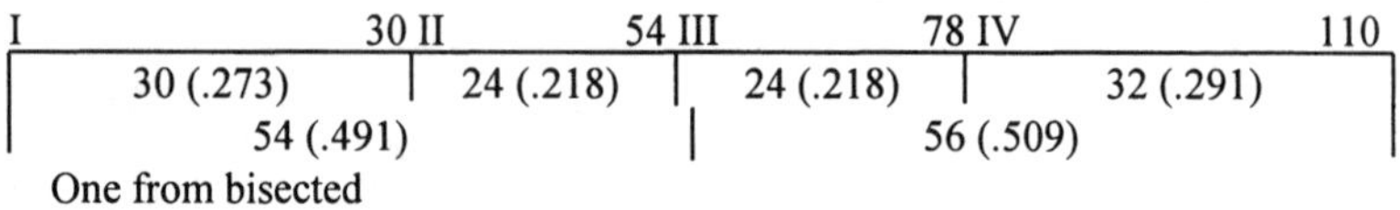

"Sirènes" from the *Nocturnes* (1892–99) is divided at one from GS^2, as shown in figure 7.10.[18]

Figure 7.10. Debussy's *Nocturnes* "Sirènes" (Parks, Madden)

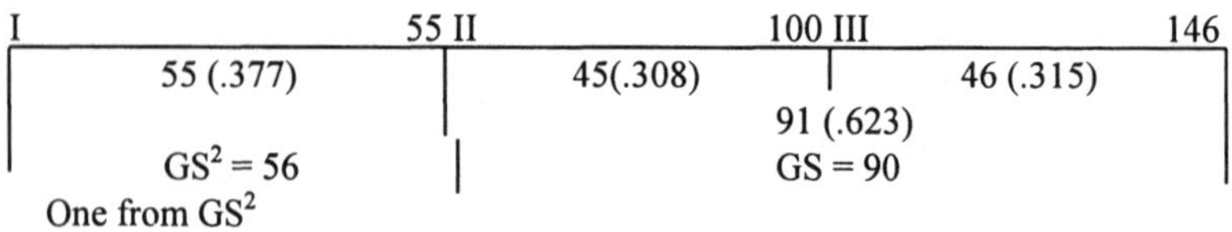

Parks' discussion of the first movement of Debussy's *La Mer* (1903–5) will be combined with Howat's for clarification because neither makes complete sense without the other.[19] Entitled "De l'aube à midi sur la mer" ("From dawn to noon on the sea"), the movement does not divide into phi proportions when analyzed in measures, as shown in figure 7.11a.

Figure 7.11a. Debussy's *La Mer* Mvt I "De l'aube à midi sur la mer" (Parks, Madden) (Measures)

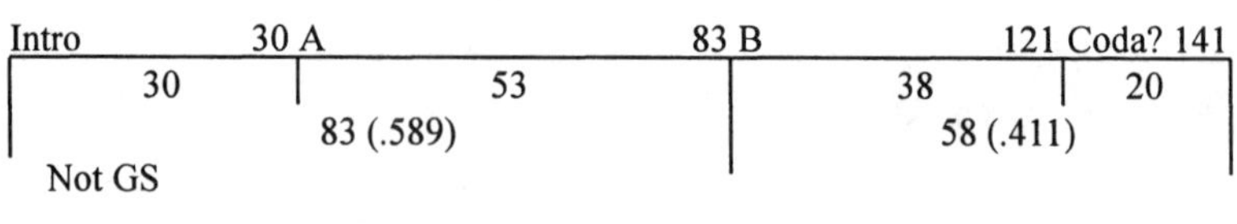

However, according to Howat, when adjusted for tempos (always risky, but in this case successful) and counted in units of 3/8 meter, the division is close to phi proportions, as shown in figure 7.11b.

Figure 7.11b. Debussy's *La Mer* Mvt I "De l'aube à midi sur la mer" (Parks, Howat) (Units)

Intro	65 A	172 B	320 Trans	380 Coda	454
65	107	148	60	74	
172 (.379)		282 (.621)			
GS^2 = 173		GS = 281			

One from GS^2

Parks left the final two movements of *La Mer* to Howat.[20] They are far less convincing. Replete with seemingly unnecessary information in an attempt to show all levels of analysis at once (dynamics, tonalities, motives, and sections), they are scrambled and hard to unravel.

For this book, just the sectional level will be extracted, which sometimes works out to GS and FN. This is not to say that other levels are not important and interesting, but they do not fit the criteria of this book. In addition, Howat's analysis could benefit from a clear separation of the levels, for one can study an analysis for some time before realizing that it is not on the assumed level.

The second movement of *La Mer*, "Jeux de vagues," appears to be roughly in sonata form: it has what Howat called a recapitulation, beginning after measure 162 (his unit 170). This is shown in figure 7.12a.[21]

Figure 7.12a. Debussy's *La Mer* Mvt II "Jeux de vagues" (Sectional level) (Howat) (Units)

	170 Recap	269
170 (.632)	99 (.368)	
GS = 166	GS^2 = 103	

Not GS

Although Howat did not say that this movement was divided at GS on the sectional level—and indeed it is not if one accepts his adjustments for changes of tempo—it does divide at one measure from GS if the adjustments are ignored and the measures used instead, as seen in figure 7.12b. This is opposite to what we did in the first movement.

Figure 7.12b. Debussy's *La Mer* Mvt II "Jeux de vagues" (Sectional level) (Madden) (Measures)

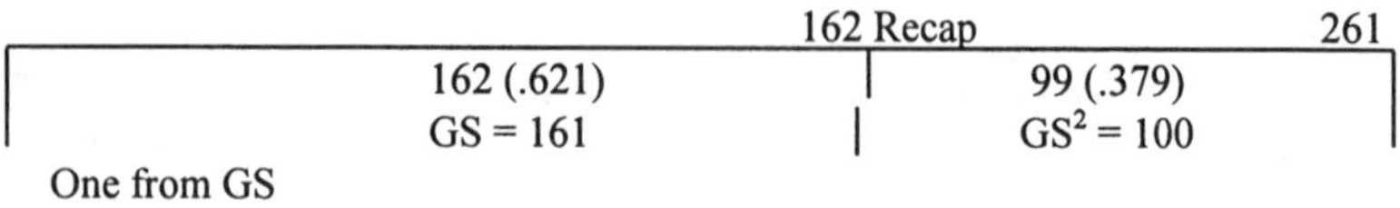

The introduction to the third movement of *La Mer* ("Dialogue du vent et de la mer") consists of a sequence of Fibonacci numbers and divides at GS^2, as shown in figure 7.13.[22]

Figure 7.13. Debussy's introduction to *La Mer* Mvt III "Dialogue du vent et de la mer" (Howat, Madden)

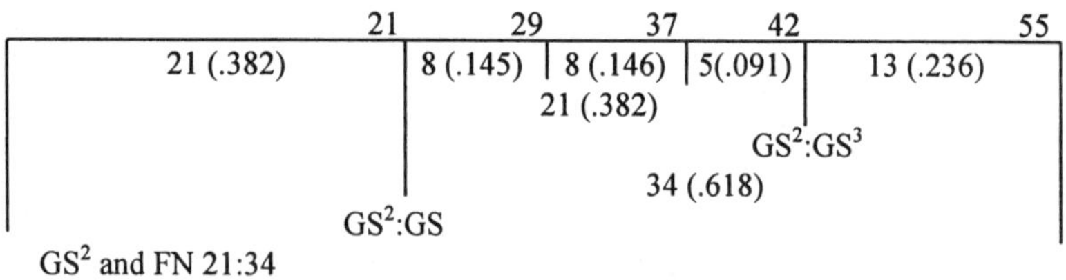

Howat provided a fascinating illustration showing an approximately logarithmic curve related to these Fibonacci numbers. The curve is similar to that shown in figure 1.8 herein, except that the quarter-circles with Fibonacci radii were apparently replaced by half-circles with Fibonacci diameters 55, 34, 21, 13, and 8. We are not certain how Howat's curve was drawn; we had to make a few adjustments to get a result reasonably close to his. Figure 7.14 shows the result.[23]

Obviously, the measure counts as found in the music (21, 8, 8, 5, 13) could not have established the curve, but rather the curve had to be drawn first, and the order of the durations extracted from the zero crossings. The spiral that Debussy saw in the water and drew on the page thus dictated the order of the durations.

As a whole, the movement is not divided at GS, but rather, the development roughly bisects it, as shown in figure 7.15. Without the introduction and coda, the piece is exactly bisected. The location of the coda, however, is in doubt; it could be at 244, 253, or even 269, as we believe.[24]

Figure 7.14. Approximately logarithmic curve found in the introduction to Debussy's *La Mer* Mvt III "Dialogue du vent et de la mer" (Howat, Madden)

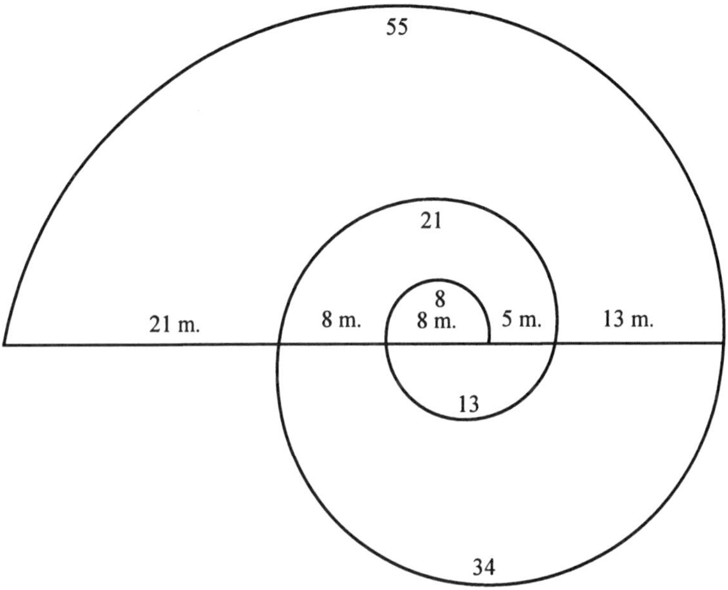

Figure 7.15. Debussy's *La Mer* Mvt III "Dialogue du vent et de la mer" (Howat, Madden) (Measures)

Intro	55 Expo	156 Devel	210 Recap	253? 257 Coda 269? 292
55 (.188)	101 (.346)	54 (.185)	47 (.161)	35 (.120)
156 (.534)		101 (.346)		

Not GS. Without introduction and coda it is bisected

Howat went to great trouble to combine the second and third movements in order to show a division at GS.[25] This is the least convincing part of his analysis. First, he appropriated the last 47 *units* of the second movement (after the dynamic peak of unit 222) to combine with the first 55 *measures* of the third movement (at a different tempo) and left off the 35 measure coda of the third movement in order to show a division at GS after measure 55 of the third movement. Neither overlapping movements, nor mixing units and measures or dynamics and sections, nor omitting the coda can be regarded as valid for this book. Thus, attempting to prove a golden proportional structure of the third movement of *La mer* is futile.

To summarize the preceding, table 7.1 illustrates Debussy's use of GS in his formal structures. In some cases, other musical parameters are referred to tangentially if seemingly important; however, it is the formal structures that are deemed important for this book.

Table 7.1. Summary of the Debussy studies

Piece	At GS or GS^2	1 meas.	Bi-sected	FN	Notes
Spleen	.382/.618		.500	13:8:13	
Mouvement		.375/.625	.500	3:2:3	Howat
					.373, Mad.
Reflets	.617				How., dyn.
			.500		How., ton.
					.362, Mad.
Clair de Lune				LN, FN	Howat
		.361	.500		Madden
L'isle joyeuse	.380				
Nuits d'étoiles		.626		.001 fm 5:3	Parks
	.615				Madden
Mandoline	.386				
String Quartet	.381				
Faun					.491
Sirènes		.377			
La Mer I					.589 unadj
		.379			Adj (units)
La Mer II					.632 unadj
		.621			Adj (meas)
La Mer III Int	.382			21:34	Log. spiral
La Mer III					.534, Mad.

5/14 = 35.7% at GS or GS^2
9/14 = 64.3% within one measure
FN used three times

In conclusion, several of Debussy's works are clearly organized in a variety of golden section structures, including Fibonacci numbers in *Spleen,*

Mouvement, and the introduction to *La mer* Mvt III; Lucas numbers (possibly) in *Clair de Lune*; balanced structures in *Spleen* and *Mouvement*; and unbalanced GS in *Reflects dans l'eau, L'isle joyeuse, Nuits d'étoiles, Mandoline, Sirènes*, *String Quartet*, and *La Mer* Mvts I and II.

The sectional structures seen here were only sometimes brought about by the long-short-long or short-long-short measure counts expected from the discussion near the beginning of this chapter. In fact, Debussy's structures are more often lop-sided, such as GS^2:GS *or* GS:GS^2, rather than both GS^2:GS *and* GS:GS^2.

Perhaps the most interesting structure illustrated by Howat was the Fibonacci spiral, which Debussy may (or not) have actually used in his planning.[26] Even if he did not draw it, a thorough understanding of spirals could have resulted in the structure seen in the music. Or, he could have merely sketched it and deduced the required measures. In any event, it seems likely that a good knowledge of spirals and the counting of measures went hand in hand.

It is interesting that Debussy used GS from the beginning of his career. *Nuits d'étoiles*, written in 1880—when he was eighteen—is already divided at one measure from GS, or *at* GS in one interpretation, while *Mandoline*, composed in 1882, is divided exactly at GS.

Once again, the question arises, Where did Debussy learn of the golden proportion? Were other composers using it at the time? There is conflicting evidence regarding the symbolists' knowledge of it, and few composers used it after Chopin.

Since, unfortunately, this book cannot contain thorough studies of any composer, there is no certainty about Debussy's other music. The question of whether he used phi in it will have to be left for other researchers. Obviously there are many pieces of Debussy omitted here because they were not studied or because they did not use phi. Numerous analyses in Parks did not show obvious evidence of phi.

The importance of GS in the dynamic and other structures as contrasted with the formal structures should be studied more thoroughly. Howat and Parks have made a good start, but the topic needs more study than can be provided in this book because of its bias toward formal structures.

Bartók

Since Béla Bartók was nearly a generation younger than Debussy and was trained in a different part of Europe, his exposure to the golden proportion must have differed from Debussy's. Yet, the large amount of material written on the subject makes it evident that he used phi to some extent. Consequently, an intriguing question is, How and when did he start using it? Where did he get the idea, and what did it mean for him? Were other composers of his circle using it? Did Debussy have an influence? How pervasive was his use of it? Whether he used it extensively and accurately will be discussed in this section.

Ernő Lendvai was the principal theorist who brought Bartók's use of phi to the attention of musicians. Several writers have commented on Lendvai's work, often negatively, primarily citing counting and procedural errors. Howat commented often enough that Lendvai felt compelled to write a defense.[27]

It should be clearly stated at the outset that the bulk of Lendvai's work was related to pitch structure, rather than form, although the criticisms have been primarily about form. Since *Fib and Phi* is about form, the pitch part of Lendvai's work will be postponed for another volume.

The first piece analyzed here is probably the best known in this debate, the *Music for String Instruments, Percussion, and Celesta.* The first movement was touted by Lendvai, Howat, Solomon, and Lowman as clearly divided at GS, which it nearly is if only the dynamics are considered (running from ***pp*** at the beginning to ***fff*** after measure 55 to ***ppp*** at the end), as shown in figure 7.16a.[28]

Figure 7.16a. Bartók's *Music for String Instruments, Percussion, and Celesta* Mvt I (Dynamics) (Lendvai, Howat, Madden) (Corrected)

	55		89? 88
55 (.625) GS = 54	5:3	33 (.375) GS^2 = 34	

One from GS and 11FN 5:3, but only in the dynamic structure

Lendvai was rightly criticized for adding a measure to make a full 89 so that the division would be at .618, but wrongly accused of distorting the analysis. In either case, the proportion is close enough. The real reason it has to be rejected for this book is that the golden proportion division is produced by the dynamic structure and not the formal structure, which is shown in figure 7.16b. The measure after 64 marks the reversal of the fugue, which is a much stronger structural point than the dynamic climax.

Figure 7.16b. Bartók's *Music for String Instruments, Percussion, and Celesta* Mvt I (Formal structure) (Madden)

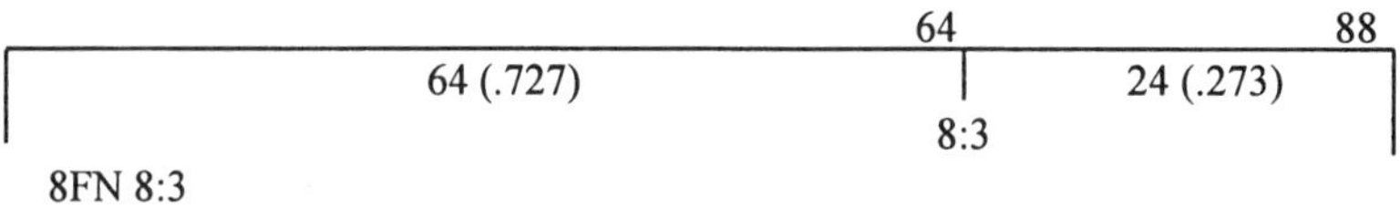

Howat made an interesting comparison between this piece and the pieces by Debussy "Reflects dans l'eau" from *Images* and by Ravel "Oiseaux tristes" from *Miroirs*.[29] They have very similar dynamic structures, as shown in figure 7.16c (certain details excised).

Figure 7.16c. A comparison of Debussy's "Reflects dans l'eau" from *Images*, Bartók's *Music for String Instruments, Percussion, and Celesta* Mvt I, and Ravel's "Oiseaux tristes" from *Miroirs* (Dynamics) (Howat, Madden)

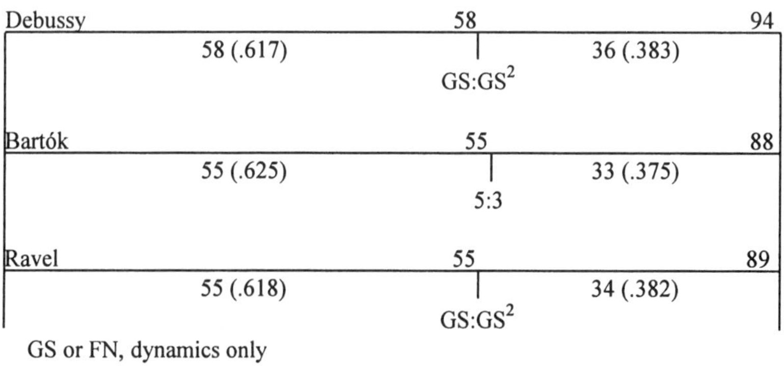

Presumably the point was to imply that these composers were inspired by each other (and that they used phi knowingly). Unfortunately, the illustration omitted to make clear that Ravel's piece consists of an additional thirty-two measures. (Overall, it is not a clear structural use of GS, as will be shown in the section on Ravel.)

The third movement of the *Music for String Instruments, Percussion, and Celesta* has an altogether different structure from the first movement. Although Lendvai's measure counts were not correct because of changing meters, recounting in units shows that this piece is without doubt divided structurally at GS^2, as shown in figure 7.17.[30]

Figure 7.17. Bartók's *Music for String Instruments, Percussion, and Celesta* Mvt III (Madden) (Units of 3/8)

76	136	194	285	325	357
76	60	58	91	40	32
136 (.381)		221 (.619)			

GS²:GS

GS²

Lendvai made his analysis of the first movement of the *Sonata for Two Pianos and Percussion* confusing by leaving out the first measure in order to show a phi proportion in measures 2 through 17.[31] The first beat of measure 18 was also missing, thus compounding the confusion. Howat criticized him for this and got a defensive response. Lendvai apparently thought that the introductory first measure and the first beat of measure 18 would throw the phi proportion off, but they would not, as shown in figure 7.18a, corrected and counted in units rather than measures.

Figure 7.18a. Bartók's *Sonata for Two Pianos and Percussion* Mvt I, first 18 measures (Madden) (Units of 3/8)

3	31	50
3	28	19 (.380)
31 (.620)		

GS:GS²

GS

Lendvai did not really show whether this movement was divided into phi proportions, but rather that the substructures were balanced similarly to those of Debussy and that they could often be related to fib or phi. Lowman did analyze the first movement, finding that it was divided after measure 273, which is one measure from GS, as shown in figure 7.18b.[32]

Figure 7.18b. Bartók's *Sonata for Two Pianos and Percussion* Mvt I (Lowman) (Measures)

Intro 31 Expo	132 Devel?	273 Recap 331		432 443	
31	101	141	58	101	11
273 (.616)			170 (.384)		
GS = 274			GS² = 169		

One from GS

The third movement is neither fib nor phi, as shown in figure 7.19, despite some use of FN in balanced substructures (not shown).

Figure 7.19. Bartók's *Sonata for Two Pianos and Percussion* Mvt III (Madden)

133 | 247 Recap? | 420

133 (.317) | 114 (.271) | 173 (.412)

247 (.588)

Not fib or phi

The *Piano Sonata* Mvt I (1926) was analyzed by Lendvai and Griffiths. Lendvai called it "an exact *bridge form*," which is shown in the central column in table 7.2 with our analysis at right.[33]

Table 7.2. Structure of Bartók's *Piano Sonata* Mvt I

Bar 1	Principle theme	A
Bar 44	Secondary theme I	B
Bar 55	Secondary theme II	C
Bar 76	CENTRAL THEME	D
Bar 135	Secondary theme II	E
Bar 155	Secondary theme I	B
Bar 176	Principal theme	A
Bar 211		D
Bar 232		Coda

Bar 135 is not related to Lendvai's Secondary theme II, so this cannot be an exact bridge form. However, the Central Theme could be a development and bar 176 a recapitulation, but that would not be phi either.

Lendvai found elements of GS in the recapitulation (measure 176) and the coda (measure 232), but the piece overall is not divided at GS, but exactly bisected in measures and nearly so in units. See figure 7.20.[34]

Figure 7.20. Bartók's *Piano Sonata* Mvt I (Measures and eighth notes)(Madden)

meas.	43	54	75	134	154	175	210	232	268
units	176	224	309	547	629	721	866	955	1098
	43	11	21	59	20	21	35	22	36

134 (.500) | 134 (.500)

1:1

Bisected in measures and two from bisected in units

Griffiths claimed that the piece is 1062 quavers (eighth notes) long divided at 656 (GS). The origin of these incorrect numbers is unknown.[35]

Bartók's *Fourth String Quartet* (1928) was mentioned briefly by Lendvai, who used fib and phi at his convenience, with mixed results. For example, he noted a proportion of 30/48 (= .625) in the exposition of the first movement although phi does not appear anywhere else in the movement, as shown in figure 7.21.[36]

Figure 7.21. Bartók's *Fourth String Quartet* Mvt I (Madden) (Measures)

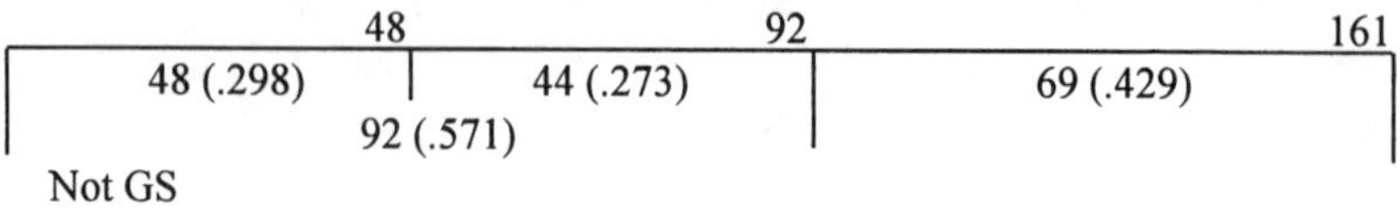

In Movement II, Lendvai noted a proportion 73/188 (= .388) for the partial movement making up the scherzo and trio, although the movement as a whole is not phi, as shown in figure 7.22.

Figure 7.22. Bartók's *Fourth String Quartet* Mvt II (Madden) (Measures)

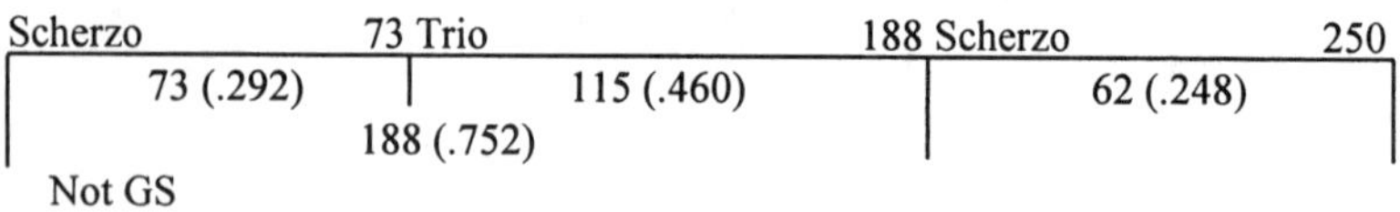

In Movement III, Lendvai mentioned sections falling at Fibonacci numbers 5, 13, 21, 34, and 55. This is correct, except that 55 is not clearly notable as a structural point and the conclusion (measure 71) is not a Fibonacci number, as shown in figure 7.23.

Figure 7.23. Bartók's *Fourth String Quartet* Mvt III (Lendvai, Madden) (Measures)

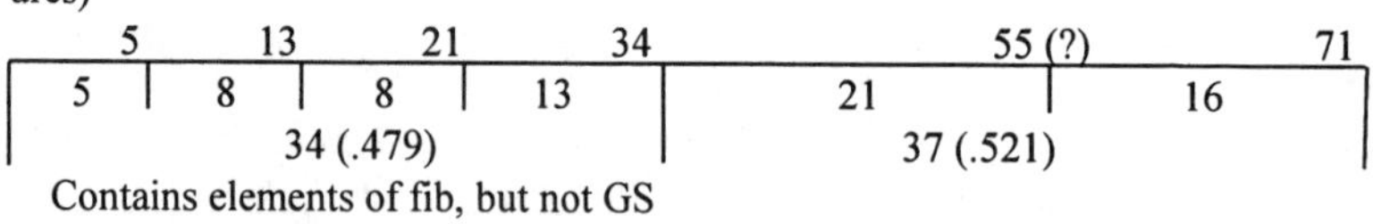

Lendvai mentioned a GS of 14/36 (= .389) in the recapitulation of Movement IV—incorrectly, because measures 88–102–124 should be 15/37 (= .405)—and phi appears nowhere else. See figure 7.24.

Figure 7.24. Bartók's *Fourth String Quartet* Mvt IV (Lendvai, Madden) (Measures)

64 87 (102) 124

64 (.516) | 23 (.186) | 37 (.298)

60 (.484)

Not GS

In Movement V, Lendvai referred to a "turning point determined by the 238:147:91 ratio." This appears to refer to the proportions of the exposition and development, which would be correct, but again, it is a partial movement. See figure 7.25a.

Figure 7.25a. Bartók's *Fourth String Quartet* Mvt V (Lendvai) (Measures)

Expo 147 Devel 238 Recap 392

147 (.375) | 91 (.232) | 154 (.393)

245 (.625)

3:5

FN 3:5

In addition, measure 151 is a more likely structural point and gives better proportions, which are one from GS^2, as shown in figure 7.25b.

Figure 7.25b. Bartók's *Fourth String Quartet* Mvt V (Madden) (Measures)

Expo 151 Devel 238 Recap 392

151 (.385) | 87 (.222) | 154 (.393)

GS^2 = 150

241 (.615)

GS = 242

One from GS^2

Thus, Bartók's *Fourth String Quartet* can in general be considered divided in phi proportions only with considerable juggling. It should be remarked that Derek Locke found Fibonacci numbers among the total number of beats in the five movements. Although the total of 2,584 beats is a Fibonacci number, we do not see that the structures divide at GS.[37]

Bartók's *Fifth String Quartet* (1934) was taken up by Clive Pascoe, George Perle, and Karen Anne Bates. Pascoe and Perle discussed Movement I, Perle discussed II and IV, while Bates discussed Movement III. Pascoe also mentioned V, but understandably did not specify its structure.[38] Preferring not to risk an analysis of this difficult movement, we say simply that nothing happens at phi. Movements I to IV are shown in figures 7.26–29. Pascoe's placement of the recapitulation at measure 132 in Movement I appears to be incorrect.

Figure 7.26a. Bartók's *Fifth String Quartet* Mvt I (Pascoe) (Measures)

Expo | 58 Devel | 132 Recap | 176 Coda | 218

58 (.266) | 74 (.340) | 44 (.202) | 42 (.192)

132 (.606) | 86 (.394)

GS = 135 | GS^2 = 83

Not GS

Perle's analysis appears to produce a perfect division at GS, but his placement of the recapitulation at $134\frac{2}{3}$ also appears to be incorrect.

Figure 7.26b. Bartók's *Fifth String Quartet* Mvt I (Perle) (Measures)

Expo | 58 Devel | 134.7 Recap | 176 Coda | 218

58 (.266) | 76.7 (.352) | 41.3 (.189) | 42 (.193)

134.7 (.618) | 83.3 (.382)

$GS:GS^2$

GS, but disputed

Measure 125 seems a better choice, but the piece is no longer divided at phi. See figure 7.26c.

Figure 7.26c. Bartók's *Fifth String Quartet* Mvt I (Madden) (Measures)

Expo | 58 Devel | 125 Recap | 176 Coda | 218

58 (.266) | 67 (.307) | 51 (.234) | 42 (.193)

125 (.573) | 93 (.427)

Not GS

Perle's analysis of Movement II is correct, as shown in figure 7.27.

Figure 7.27. Bartók's *Fifth String Quartet* Mvt II (Perle) (Measures)

$34\frac{5}{8}$ | 56

$34\frac{5}{8}$ (.618) | $21\frac{3}{8}$ (.382)

$GS:GS^2$

GS

Bates did not find GS in Movement III. See figure 7.28.

Figure 7.28. Bartók's *Fifth String Quartet* Mvt III (Bates) (Measures)

Scherzo 66 Trio 131 Scherzo 223

66 (.296) 65 (.291) 92 (.413)

131 (.587)

Not GS

Movement IV is within one measure, as shown in figure 7.29.

Figure 7.29. Bartók's *Fifth String Quartet* Mvt IV (Perle) (Measures)

63 101

63 (.624) 38 (.376)

GS = 62 $GS^2 = 39$

One from GS and .001 from 5:3

The first movement of Bartók's *Sixth String Quartet* (1939) was discussed by James Rothwell.[39] It is counted in seconds as given by Bartók in the score, as shown in figure 7.30. There is a structural inflection after 250 (one from GS), but it is not the recapitulation.

Figure 7.30. Bartók's *Sixth String Quartet* Mvt I (Rothwell) (Seconds)

Expo 189 Devel 250 288 Recap 406

189 (.465) 99 (.244) 118 (.291)

288 (.709)

GS = 251 $GS^2 = 155$

Not GS

Bartók's piano solo *Allegro Barbaro* (1911) was mentioned by Lendvai and Kramer. These analyses are tricky because they do not give measure numbers, but counting the lowest-register F-sharp minor chords in the sequence 8, 5, 5, 3, 13 beginning at measure 50, as instructed by Lendvai, does gives this sequence, with some gaps and an extra half measure after 13. Thus the Fibonacci sequence is supported for the most part. More importantly, the piece is divided at GS overall, as shown in figure 7.31.[40]

Figure 7.31. Bartók's *Allegro Barbaro* (Madden) (Measures)

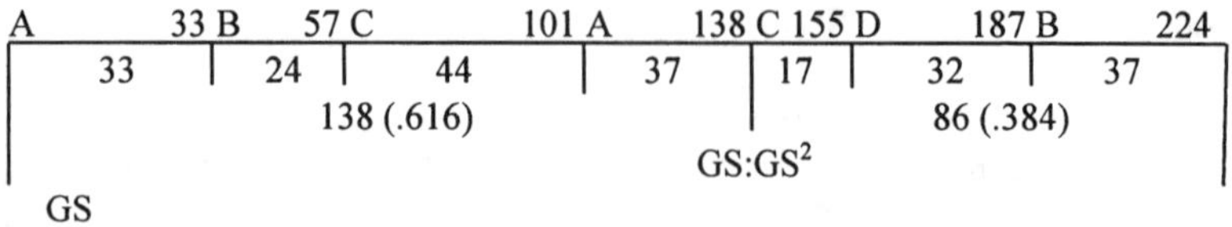

Lendvai mentioned that "Free Variations" (Number 140 of *Mikrokosmos Book VI*) has a GS structure.[41] This is correct if analyzed in measures, but not if analyzed in units, which vary greatly through many changes of meter. See figures 7.32a and b.

Figure 7.32a. Bartók's *Mikrokosmos VI no. 140* "Free Variations" (Lendvai) (Measures)

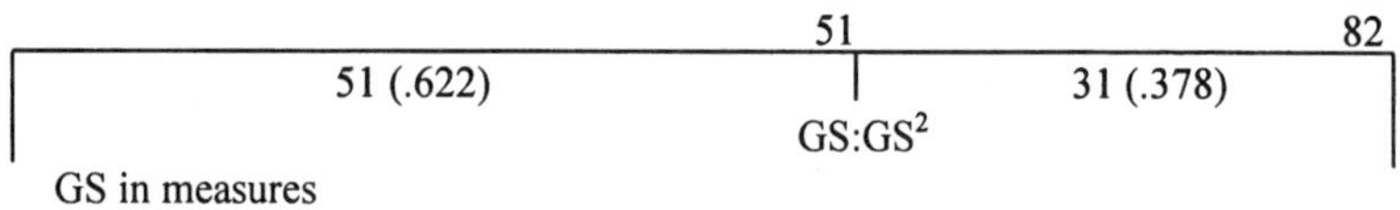

Figure 7.32b. Bartók's *Mikrokosmos VI no. 140* "Free Variations" (Madden) (Units)

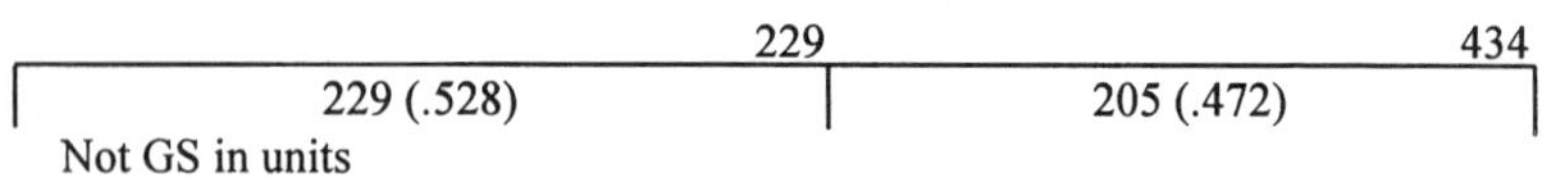

Lendvai also mentioned that "From the Diary of a Fly" (Number 142 of the *Mikrokosmos Book VI*) is divided in phi proportion.[42] This appears to be incorrect, as shown in figures 7.33a and b, although in units the division is close to 3:2.

Figure 7.33a. Bartók's *Mikrokosmos VI no. 142* "From the Diary of a Fly" (Lendvai) (Measures)

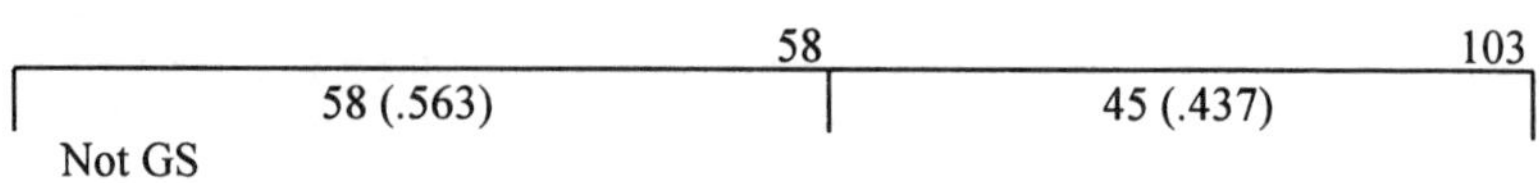

Figure 7.33b. Bartók's *Mikrokosmos VI no. 142* "From the Diary of a Fly" (Madden) (Units)

	136	225
	136 (.604)	89 (.396)
	GS = 139	$GS^2 = 86$
Not GS		

Lendvai mentioned the Stollens and Abgesangs of the first movement of the *Violin Concerto* (1938) as lying "approximately in golden section proportions," as shown in Table 7.3.[43]

Table 7.3. Structure of Bartók's *Violin Concerto* Mvt I

Exposition	Principle Theme	Stollen	55 bars
	Secondary Theme	Stollen	55 bars
Development		Abgesang	83 bars
Recapitulation	Principle Theme	Stollen	54 bars
	Secondary Theme	Stollen	54 bars
Coda		Abgesang	88 bars

This is an interesting observation, but it probably could not have entered into Bartók's thinking, as the larger sections are not in golden proportion, and, according to Bartók's own timing marks, the development begins at measure 114, throwing off the scheme. See figures 7.34a and b.

Figure 7.34a. Bartók's *Violin Concerto* Mvt I (Lendvai) (Measures)

Expo	110 Devel	193 Recap	301 Coda	389
	110 (.283)	83 (.213)	108(.278)	88 (.226)
	193 (.496)		196 (.504)	
Not GS, but close to bisected				

Figure 7.34b. Bartók's *Violin Concerto* Mvt I (Madden) (Measures)

Expo	55	114 Devel	193 Recap 247	301	308 Cadenza 343 Coda	389
	55 (.141)	59 (.152)	79 (.203)	54 (.139)	54 (.139)	88 (.226)
	193 (.496)			196 (.504)		
Not GS, but close to bisected						

In the first movement of the *Divertimento* (1939), Lendvai referred to

"563 triplet units" to arrive at GS 348.[44] However, some of the measures are not divisible by three, so this cannot be correct. Recounted, the piece is revealed to have 1681 eighth notes (560⅓ triplet units), an error of about one measure, whose GS is at 1039, which marks the end of the development. This piece is therefore divided exactly at GS, as shown in figure 7.35.

Figure 7.35. Bartók's *Divertimento* Mvt I (Madden) (Eighth notes)

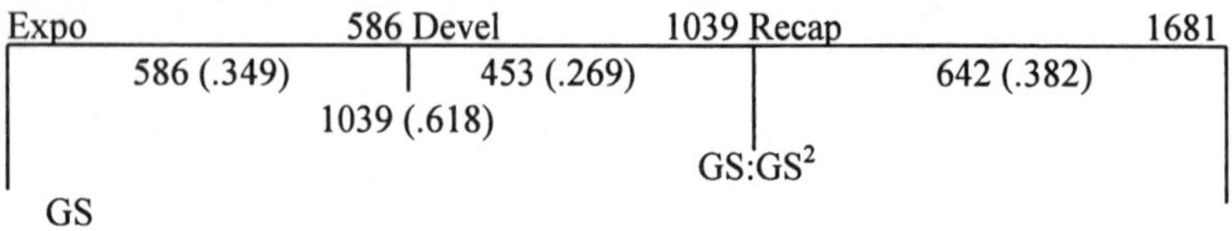

Pascoe discussed Bartók's *Rhapsody no. 1* (1928).[45] His numbers for the first movement did not reveal the golden section in this piece, and some were off by one measure. See figure 7.36.

Figure 7.36. Bartók's *Rhapsody no. 1* Mvt I (Pascoe, Madden) (Meas.) (Corr.)

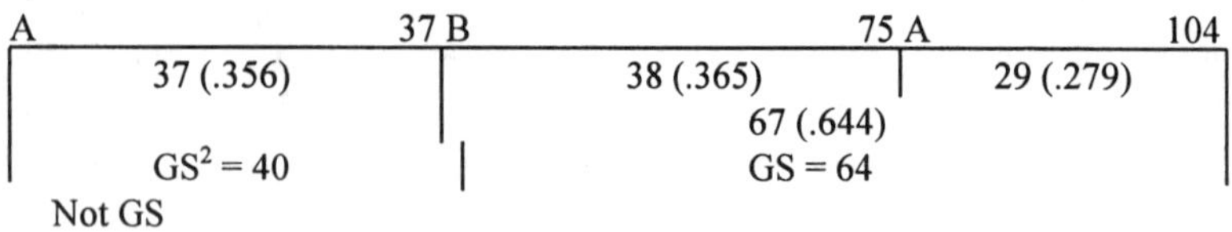

In his analysis of the second movement, Pascoe neglected to count the second system on some pages after measure 96. His measure 133 should be 150, and the total number of measures is 180. The piece is risky to analyze because of many meter and tempo changes, but neglecting these, it *is* divided at one measure from GS^2, as shown in figure 7.37. Note that part F reflects part A of the first movement.

Figure 7.37. Bartók's *Rhapsody no. 1* Mvt II (Pascoe, Madden) (Corrected)

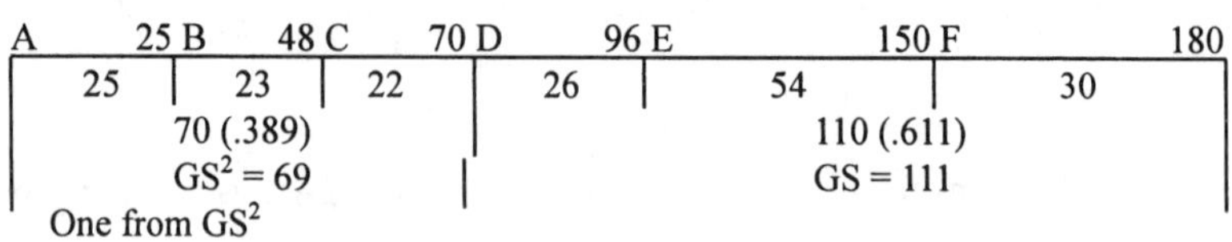

Lendvai analyzed the first movement of Bartók's trio *Contrasts* (1938).[46] This piece is divided at GS, as shown in figure 7.38, if counted in measures.

Figure 7.38. Bartók's *Contrasts* I (Lendvai, Madden) (Measures)

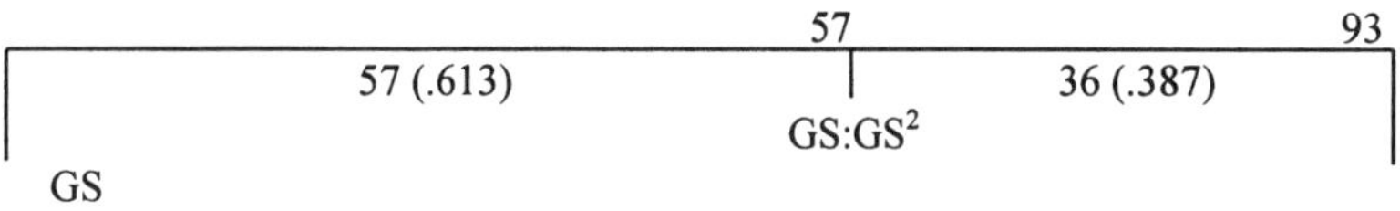

Several of the Dances in Bulgarian Rhythm found in the *Mikrokosmos Book VI* have Fibonacci numbers in the meter signatures, but they are constant throughout the individual pieces and cannot be regarded as effective usages of Fibonacci sequences.

In conclusion, some of the analysts' claims of Bartók's use of golden proportion in his structures have been proven correct. Even Lendvai, whose "controversial" study took so much criticism about "wayward arithmetic" and failure to answer the question whether Bartók used phi consciously, was not far off.[47] His courageous pioneering work should have been praised rather than treated with contempt.

Of the twenty-three pieces analyzed here, six are clearly divided at GS, four more are at one measure from GS, and two have Fibonacci proportions, leaving thirteen which cannot be accepted for inclusion as golden proportion constructions. The failure rate is higher than for Debussy, but this does not mean that previous work was not valuable; rather, that some authors were careless or made unsound assumptions.

It cannot be said that Bartók used fib or phi regularly or consistently. When he did, however, it cannot have been by accident. Seven of the pieces fall within three points of phi (.5% error) and there are few pieces to suggest that intuition played any part.

Clearly, the first movement of the *Music for String Instruments, Percussion, and Celesta* was modeled after Debussy's "Reflects dans l'eau" from *Images*, as was Ravel's "Oiseaux tristes" from *Miroirs*. All three pieces have approximately the same number of measures and all have similar dynamic structures. Also, all are otherwise not in phi proportion.

Although the reputations for the star performers in this debate have been tarnished, they were not knocked completely out, and the reputations of several others were strengthened.

A summary of Bartók is shown in table 7.4.

Table 7.4. Summary of Bartók

Piece (Abbr.)	At GS or GS^2	1 meas.	Bi-sected	FN	Notes
Music for Str. I		.625		5:3	Dynamics
				8:3	Madden
Music for Str. III	.381				Madden
Sonata 2 Pianos I		.616			Lowman
Sonata 2 Pianos III					.588 Madden
Piano Sonata I			.500		Madden
String Qt. 4 I					.571 Madden
String Qt. 4 II					.752 Madden
String Qt. 4 III					Elem. of FN
String Qt. 4 IV					.516 Madden
String Qt. 4 V				3:5	Lendvai
		.385			Madden
String Qt. 5 I					.606 Pascoe
	.618				Perle, disputed
					.573 Madden
String Qt. 5 II	.618				Perle
String Qt. 5 III					.587 Bates
String Qt. 5 IV		.624			Perle
String Qt. 6 I					.709 Roth., sec.
Allegro Barbaro	.616				Madden
Mikrokosmos 140	.622				Lendvai
Mikrokosmos 142					.563 Lendvai
Violin Conc. I					.496 Lendvai
Divertimento I	.618				Madden
Rhapsody 1 I					.356 Pascoe
Rhapsody 1 II		.389			Pascoe
Contrasts I	.613				Lendvai

Hindemith

Paul Hindemith was said by David Neumeyer to have used phi proportion in the "Fugue [No. 2] in G" from the *Ludus Tonalis* (1943).[48] Although we disagree with Neumeyer's section labels, a structural inflection is indeed found at GS, as shown in figure 7.39.

Figure 7.39. Hindemith's "Fugue in G" from *Ludus Tonalis* (Neumeyer, Madden) (Measures) (Corrected)

Stretto

11 17 23 28 46 53 (58?) 65 75

11 (.147) | 6 | 11 (.147) | 18 (.240) | 7 (.093) | 12 | 10

GS^4 | | GS^4 | GS^3 | GS^5

46 (.613) | 29 (.387)

$GS:GS^2$

GS

According to Rothwell, "Interludium (Valse)" (No. 11) from the *Ludus Tonalis* (1943) is divided in phi proportions.[49] Rather, it is one measure from GS, as shown in figure 7.40, and is probably off only because the composer truncated the decimal instead of rounding.

Figure 7.40. Hindemith's *Interludium (Valse)* (Rothwell, Madden) (Measures)

28 61 100

28 | 33 | 39 (.390)

61 (.610)

GS = 62 | $GS^2 = 38$

One from GS

Rothwell also thought that Hindemith's *Piano Sonata no. 2* (1936) Mvt I (Maessig schnell) is in phi proportion.[50] But it is two measures from GS. See figures 7.41a and b.

Figure 7.41a. Hindemith's *Piano Sonata no. 2* Mvt I (Rothwell) (Measures)

16 41 95 155

16 | 25 | 54 | 60 (.387)

95 (.613)

GS = 96 | $GS^2 = 59$

One from GS, but disputed

Figure 7.41b. Hindemith's *Piano Sonata no. 2* Mvt I (Madden) (Measures)

16 40 73? 94 116 156

16 | 24 | 54 | 22 | 40

94 (.603) | 62 (.397)

GS = 96 | $GS^2 = 60$

Not GS

Movement II was marked by Rothwell as accurate to GS^2. In this case he was right, as shown in figure 7.42.

Figure 7.42. Hindemith's *Piano Sonata no. 2* Mvt II (Rothwell) (Measures)

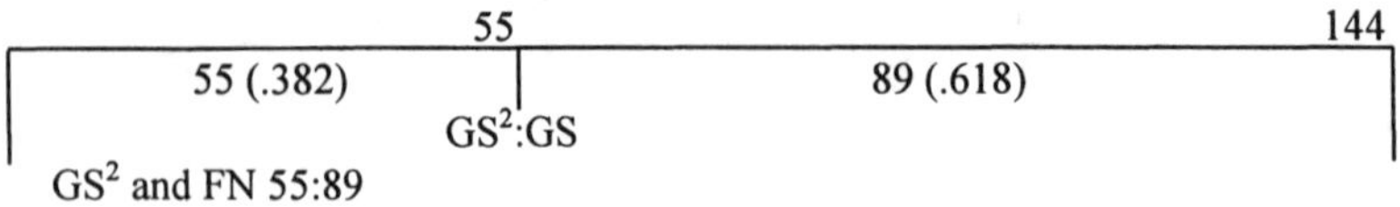

Rothwell measured Movement III in seconds and did not find GS. However, we analyzed it in measures and did find phi. See figure 7.43.

Figure 7.43. Hindemith's *Piano Sonata no. 2* Mvt III (Madden) (Measures)

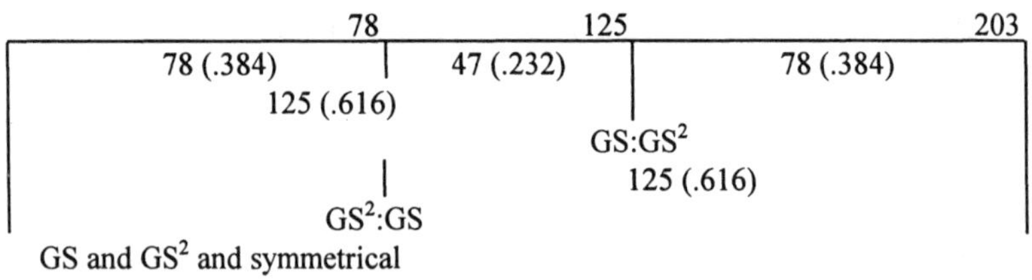

According to Pascoe, the first movement of the *Clarinet Sonata* (1939) is 505 quarter notes long with the development after 53 measures and the recapitulation after 99 measures.[51] This does not tell us anything because the units disagree. Recounted in measures, Pascoe's estimation of the development and the recapitulation does not give a division at GS. However, there are what appear to be structural events at one measure from GS^2, as shown in figure 7.44.

Figure 7.44. Hindemith's *Sonata for Clarinet* Mvt I (Madden) (Measures)

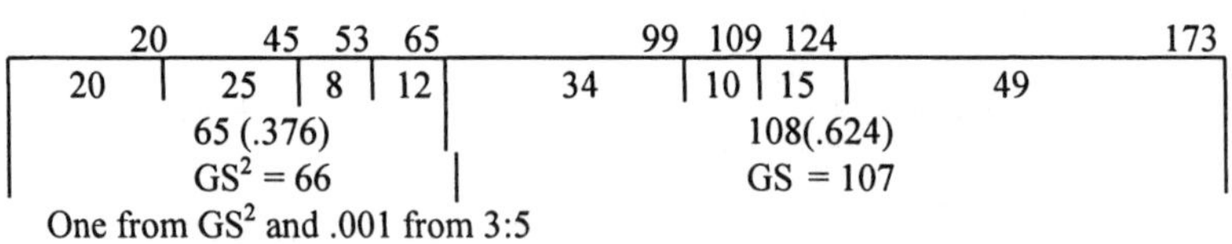

Pascoe also studied the fourth movement. This piece is not GS, as shown in figure 7.45.

Figure 7.45. Hindemith's *Sonata for Clarinet* Mvt IV (Madden) (Measures)

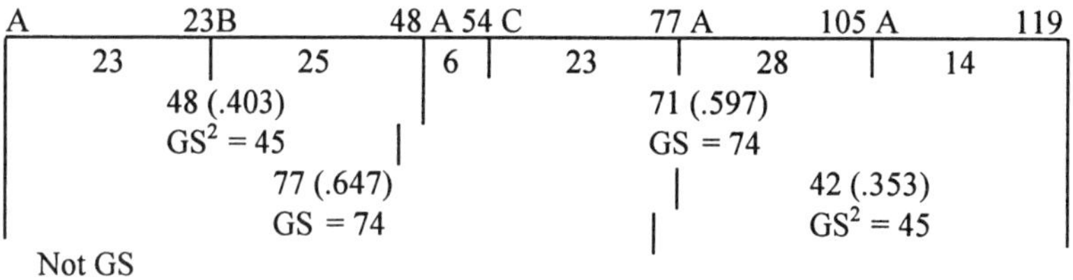

Pascoe reported that the first movement of the *Concert Music for Strings and Winds* (1930) measures 484 seconds when calculated according to the metronomic values, and is divided at 284 seconds, as shown in figure 7.46.[52]

Figure 7.46. Hindemith's *Concert Music for Strings and Winds* Mvt I (Pascoe) (Seconds) (Pascoe's numbers)

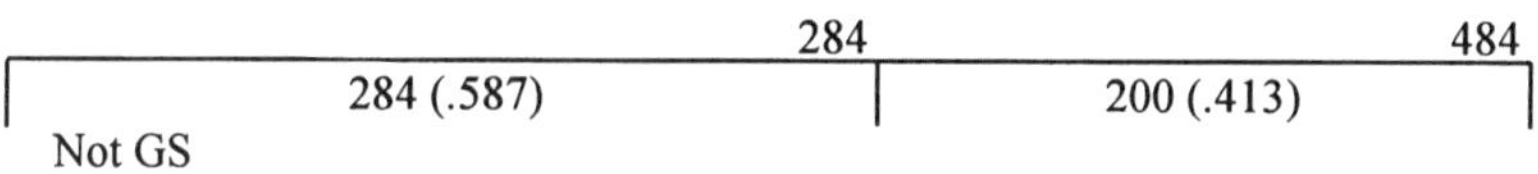

Pascoe also reported the second movement as shown in figure 7.47. Interestingly, the central section of this piece is one second from GS^2, but the rest of the piece is not GS.

Figure 7.47. Hindemith's *Concert Music for Strings and Winds* Mvt II (Pascoe) (Seconds) (Pascoe's numbers)

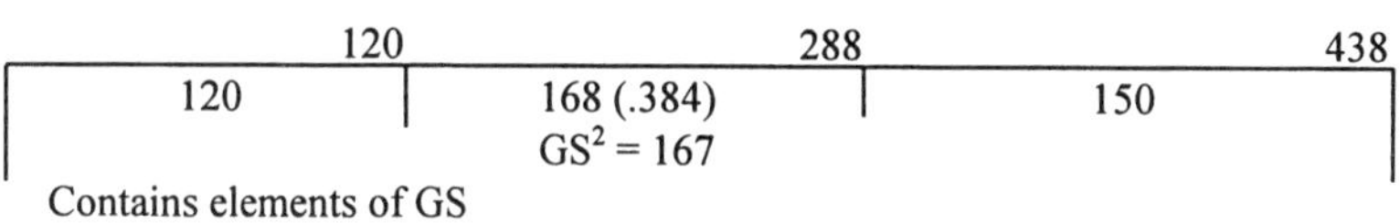

Table 7.5 summarizes Hindemith. We see that Hindemith seems to have understood phi and may have used it often.

Table 7.5. Summary of Hindemith

Piece (Abbr.)	At GS or GS^2	1 meas.	Too far	FN	Notes
Ludus Fugue G	.613				Neumeyer
Ludus Valse		.610			Rothwell
Son. No. 2 I		.613			Rothwell
			.603		Madden
Son. No. 2 II	.382			55:89	Rothwell
Son. No. 2 III	.384/.616				Madden
Son. for Clar. I		.376			.001 from 3:5
Son. for Clar. IV			.403/647		Madden
Conc. Music I			.587		Pascoe
Conc. Music II			.658		Pascoe

Prokofieff

Sergei Prokofieff's *Classical Symphony* (1916–7) has all of its movements close to GS, according to Adkins.[53] The first movement is two measures from GS if measure 130 is taken as important (a major break, but perhaps not the recapitulation), as shown in figure 7.48.

Figure 7.48. Prokofieff's *Classical Symphony* Mvt I (Madden) (Measures)

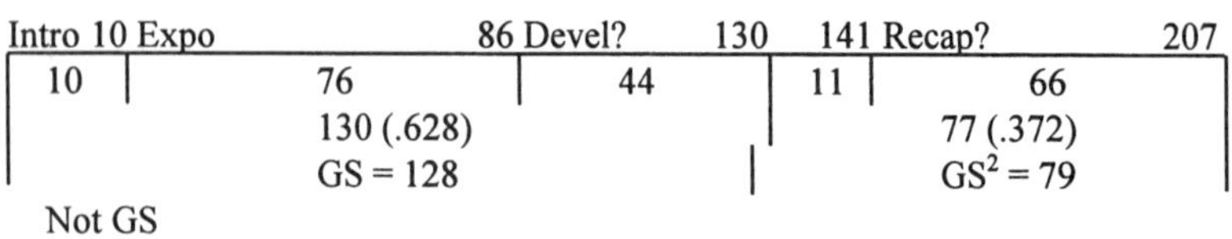

The second movement divides at one measure from GS^2, as shown in figure 7.49.

Figure 7.49. Prokofieff's *Classical Symphony* Mvt II (Madden) (Measures)

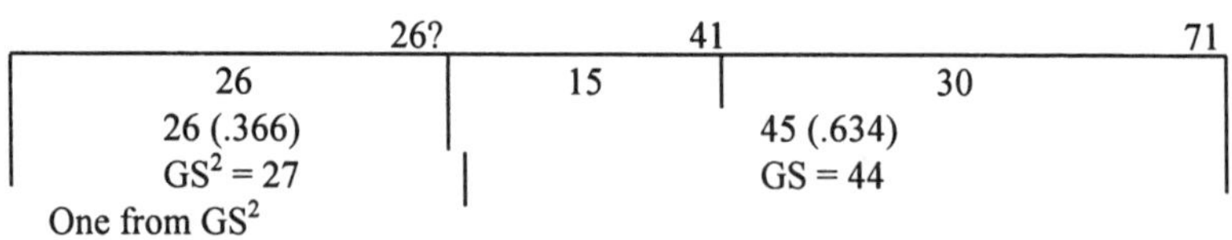

The third and fourth movements are far from GS, as shown in figures 7.50 and 7.51.

Figure 7.50. Prokofieff's *Classical Symphony* Mvt III (Madden) (Half notes)

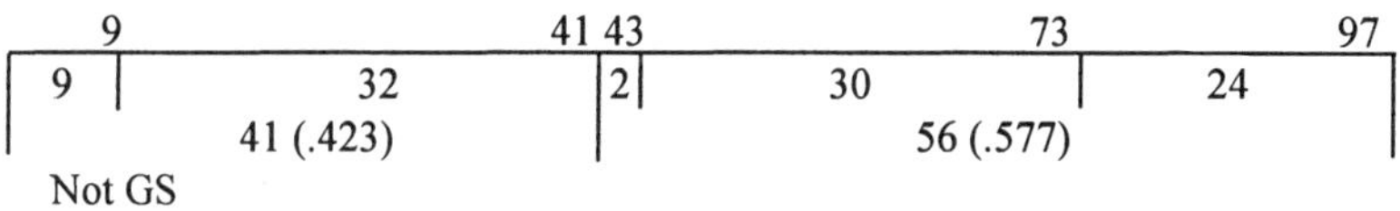

Figure 7.51. Prokofieff's *Classical Symphony* Mvt IV (Madden) (Measures)

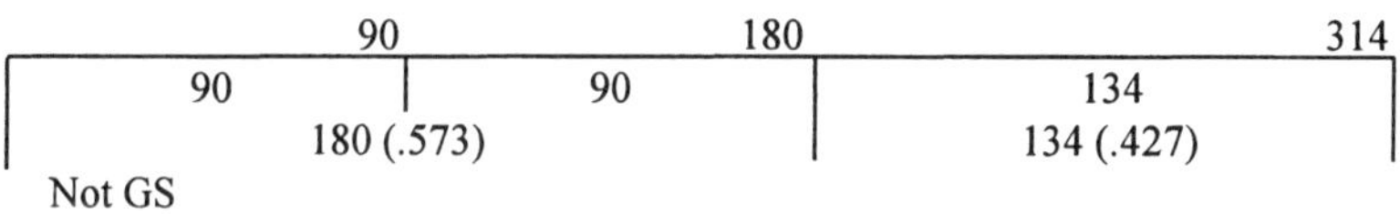

Adkins suggested that in Prokofieff's *Lieutenant Kije Suite* (1934) the first and second movements are in phi proportion.[54] They are, but the percentages for the second movement are poor. See figures 7.52 and 7.53.

Figure 7.52. Prokofieff's *Lieutenant Kije Suite* Mvt I (Madden) (Measures)

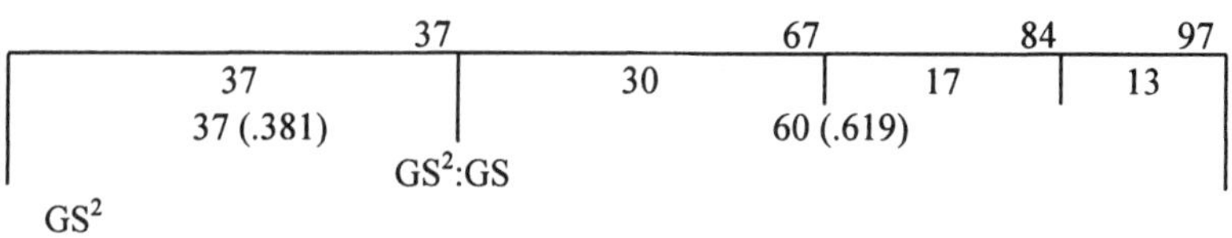

Figure 7.53. Prokofieff's *Lieutenant Kije Suite* Mvt II (Madden) (Measures)

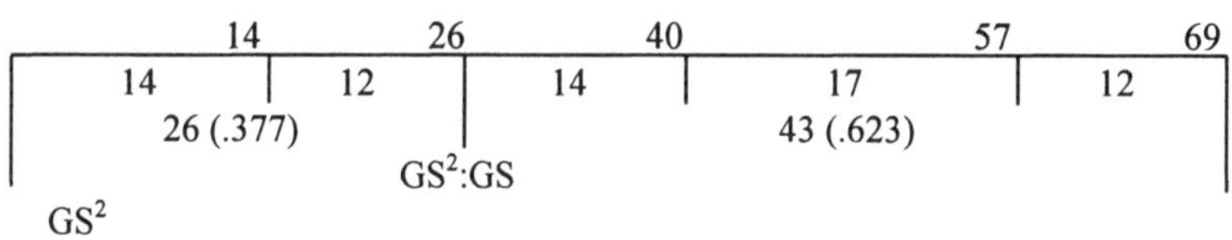

He correctly suggested that the third and fifth movements are in phi proportion as well, while the fourth is not. See figures 7.54 and 7.55.[55]

Figure 7.54. Prokofieff's *Lieutenant Kije Suite* Mvt III (Madden) (Measures)

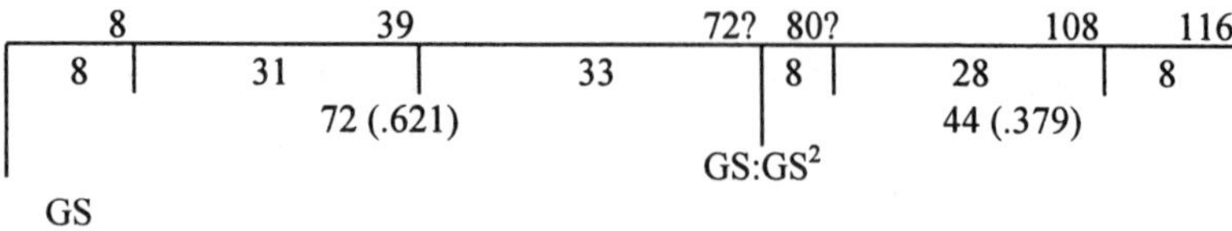

Figure 7.55. Prokofieff's *Lieutenant Kije Suite* Mvt V (Madden) (Measures)

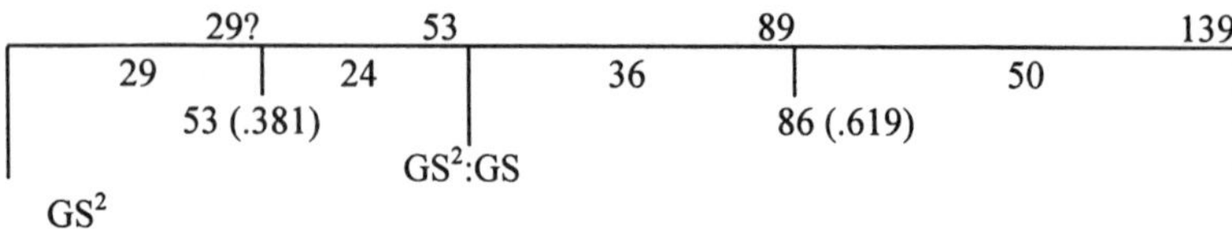

Rothwell studied two of Prokofieff's piano pieces *Penseés op. 62.*[56] Counting the first piece in beats and removing the introduction, he found phi, as shown in figure 7.56a.

Figure 7.56a. Prokofieff's *Penseés no. 1* (Rothwell) (Beats) (Without intro)

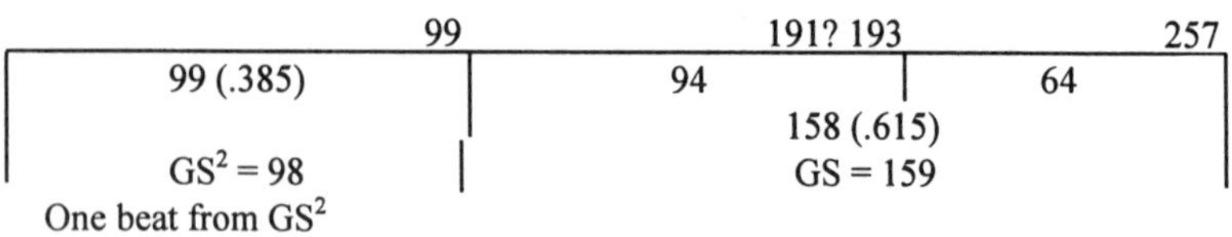

In measures, the result is rather different, as shown in figure 7.56b.

Figure 7.56b. Prokofieff's *Penseés no. 1* (Madden) (Measures) (Without intro)

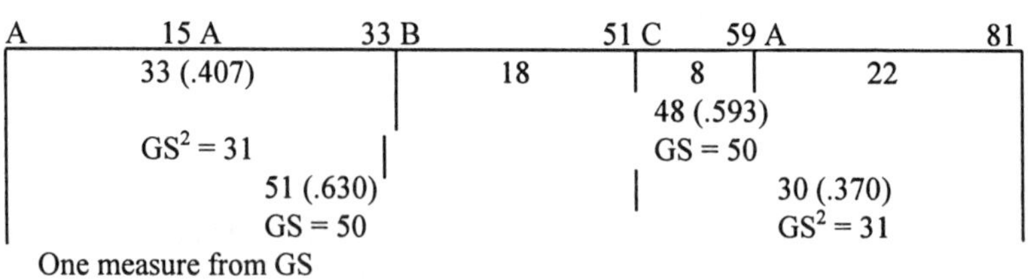

With the introduction, in beats, the result is not near GS, as correctly reported by Rothwell, and as shown in figure 7.56c.

Figure 7.56c. Prokofieff's *Penseés no. 1* (Madden) (Beats) (With introduction)

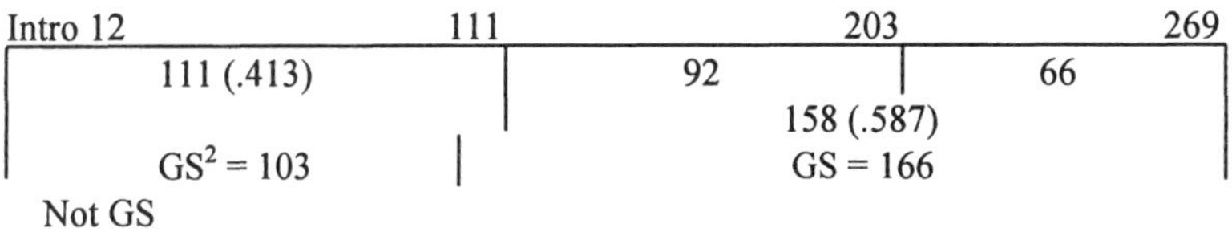

Counted in measures, it is even worse, as shown in figure 7.56d.

Figure 7.56d. Prokofieff's *Penseés no. 1* (Madden) (Measures) (With intro)

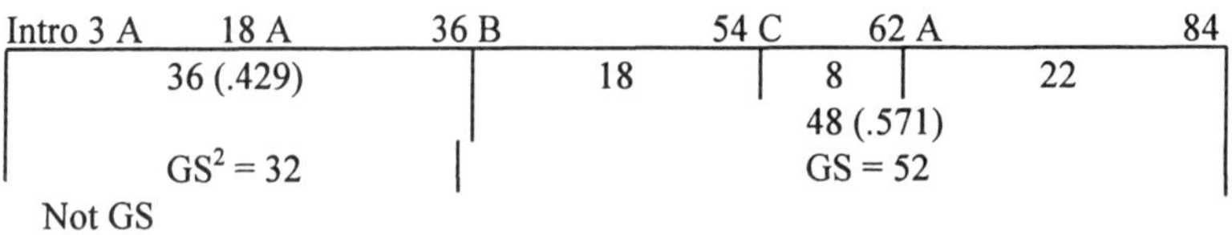

Rothwell's study of *Penseés no. 2* correctly showed a result at GS, as shown in figure 7.57a.

Figure 7.57a. Prokofieff's *Penseés no. 2* (Rothwell) (Beats)

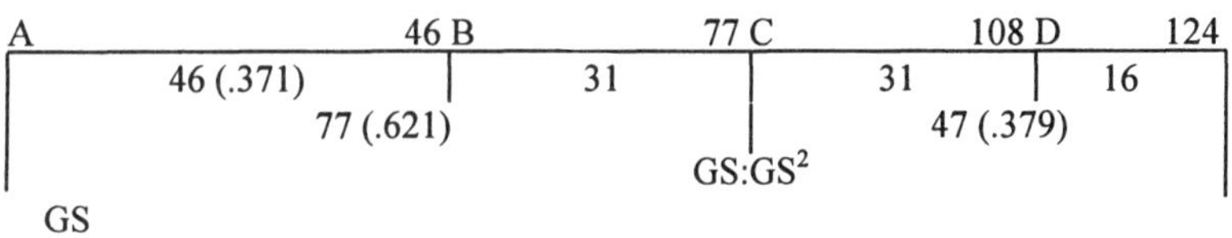

If this were counted in measures, the result would be a trisection, as shown in figure 7.57b.

Figure 7.57b. Prokofieff's *Penseés no. 2* (Madden) (Measures)

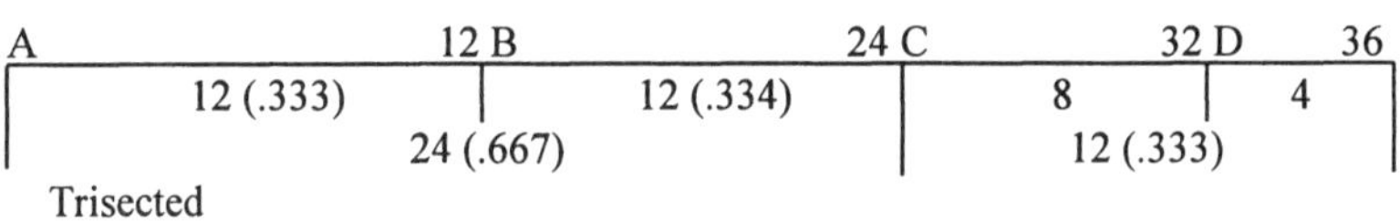

Rothwell did not report on *Penseés no. 3*. The structure is not obvious, and we do not see any structural points at GS or GS^2.

Rothwell also studied the first piece of *Op. 3* entitled "Conte" (Story).[57] Counted in beats, it is one beat from GS, as shown in figure 7.58a.

Figure 7.58a. Prokofieff's *Op. 3 no. 1* (Rothwell) (Beats)

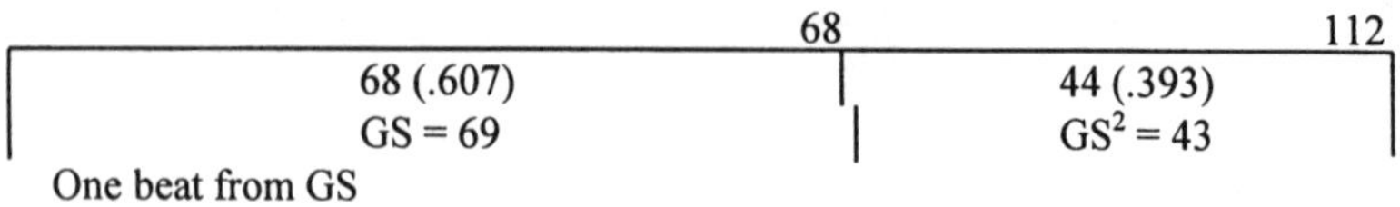

Counted in measures, it is GS, with poor percentages. See figure 7.58b.

Figure 7.58b. Prokofieff's *Op. 3 no. 1* (Madden) (Measures)

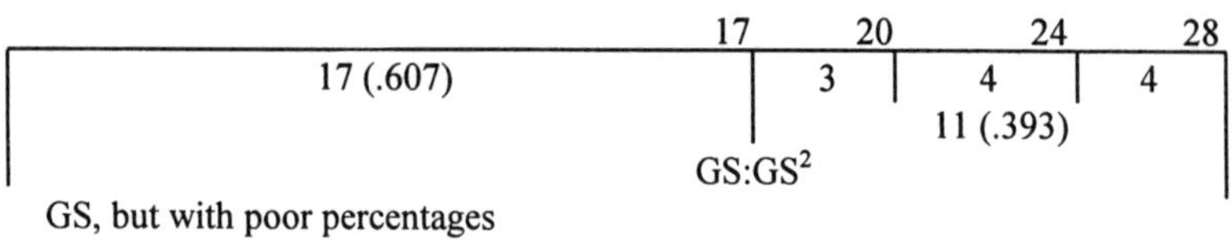

Rothwell would have been well advised to study the other three pieces in this opus, for there are some impressive results. Figure 5.59 is not GS.

Figure 7.59. Prokofieff's *Op. 3 no. 2* (Madden) (Measures)

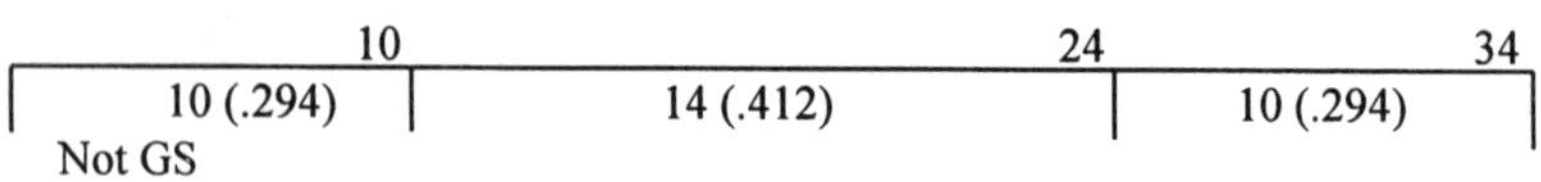

Figure 5.60 is.

Figure 7.60. Prokofieff's *Op. 3 no. 3* (Madden) (Measures)

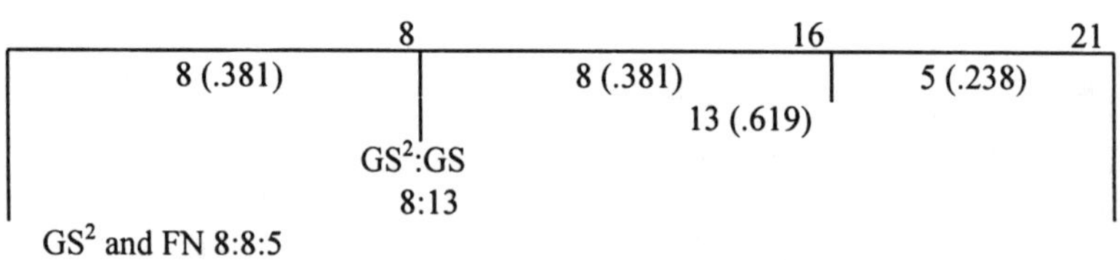

Figure 7.61. Prokofieff's *Two Pieces op. 3 no. 4* (Madden) (Measures)

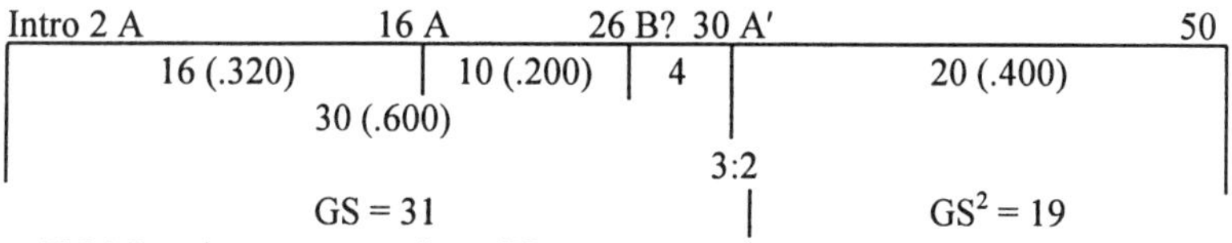

FN 3:2 and one measure from GS

Table 7.6 summarizes Prokofieff. It looks like he knew about phi.

Table 7.6. Summary of Prokofieff

Piece (Abbr.)	At GS or GS^2	1 unit	Bi/Tri-sected	Too far	FN	Notes
Class Symph I				.628		Measures
Class Symph II		.366				Measures
Class Symph III				.423		Measures
Class Symph IV				.573		Measures
Lieut Kije I	.381					Measures
Lieut Kije II	.377					Measures
Lieut Kije III	.621					Measures
Lieut Kije V	.381					Measures
Penseés 1		.385				Roth., beats
		.630				Mad., meas
				.413		Mad., beats
				.429		Mad., meas
Penseés 2	.621					Roth., beats
			.667			Mad., meas
Penseés 3						Not analyzed
Op. 3 no. 1		.607				Roth., beats
	.607					Mad., meas
Op. 3 no. 2				.294		Mad., meas
Op. 3 no. 3	.381				8:13	Mad., meas
Op. 3 no. 4		.600			3:2	Mad., meas

Satie

Erik Satie was an idiosyncratic musician and close friend of Debussy, and made his living playing piano at the Chat Noir, a cabaret and hotbed of arcane ideas frequented by many of the odder personalities in Paris. He was closely acquainted with Rosicrucian Grand Master and proprietor of the annual—and hugely successful—Salon de la Rose+Croix fine arts exhibits, Joséphin Péladan, and was its official composer for the first exhibit in 1892. Thus, it can be assumed that he, too, was a Rosicrucian which leads to the provocative question of this order's possible influence.[58]

Courtney S. Adams reported an extensive study of Satie's use of phi.[59] He found that Satie did use it in two distinct periods of his musical career, 1887–92 and 1914–19, but not in the rest of his output. Adams' work was thorough and correct. He reported the relationship of the three movements of Satie's *Trois Gymnopédies* (1888?) as shown in figure 7.62.

Figure 7.62. Satie's *Trois Gymnopédies* (Adams) (Measures)

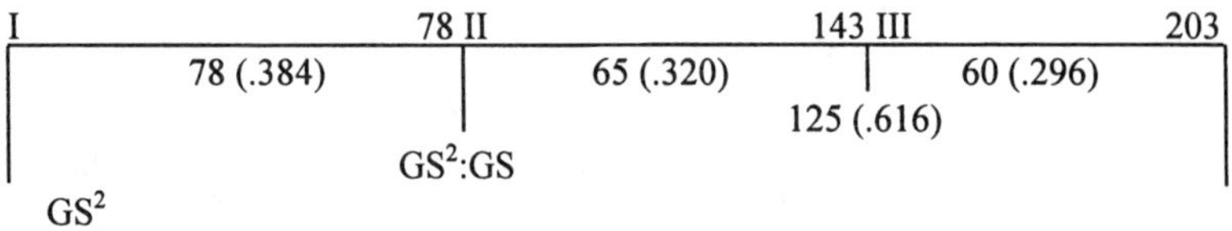

Adams found that, although *Les Trois Valses distinguées du précieux dégôuté* (1914) has neither time signature nor bar lines, it can be interpreted as 3/4 time and is divided at GS. See figure 7.63.

Figure 7.63. Satie's *Les Trois Valses distinguées du précieux dégôuté* (Adams)

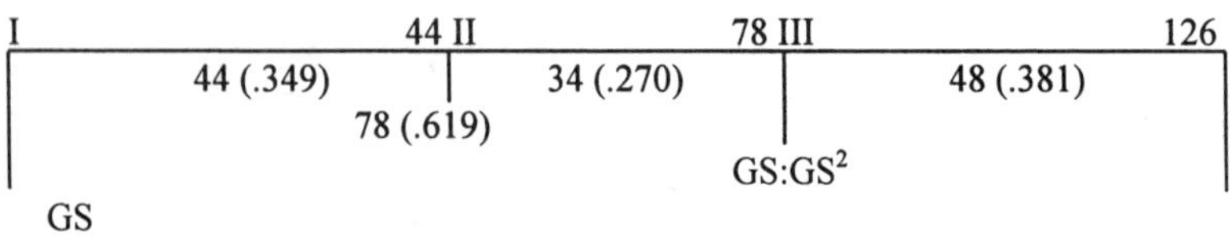

Adams' interpretation of the relationship of the three movements of *Avant Dernières Pensées* (1915) is GS, as shown in figure 7.64.

Figure 7.64. Satie's *Avant Dernières Pensées* (Adams) (Measures)

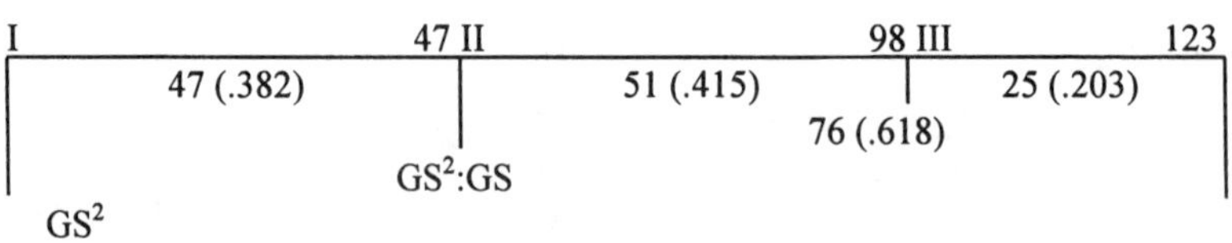

He also found GS in the first three *Nocturnes* (1919). See figure 7.65.

Figure 7.65. Satie's *Nocturnes I–III* (Adams) (Measures)

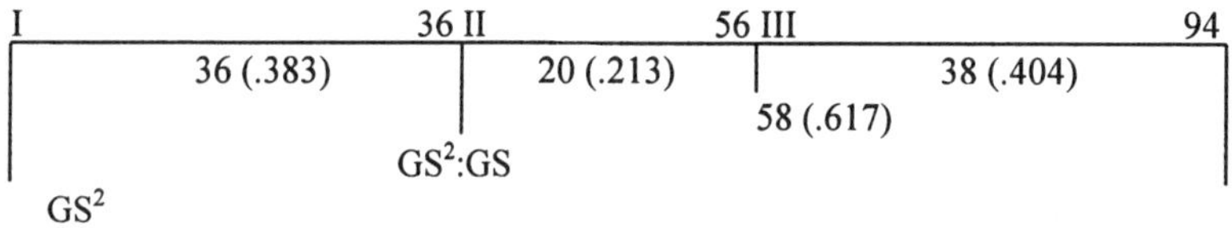

The pieces in the *Trois sonneries de la Rose+Croix* (1891) do not show GS overall, but the first one, entitled "Air de l'Ordre," is divided exactly at GS, as shown in figure 7.66.[60] Neither the second piece "Air du Grand Maître" nor the third "Air du Grand Prieur" is divided at GS, as shown in figure 7.67–8. Gillmor stated correctly the durations of 68 and 105 and claimed that this is a golden section. However, it is two measures away.

Figure 7.66. Satie's "Air de l'Ordre" from *Trois sonneries de la Rose+Croix* (Gillmor) (Beats)

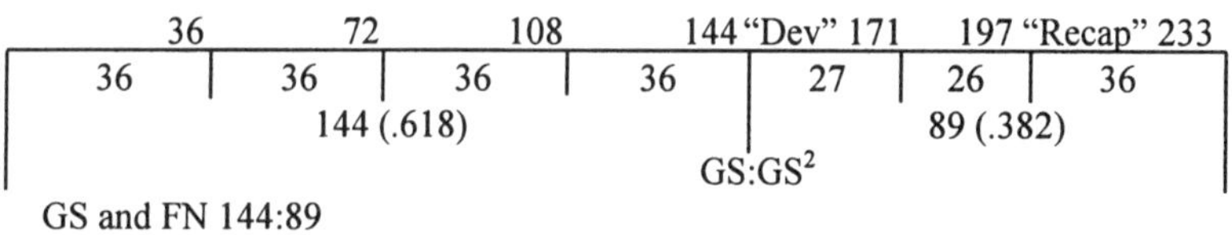

Figure 7.67. Satie's "Air du Grand Maître" from *Trois sonneries de la Rose+Croix* (Madden) (Beats)

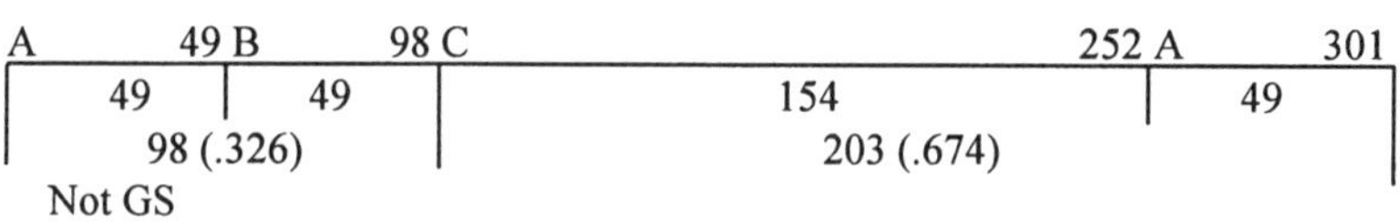

Figure 7.68. Satie's "Air du Grand Prieur" from *Trois sonneries de la Rose +Croix* (Gillmor, Madden) (Beats)

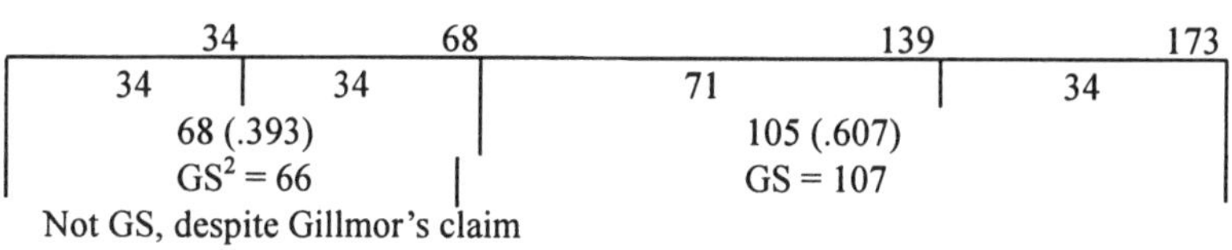

Adams also examined a group of pieces that fall into the so-called Rose+Croix style but did not give measure counts. Consequently, the following analyses are ours.

Le Fils des étoiles (1892) has three movements, only one of which is divided at GS, as shown in figures 7.69–71.

Figure 7.69. Satie's "La Vocation" from *Le Fils des étoiles* (Madden) (Beats)

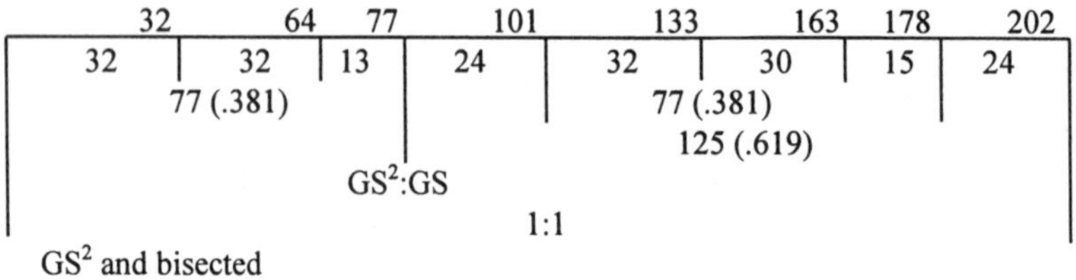

Figure 7.70. Satie's "L'Initiation" from *Le Fils des étoiles* (Madden) (Beats)

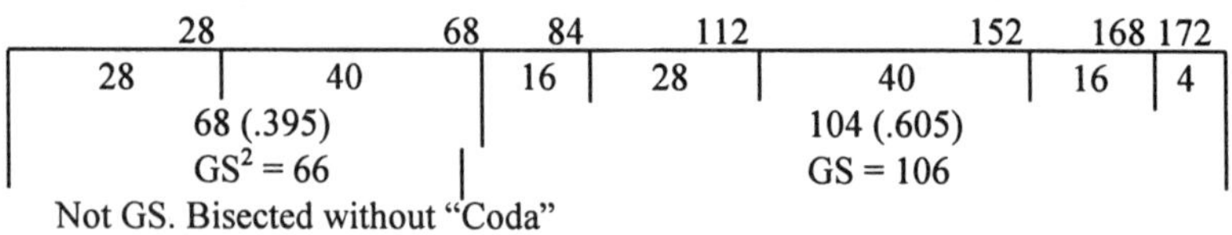

Figure 7.71. Satie's "L'Incantation" from *Le Fils des étoiles* (Madden) (Beats)

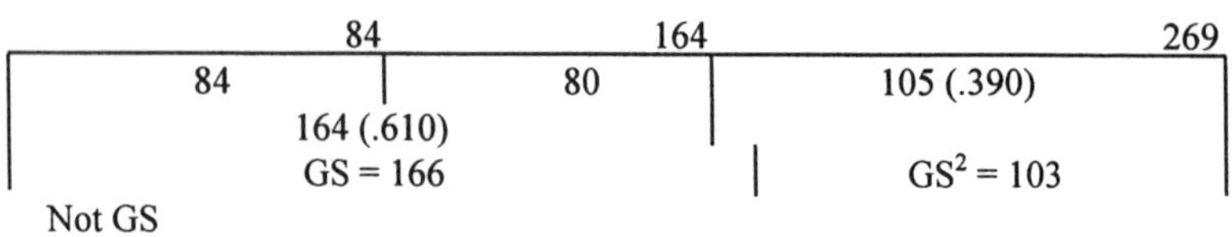

Prélude de la porte héroïque du ciel (1894) divides at two beats from GS, as shown in figure 7.72.

Figure 7.72. Satie's *Prélude de la porte héroïque du ciel* (Madden) (Beats)

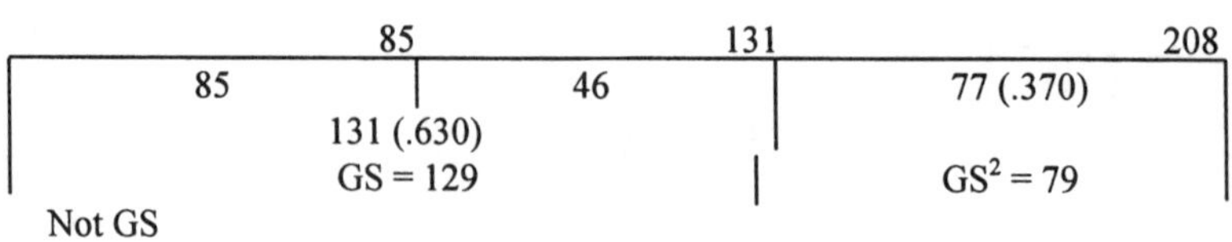

Table 7.7 summarizes Satie. We see that the proportions were quite exact in six of the eleven pieces reported here. By no means, however, did he use phi in all his music, and he used Fibonacci numbers only once in this study.

Table 7.7. Summary of Satie

Piece (Abbr.)	At GS or GS^2	1 meas.	Too far	Bi-sected	FN	Other
3 Gymnopedies	.384					
Les 3 Valses	.619					
Dern Pensees	.382					
3 Nocturnes	.383					
3 Sonneries I	.618				144:89	
3 Sonneries II			.326			
3 Sonneries III			.393			
Fils des etoiles I	.381			.500		
Fils des etoiles II			.395			
Fils des etoiles III			.610			
Porte héroïque			.630			

As with Debussy, the question remains, Where did this come from? The Rosicrucian connection is by no means certain, since GS was not confined to the so-called Rose+Croix period.

Schoenberg

Arnold Schoenberg seemed to disavow the use of phi when he wrote, "I do not believe in the Golden Section. At least I do not believe it is the sole formal law governing our perception and sensations of beauty, rather, at most, one among many."[61] Of course, he was right that phi is not the sole acceptable structural principle, but Schoenberg did use phi on a few occasions.

One of these occasions was noted by Wardell, who stated that *Piano Piece op. 33a* (1929) had attacks in the ratio 8:5:3, i.e., "The first phrase . . . has eight attacks. . . . Schoenberg here deliberately breaks the 8 attacks into 5 and 3 by placing the dynamic sign < > at the fifth attack."[62] Wardell made similar statements about measures three, four, and five. It can be debated whether this construction is five plus three or four plus four and the argument is weak. It is rejected for this book.

Wardell's discussion of measures 14 and 15 (16 is similar) is stronger. Here there are durations of five and three eighth notes. This is by no means a common construction in this piece, and is most likely incidental.

However, Schoenberg did place a rather prominent sectional break after 60 quarter note units (middle of measure 16), which would be GS^2, as shown in figure 7.73.

Figure 7.73. Schoenberg's *Piano Piece op. 33a* (Madden) (Quarter notes)

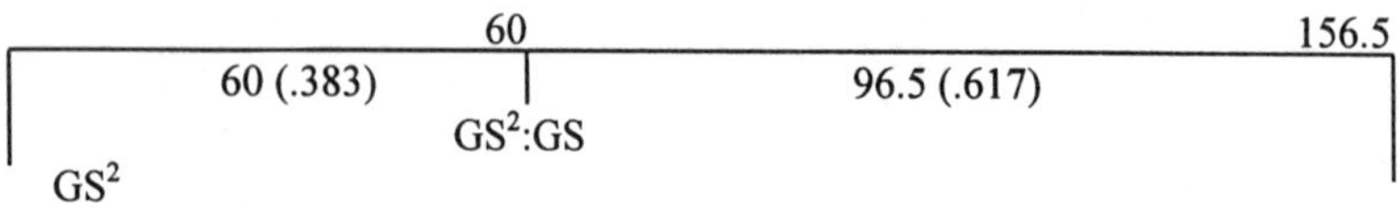

Pascoe discussed Movement I of *Five Pieces for Orchestra op. 16* using the *Eulenburg Study Scores.*[63] See figure 7.74. Pascoe's measure numbers differ from those in the Henmar Press edition reprinted in Morgan's *Anthology* (dated here as 1909). Pascoe had the "climax" (score-number 10) after measure 80, whereas Morgan's score has it after measure 78. Either interpretation is one measure from GS.

Figure 7.74. Schoenberg *Five Orchestral Pieces op. 16* Mvt I (Pascoe, Madden)

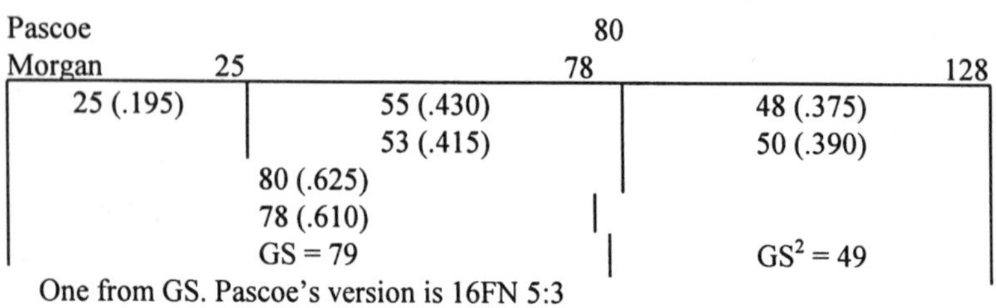

Pascoe also discussed Movement V but nothing much happens in this analysis.[64] In Morgan's discussion, Section III begins at measure 414 (measure 84 of the movement). This is after 83, one measure from GS, as shown in figure 7.75.

Figure 7.75. Schoenberg's *Five Orchestral Pieces op. 16* Mvt V (Morgan, Madden)

I — 40 II — 83 III — 136

40 (.294) | 43 (.316) | 53 (.390)

83 (.610)

GS = 84 | $GS^2 = 52$

One from GS

Pascoe discussed the first movement of the *Quintet for Wind Instruments op. 26* (1924).[65] Of 227 measures, he placed the development after measure 71 and the recapitulation after measure 127. See figure 7.76a.

Figure 7.76a. Schoenberg's *Quintet for Wind Instruments op. 26* Mvt I (Pascoe)

Expo? — 71 Devel — 127 Recap — 227

71 (.313) | 56 (.247) | 100 (.440)

127 (.560)

Not GS

We found the development to begin after measure 75 (or even 82, as given in the notes of the score) if the repeat is not taken. See figure 7.76b.

Figure 7.76b. Schoenberg's *Quintet for Wind Instruments op. 26* Mvt I (Madden)

Expo? — 75 Devel 82? — 127 Recap — 227

75 (.330) | 52 (.230) | 100 (.440)

127 (.560)

Not GS

However, we found that if the repeat is taken, the piece is trisected, as shown in figure 7.76c.

Figure 7.76c. Schoenberg's *Quintet for Wind Instruments op. 26* Mvt I (Madden) (With repeat)

Expo? — 73 Rpt — 148 Devel — 200 Recap — 300

73 (.244) | 75 (.250) | 52 (.173) | 100 (.333)

200 (.667)

Trisected and one from bisected

Pascoe also analyzed Movement IV. He placed the return of section A after 222. If correct, this would be GS, as shown in figure 7.77a.

Figure 7.77a. Schoenberg's *Quintet for Wind Instruments op. 26* Mvt IV (Pascoe)

222 359

222 (.618) 137 (.382)

GS:GS^2

GS but incorrect

However, it is incorrect. Part A returns after measure 225, which is 3 measures from GS. Interestingly, there is a grand pause at 179, which is one half-measure from bisected. See figure 7.77b.

Figure 7.77b. Schoenberg's *Quintet for Wind Instruments op. 26* Mvt IV (Madden)

179 225 359

179 (.499) 46 (.128) 134 (.373)

225 (.627)

GS = 222 GS^2 = 137

Not GS but one-half measure from bisected

Webster gave GS proportions for three of Schoenberg's string quartets.[66] For the first movement of *String Quartet no. 2* (1908), of 233 measures length, he placed the development at measure 92 and the recapitulation at measure 159 which, if correct, would be three measures from GS^2, as shown in figure 7.78a.

Figure 7.78a. Schoenberg's *String Quartet no. 2* Mvt I (Webster)

92 159 233

92 (.395) 67 (.287) 74 (.318)

141 (.605)

GS^2 = 89 GS = 144

Not GS

However, we found a strong structural point at measure 89, thus dividing the piece at GS^2. See figure 7.78b.[67]

Figure 7.78b. Schoenberg's *String Quartet no. 2* Mvt I (Madden)

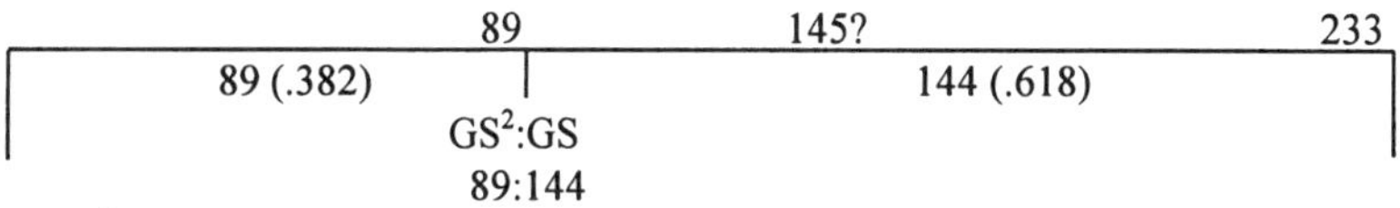

GS^2 and FN 89:144

The second movement was given by Webster as shown in figure 7.79. This result is FN 3:2, but there is nothing near GS.

Figure 7.79. Schoenberg's *String Quartet no. 2* Mvt II (Webster)

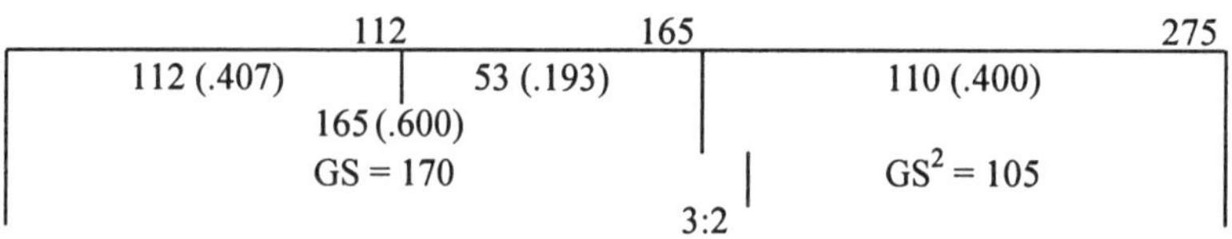

FN 3:2, but not GS

In Movement III, we found a division at one from GS. See figure 7.80.

Figure 7.80. Schoenberg's *String Quartet no. 2* Mvt III (Madden)

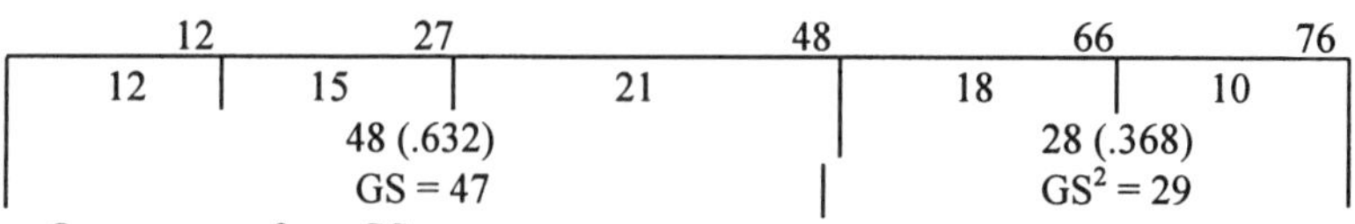

One measure from GS

For the first movement of *String Quartet no. 3* (1927), Webster gave the values shown in figure 7.81a. It is not GS.

Figure 7.81a. Schoenberg's *String Quartet no. 3* Mvt I (Webster)

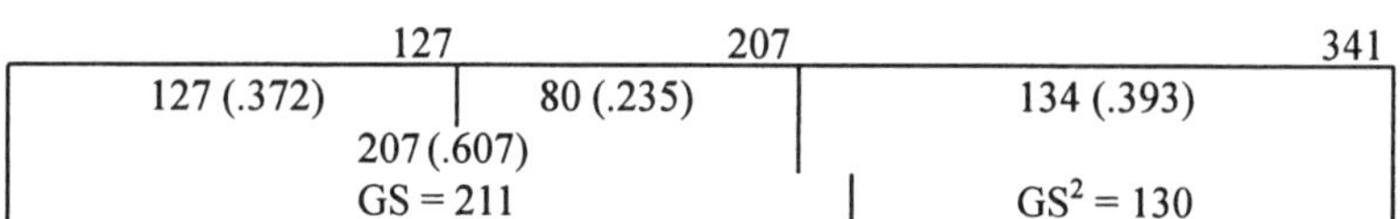

Not GS

However, we found it to be exactly GS. See 7.81b.

Figure 7.81b. Schoenberg's *String Quartet no. 3* Mvt I (Madden)

94 211? 341

94 (.276) 117 (.343) 130 (.381)

211 (.619)

$GS:GS^2$

GS

Webster's values for the first movement of *String Quartet no. 4* (1936) were as shown in 7.82. Webster was right that this is not GS.

Figure 7.82. Schoenberg's *String Quartet no. 4* Mvt I (Webster)

165 284

165 (.581) 119 (.419)

GS = 175 GS^2 = 109

Not GS

Webster's values for Movement III were as shown in figure 7.83a. This is one measure from GS.

Figure 7.83a. Schoenberg's *String Quartet no. 4* Mvt III (Webster)

55 90

55 (.611) 35 (.389)

GS = 56 GS^2 = 34

One measure from GS, but disputed

However, we found it to be more like in figure 7.83b.

Figure 7.83b. Schoenberg's *String Quartet no. 4* Mvt III (Madden)

16 50 66 90

16 34 16 24

50 (.556) 40 (.444)

GS = 56 GS^2 = 34

Not GS

Table 7.8 summarizes Schoenberg. Although we disagreed with some of the other analysts, we found that seven of these eleven pieces are within tolerance.

Table 7.8. Summary of Schoenberg

Piece (Abbr.)	At GS or GS^2	1 meas.	Too far	Tri-sected	FN	Notes
Piano Piece op. 33a	.383					
Five Orch op. 16 Mvt I		.625			5:3	Pasc
		.610				Mad
Five Orch op. 16 Mvt V		.610				Morg
Quintet Wind Mvt I			.560			Pasc
			.560			Mad
				.667		W/rpt
Quintet Wind Mvt IV	.618					Pasc
			.627			Mad
Str Qt no. 2 Mvt I			.395			Webst
	.382				89:144	Mad
Str Qt no. 2 Mvt II			.600		3:2	Webst
Str Qt no. 2 Mvt III		.632				Mad
Str Qt no. 3 Mvt I			.607			Webst
	.619					Mad
Str Qt no. 4 Mvt I			.581			Webst
Str Qt no. 4 Mvt III		.611				Webst
			.556			Mad

Stockhausen

Karlheinz Stockhausen did not make *Klavierstück IX* (1954–5) easy to analyze. Jonathan Kramer reported his struggle with this piece in an article and a book, with somewhat more success than we had.[68] Stockhausen used Fibonacci numbers to structure the meters in parts of the piece and part of a sequence of "Fibonacci differences" as in the following

Sequence	1 3 6 11 19 32 53 87 142
Differences	2 3 5 8 13 21 34 55

to structure other parts, but there are numerous anomalies, such as non-Fibonacci meters and changes from eighth-note to sixteenth note meters seemingly at random. Table 7.9 shows these meters section by section.

Table 7.9. Meter signatures used by Stockhausen in *Klavierstück IX*, by section (marked by heavy barlines in the score) (Each box is one measure)

142/8	87/8																
42/8																	
13/8	2/8	21/8	8/8	1/8	3/8	8/8	1/8	5/8	13/8	2/8	5/8	3/8					
21/8	3/8	1/8	13/8	2/8	8/8	5/8	13/8	3/8	2/8	8/8	5/8	1/8	8/8	3/8	5/8	1/8	2/8
4/8																	
2/8	3/8	1/8	2/8	3/8	2/8	1/8	2/8	1/8	1/8								
8/8	8/8	8/8	5/8	8/8	13/8	8/8	3/8	1/8	8/8	13/8	3/8	10/8	5/16	5/8	8/8	10/8	13/8
3/16	3/8	7/16															
4/8																	
1/8	4/8	7/8	4/8	5/16													
4/8																	
3/16	4/8	3/16	4/8	13/16													
8/8																	
1/8	5/16	3/8															
8/16																	
1/8	3/16																
4/8																	
4/8																	
4/8																	
5/8																	
13/8	5/8	8/8	3/8														
8/8	1/8	5/8	3/8	2/8	5/8	2/8	3/8	1/8	3/8	1/8	2/8	2/8	1/8	1/16			

8/8	3/8	5/8															
1/8	6/8	3/8	11/8	10/8													
3/8	5/8	21/8	13/8	8/8	2/8	1/8											
2/8	3/8	1/8	5/8	8/8													
3/8	2/8	1/8															
1/8																	
2/8	1/8																
1/8	3/8	2/8	5/8														
1/8	2/8	5/8	3/8	8/8	13/8												
5/8	1/8	2/8	21/8	3/8	13/8	8/8	34/8										
34/8																	

Since the non-Fibonacci and sixteenth note meters occur in the middle of the piece, perhaps this section is intended to be a development. If so, an exposition of sorts could be said to begin after measure 3, in which only meters 1/8, 2/8, 3/8, 5/8, 8/8, 13/8, and 21/8 are used. These meters plus 34/8 are again used exclusively in the "Coda" after measure 116.

After measure 34 (marking the development?) there is coincidentally a section break and eighth note and sixteenth note meters appear. After measure 63, meters as diverse as 1/16, 3/16, 5/16, 7/16, 8/16, 13/16, 4/8, 6/8, 7/8, 10/8, and 11/8 appear. How are these to be explained?

As for one of Kramer's open questions, we found that the meters of the first two measures in the Introduction (142/8 and 87/8) can be explained as the last two numbers of the Fibonacci difference sequence shown above. The meter of the third measure (42/8), however, remains unexplained. In addition, the reason for the use of the difference sequence as opposed to the Fibonacci sequence at this point is not clear.

Each of the sections is at a different tempo. They alternate in metronome markings of ♪ = 60 and ♪ = 160, which are ratios 3:8 and 8:3, except in the coda, which has a metronome marking of ♪ = 120. This is inconsistent, for 60:120:160 = 3:6:8, which of course is not Fibonaccian, whereas 60:100:160 = 3:5:8 would be.[69]

Kramer cited a reference to an analysis by Stockhausen of *Zyklus*, but the piece has no durational information and does not seem susceptible to proportional analysis.[70]

Kramer also reported an analysis of *Adieu* (1966) for wind quintet.[71] This piece does indeed use Fibonacci numbers to determine durations in a mostly straightforward fashion. There are five sections marked by general pauses. The first section is fairly slow, consisting of durations of 8, 13, 21, 55, and 89 seconds.

The longer second section begins slowly using durations of 34, 55, and 144 and accelerates greatly using durations of 2, 3, 5, and 8 in seemingly random order. The third, and shortest, section uses durations of 1, 2, 3, and 5. The fourth section uses durations of 5, 8, and 13, while the fifth section uses durations of 21 and 34.

Three sections (III, IV, and V) and four subsections of I and II combine to durations equal to Fibonacci numbers. Hence the smaller Fibonacci sections build up into larger Fibonacci sections. Nevertheless, the piece does not divide at or near GS. Several levels of Fibonacci subsections overlap section boundaries and combine into four units of Fibonacci number 144, totaling 576, the length of the piece. Since the sections themselves do not always add to Fibonacci numbers, we are led to wonder if the general pauses marking the sections are a subterfuge to reduce our perception of this structure as a double bisection.

These details are shown in figure 7.84.

Figure 7.84. Stockhausen's *Adieu* (Madden) (Units)

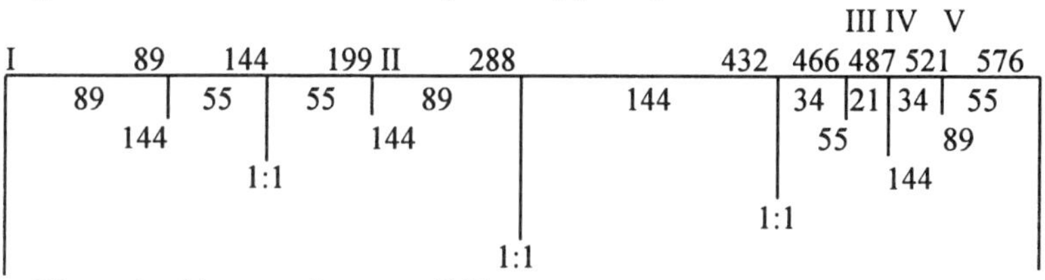

Bisected, with many elements of FN

David Cope showed a portion of *Stop* (1965) that uses Fibonacci durations throughout.[72] The piece is divided so that all adjacent measures relate as phi. They run the gamut from GS^5 (.090) to ϕ^5 (11.089), as shown in table 7.10. Notice that the one-unit anomaly in measure 19 is made up in measure 22.

Table 7.10. Summary of Stockhausen's *Stop* (Units)

Measures	Ratio	Dec. No.	GS or phi	Proximity to ϕ
1-2	8:55	0.145	GS^4	Good
2-3	55:5	11.000	ϕ^5	Good
3-4	5:5	1.000	1:1	Good
4-5	5:5	1.000	1:1	Good
5-6	5:13	0.385	GS^2	Good
6-7	13:5	2.600	ϕ^2	Poor
7-8	5:5	1.000	1:1	Good
8-9	5:13	0.385	GS^2	Good
9-10	13:8	1.625	ϕ	Fair
10-11	8:8	1.000	1:1	Good
11-12	8:13	0.615	GS	Good
12-13	13:8	1.625	ϕ	Fair

Measures	Ratio	Dec. No.	GS or phi	Proximity to ϕ
13-14	8:5	1.600	ϕ	Poor
14-15	5:13	0.385	GS^2	Good
15-16	13:5	2.600	ϕ^2	Poor
16-17	5:21	0.238	GS^3	Good
17-18	21:34	0.618	GS	Good
18-19	34:35	0.971	1:1	Poor
19-20	35:5	7.000	ϕ^4	Poor
20-21	5:5	1.000	1:1	Good
21-22	5:54	0.092	GS^5	Poor
22-23	54:5	10.800	ϕ^5	Poor
23-24	5:5	1.000	1:1	Good
24-25	5:21	0.238	GS^3	Good
25-26	21:5	4.200	ϕ^3	Poor
26-27	5:21	0.238	GS^3	Good
27-28	21:8	2.625	ϕ^2	Fair
28-29	8:13	0.615	GS	Good
29-30	13:21	0.619	GS	Good
30-31	21:5	4.200	ϕ^3	Poor
31-32	5:8	0.625	GS	Fair
32-33	8:5	1.600	ϕ	Poor
33-34	5:5	1.000	1:1	Good
34-35	5:8	0.625	GS	Fair
35-36	8:8	1.000	1:1	Good
36-37	8:5	1.600	ϕ	Poor
37-38	5:5	1.000	1:1	Good
38-39	5:13	0.238	GS^3	Good
39-40	13:8	1.625	ϕ	Fair
40-41	8:34	0.235	GS^3	Good
41-42	34:5	6.800	ϕ^4	34:8 optional

Thus Stockhausen used fib differently than any of the other composers studied so far and did not divide his structures into GS proportions. The Fibonacci numbers must have been stimulating to him because he used them differently each time.

Stravinsky

Igor Stravinsky is surprisingly well-represented in the golden proportion literature. Kramer discussed four of Stravinsky's pieces.[73] Using section-like "moment" structures, he meticulously worked out the timings in seconds from the metronome markings. They will be used here.

To begin with the earliest piece that Kramer treated, *Les Noces* (1921–1923) is almost exactly bisected, as shown in figure 7.85.

Figure 7.85. Stravinsky's *Les Noces* (Kramer) (Seconds)

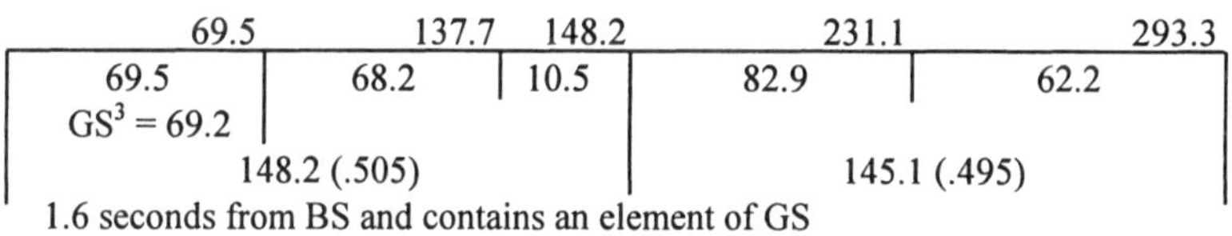

Kramer's centerpiece was Stravinsky's *Symphonies of Wind Instruments* (rev. 1947). Kramer stated that the ratios of adjacent moments were within 7% of the ratio 3:2, or 1.5; thus they ranged from 1.4 to 1.6 or so, including phi (1.618). The relationships needed for our purposes are shown in figure 7.86. The piece divides almost exactly at GS.[74]

Figure 7.86. Stravinsky's *Symphonies of Wind Instruments* (Kramer) (Seconds)

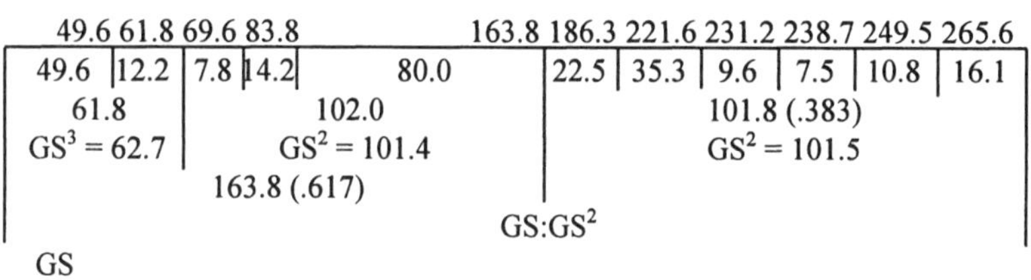

Sonata for Two Pianos (1944) is equally revealing. The timings for the three movements are as shown in figure 7.87. This is FN 2:3.

Figure 7.87. Stravinsky's *Sonata for Two Pianos* (Kramer) (Seconds)

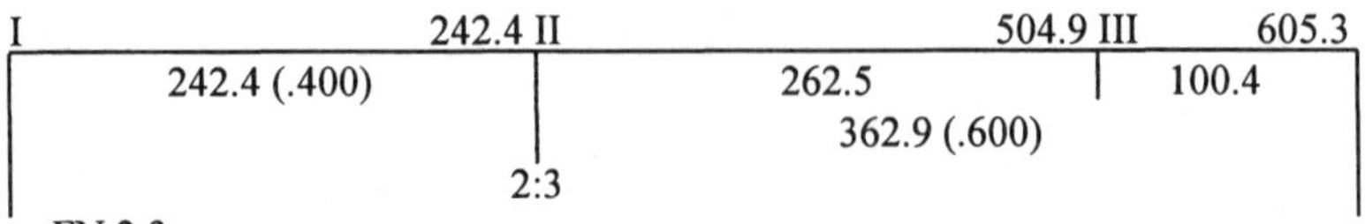

FN 2:3

Agon (1954–7) is only .8% from GS^2, as shown in figure 7.88.

Figure 7.88. Stravinsky's *Agon* (Kramer) (Seconds)

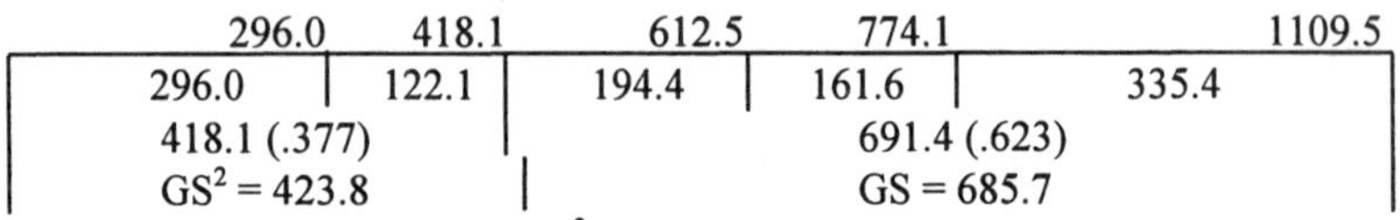

Less than one percent from GS^2

Rothwell correctly stated that the first movement of Stravinsky's *Octet for Wind Instruments* (1923) is constructed in phi proportions.[75]

Figure 7.89. Stravinsky's *Octet for Wind Instruments* (Rothwell) (Seconds)

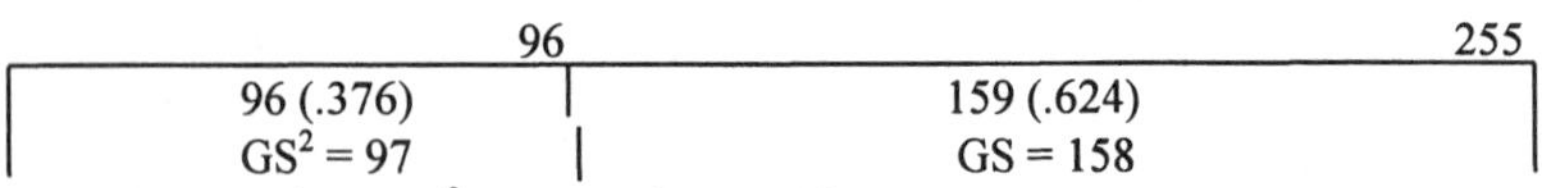

One second from GS^2 and .001 from FN 3:5

Pascoe suggested that the first movement of the *Concerto in D* (1931) has 2000 eighth-note units when converted according to the metronomic markings.[76] However, he erred by counting two slower sections of 26 and 54 units with the faster units.[77] The correct total is 2040 units. Either way, the structure is not GS. See figure 7.90.

Figure 7.90. Stravinsky's *Concerto in D* Mvt I (Madden (Eighth-notes) (Corrected)

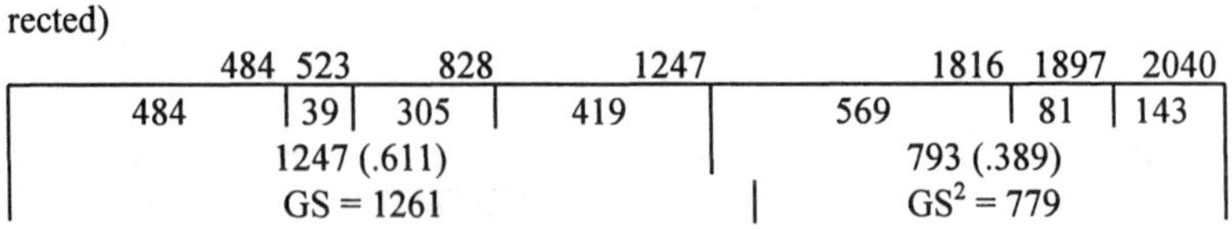

Percentages are close, but greater than one percent from GS

He also reported that movement III is 119 measures long with Section C beginning after measure 73. However, that should be measure 76, which does not mark GS, as shown in figure 7.91.

Figure 7.91. Stravinsky's *Concerto in D* Mvt III (Madden) (Measures) (Corrected)

41 60 76 93 119

41 | 19 | 16 | 17 | 26

76 (.639) 43 (.361)

GS = 74 $GS^2 = 45$

Not GS

Three Pieces for String Quartet was mentioned by Watkins as appropriate for this study.[78] Movement I is a one-part form and is not divided at GS. Watkins' argument for Fibonacci control escapes us. Movements II and III are also not divided at GS. See figures 7.92–3.

Figure 7.92. Stravinsky's *Three Pieces for String Quartet* Mvt II (Eighth notes)

70 93 138 165 215 246 280

70 | 23 | 45 | 27 | 50 | 31 | 34

165 (.589) 115 (.411)

GS = 173 $GS^2 = 107$

Not GS but two from bisected

Figure 7.93. Stravinsky's *Three Pieces for String Quartet* Mvt III (Eighth notes)

12 38 46 73 81 107 115 131 179 229

12 | 26 | 8 | 27 | 8 | 26 | 8 | 16 | 48 | 50

131 (.572) 98 (.428)

GS = 142 $GS^2 = 87$

Not GS

In Memoriam Dylan Thomas was also suggested by Watkins.[79] Movement I is indeed divided at GS, as shown in figure 7.94 and movement II is close, as shown in figure 7.95. Movement III is not.

Table 7.11 summarizes Stravinsky. At first it seemed that Stravinsky may have been a heavy user of phi, but this table shows that his usage was sporadic.

Figure 7.94. Stravinsky's *In Memoriam Dylan Thomas* Mvt I (Madden) (Quarter notes)

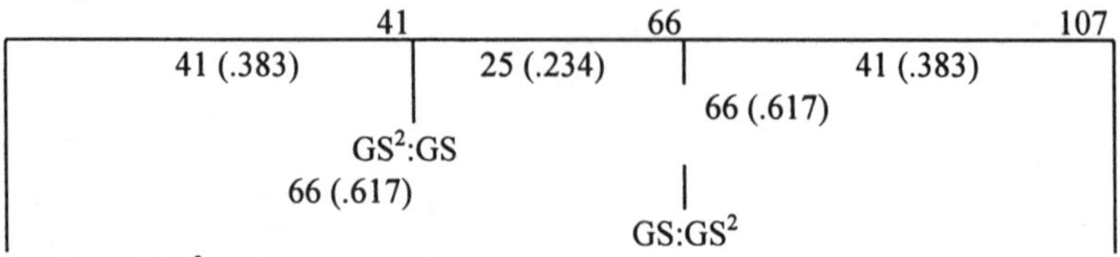

GS and GS^2 and symmetrical

Figure 7.95. Stravinsky's *In Memoriam Dylan Thomas* Mvt II (Madden) (Simplified)

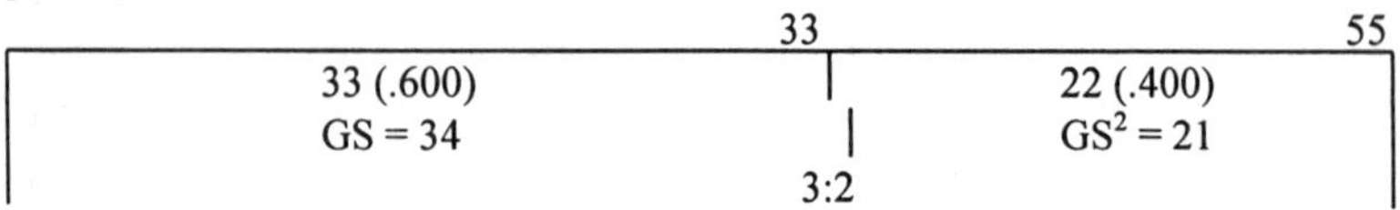

One from GS and FN 3:2

Figure 7.96. Stravinsky's *In Memoriam Dylan Thomas* Mvt III (Madden)

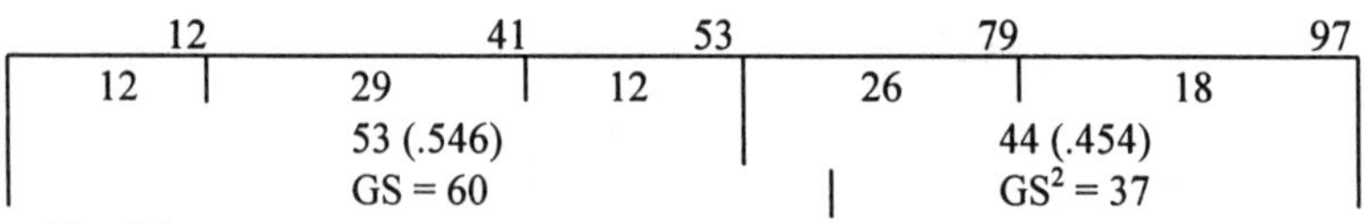

Not GS

Table 7.11. Summary of Stravinsky

Piece (Abbr.)	GS/GS^2	1 meas.	Bi-sected	Too far	FN	Notes
Les Noces				.505		
Sym Wind Instr	.617					
Son Two Pianos				.400	2:3	
Agon		.377				5 sec.
Oct. Wind Instr		.376				1 sec.
Conc. in D I				.611		15 sec.
Conc. in D III				.639		2 meas.
3 Piece Str Qt II				.589		
3 Piece Str Qt III				.572		
In Memoriam I	.383/.617					
In Memoriam II		.600			3:2	
In Memoriam III				.546		

Varèse

Edgard Varèse may have used phi in *Hyperprism* (1923). *Fractals in Music* eagerly favored this view, but a reassessment of the piece in view of its notation questions this.[80] The first measure, for example, is counted as measure 0 by its publisher, yet it is meticulously written out. This is followed later in the piece by two measures of 7/4 meter (measures 84 and 85), thus combining measures of 3/4 and 4/4. Of course, it is legitimate to do this, but there seems to be no compelling reason other than to shorten the count.

Did Varèse fudge? Or, if the publishers did why would they? Does it matter? In either case, the piece—as published—divides at one measure from GS. The alternative analyses starting at measure 0 and ending at measure 89 versus starting at 1 and ending at 90 are shown in figure 7.97.

Figure 7.97. Varèse's *Hyperprism* (Madden) (Measures)

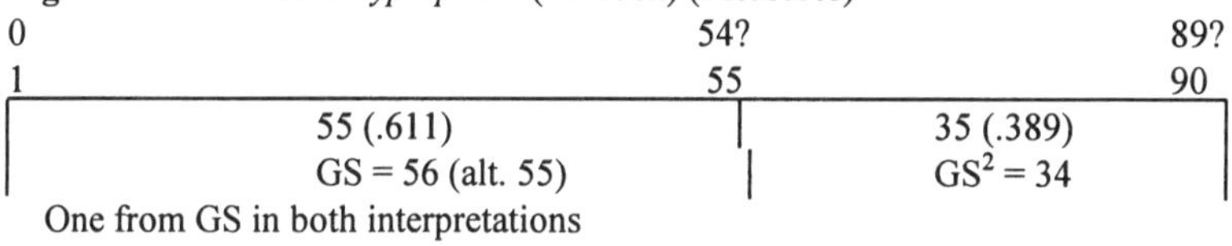

In *Offrandes* (1921), the division is at an elision at or near GS (32 or 33), although the percentages are poor, as shown in figure 7.98.[81]

Figure 7.98. Varèse's *Offrandes I* (Madden) (Measures)

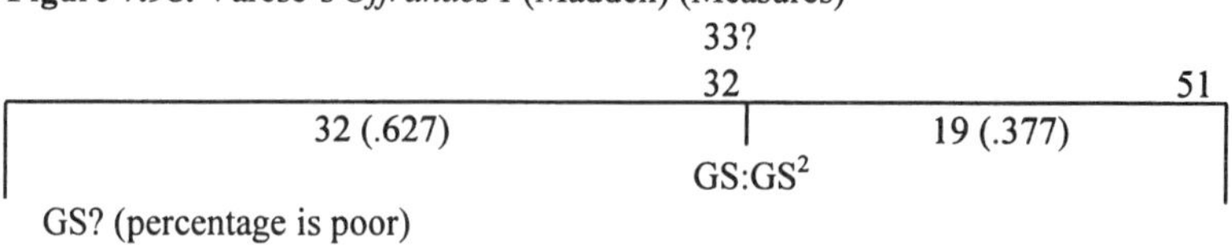

Offrandes II is divided exactly at GS and FN, as shown in figure 7.99.

Figure 7.99. Varèse's *Offrandes II* (Madden) (Measures)

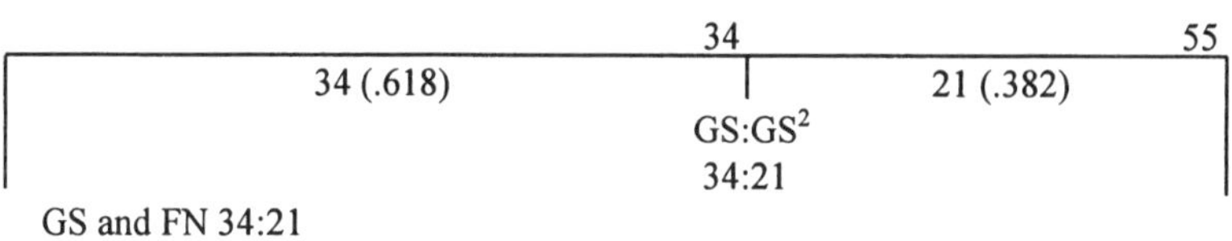

In *Intégrales* (1925) there is ambiguity of another kind. The event after 126 is strong but not GS; the events between measures 136 and 140 are weak and one or two measures from GS. However, 140 is 28FN 5:3. Thus Varèse might have known about Fibonacci numbers. See figure 7.100.[82]

Figure 7.100. Varèse's *Intégrales* (Madden) (Measures)

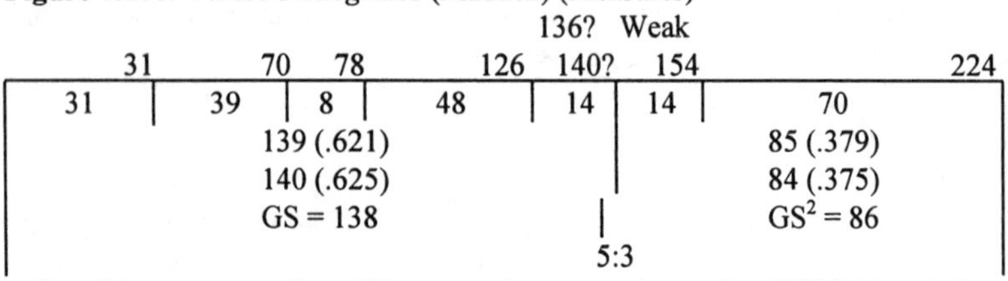

Possibly one or two from GS on a weak structural point but 28FN 5:3 at 140

In *Arcana* (1927), ambiguity also arises. Phi does not appear near any clear structural point, although there may be a structural inflection in measure 171. See figure 7.101.[83]

Figure 7.101. Varèse's *Arcana* (Madden) (Measures)

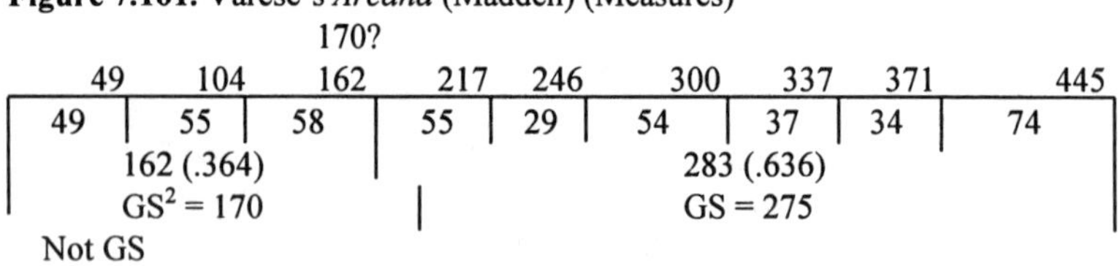

Equatorial (1934), is divided almost exactly at GS. See figure 7.102.[84]

Figure 7.102. Varèse's *Equatorial* (Madden) (Measures)

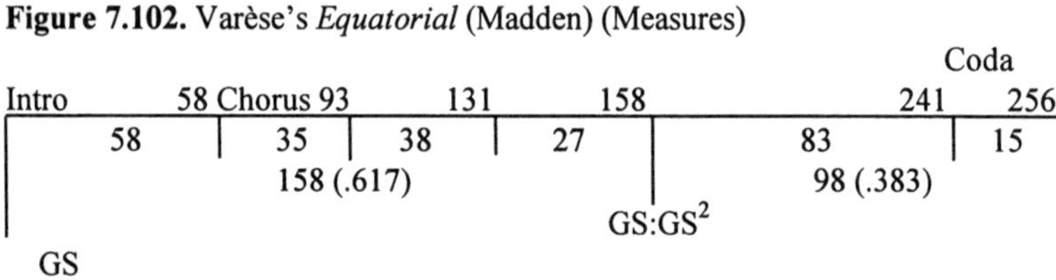

Déserts (1954) has been avoided here because of the ambiguity introduced by the lack of durations for the electronic tape sections. Moreover, calculating the structures omitting the tape interpolations does not suggest the presence of phi.[85]

Varèse is summarized in table 7.12.

Table 7.12. Summary of Varèse

Piece	GS/GS^2	1 meas.	Too far	FN	Notes
Hyperprism		.611			
Offrandes I	.627				Poor %
Offrandes II	.618			34:21	Strong
Intégrales		.621			Weak
			.625	5:3	Weak
Arcana			.364		
Equatorial	.617				
Déserts					Not anal.

3/6 = 50.0% at GS
5/6 = 83.3% within one measure
FN used twice

In conclusion, five of the six pieces by Varèse analyzed here divide within one measure of GS and FN is used twice. Except for *Hyperprism* (which may have been a simple miscount), they leave no doubt that Varèse knew that he was doing this. Nevertheless, it was not so important to him that he would use phi in every piece.

Minor Users

Barber

Samuel Barber's *Piano Sonata* (1949–50) Movement I is in phi proportion, according to Rothwell.[86] His count was 620 beats with a division after 384, making, as he wrote, a percentage figure of .618. (We nit-pick this and say that it should be .619.) That would be one beat from GS (.383). Our count differs at 616.5, a difference of about one measure, with the division after 385, which is about four beats from GS (381) and very close to FN 5:3. This is shown in figures 7.103a–b.

Figure 7.103a. Barber's *Piano Sonata op. 26* Mvt I (Rothwell) (Beats)

384 620

384 (.618) | 236 (.382)

GS = 383 | GS^2 = 237

One beat from GS, but miscounted

Figure 7.103b. Barber's *Piano Sonata op. 26* Mvt I (Madden) (Beats)

385 616.5

385 (.624) | 231.5 (.376)

GS = 381 | GS^2 = 235.5

Four beats from GS (one measure) and .001 from 5:3

Berg

Alban Berg's *Lyric Suite* (1925–6) Movements I and VI are divided in phi proportions, according to Pascoe. In Movement I Pascoe's division measured in beats is very close to bisected, as shown in figure 7.104a.

Figure 7.104a. Berg's *Lyric Suite* Mvt I (Pascoe) (Beats)

125 248

125 (.504) | 123 (.496)

One beat from bisected

Morgan's analysis, in measures, is approximately the same. See figure 7.104b.[87]

Figure 7.104b. Berg's *Lyric Suite* Mvt I (Morgan) (Measures)

B-II B-III

I 12 II 15 III 22 IV 35 B-I 48 51 53 B-IV 69

12 | 3 | 7 | 13 | 13 | 3 | 2 | 16

35 (.507) | 34 (.493)

GS = 43 | GS^2 = 26

One half-measure from bisected

Mvt VI has too many tempo and meter changes to risk an analysis.

Berg's *Altenberg Lieder no. 5* (1912), however, is one from GS, as shown in figure 7.105.[88]

Figure 7.105. Berg's *Altenberg Lieder no. 5* (Madden) (Beats and measures)

Beats	137	220
Measures	35	55
137 (.623)	83 (.377)	
35 (.636)	20 (.364)	
GS = 136	GS^2 = 84	
GS = 34	GS^2 = 21	

One beat and one measure from GS

Cage

In discussing John Cage's music, Reginald Smith Brindle and Michael Richard Rogers stated, "Between the pre-war *Construction in Metal* and *Music of Changes* (1951), the structure of several works is based on a number of measures having a square root, so that the large lengths have the same relation within the whole as the small lengths have within the larger units."[89] This seems to imply phi proportion, but the only way this could happen is for the square roots to be Fibonacci numbers, as shown in table 7.13. Square roots 13 and 21 are in phi proportion and 20, 18, 16, and 15 are fairly close. Obviously, 19, 17, and 14 are far from phi.

Table 7.13. Proportions of square roots appropriate to the music of Cage

Square	Square Root	Nearest ratio to phi						
441	21	8:13	=	8/21	=	.381		
400	20	8:12	=	8/20	=	.400	=	FN 2:3
361	19	7:12	=	7/19	=	.368		
324	18	7:11	=	7/18	=	.389		
289	17	7:10	=	7/17	=	.412		
(289)	(17)	6:11	=	6/17	=	.353		
256	16	6:10	=	6/16	=	.375	=	FN 3:5
225	15	6:9	=	6/15	=	.400	=	FN 2:3
196	14	5:9	=	5/14	=	.357		
169	13	5:8	=	5/13	=	.385		

The nearest Cage came to phi in the *Piano Works 1935–1948* was in the piece entitled *In a Landscape* (1948), which is constructed of 15 sections of 15 measures.[90] The only possible phi divisions would come after measure 90 or 135, as shown in figure 7.106.

Figure 7.106. Cage's *In a Landscape* (Madden) (Measures)

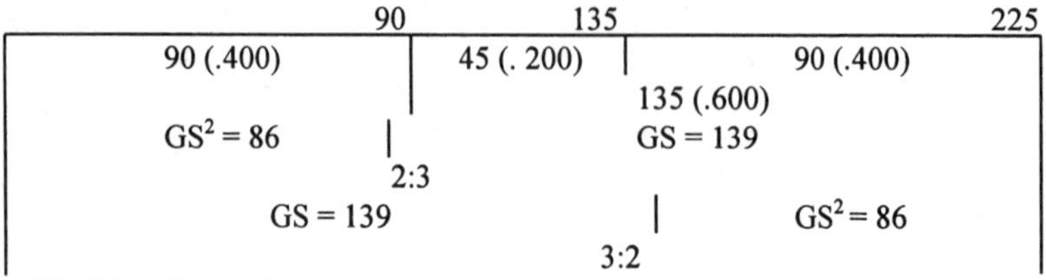

FN 2:3, 3:2, or 2:1:2

Double Music (1941) by John Cage and Lou Harrison is 200 measures long and shows no obvious divisions at phi.[91]

First Construction in Metal (1939) is constructed of 16 units of beats grouped as 4-3-2-3-4, plus a coda of 2-3-4 beats for a total of 265.[92] This piece has a similar problem as above, with the additional problem of the coda. To be even a Fibonacci piece, the coda would have to be removed, as shown in figure 7.107.

Figure 7.107. Cage's *First Construction in Metal* (Madden) (Measures)

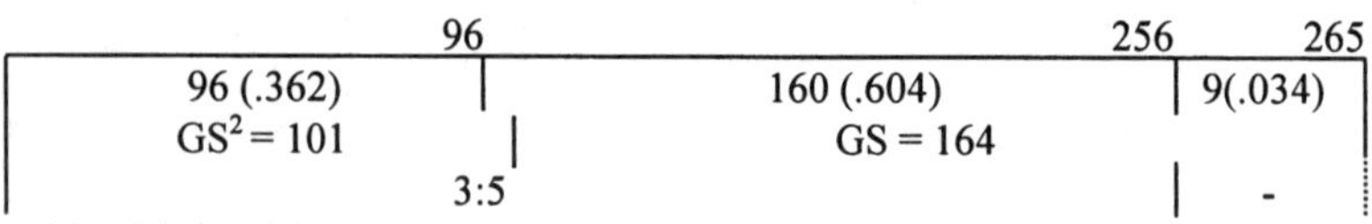

Not GS, but 96:160 is FN 3:5

Thus Cage's use of phi proportions seems to be only coincidental to his use of another structural system. There is no real evidence that he knew of phi or would have been interested in it.

Carter

Elliott Carter's *String Quartet no. 2* (1959) Movement III is in phi proportion according to Hoffman.[93] See figure 7.108a. (Number 286 is the beginning measure number in the score, which is numbered continuously from movement to movement.)

Figure 7.108a. Carter's *String Quartet no. 2* Mvt III (Hoffman) (Measures)

286 | 340 Climax | 373 Cadenza | 408 Close | 426
55 (Dynamic) | 88 | 123 | 141

55 (.625) GS = 54 of 88	33 (.375) GS^2 = 33 of 88	35 (.660) GS = 33 of 53	18 (.340) GS^2 = 20 of 53

5:3

88 (.624) GS = 87	53 (.376) GS^2 = 54

One measure from GS and .001 from 5:3

To begin, Hoffman's total count is off by one measure, and his first structural point (340) is dynamic. This brings him to one measure from GS. We found that the piece is in fact divided at GS. See 7.108b.

Figure 7.108b. Carter's *String Quartet no. 2* Mvt III (Madden) (Measures)

286 | 320 | 373 Cadenza | 408 Close | 427
35 | 88 | 123 | 142

35 (.247)	53 (.373)	35 (.247)	19 (.133)

88 (.620)	54 (.380)

GS:GS^2

GS

Rogers thought that Carter's *Eight Studies and a Fantasy for Woodwind Quartet* (1952) Movement VII is divided at GS.[94] See figure 7.109a.

Figure 7.109a. Carter's *Eight Studies and a Fantasy for Woodwind Quartet* Mvt VII (Rogers) (Measures) (Dynamic)

19 | 31

19 (.613)	12 (.387)

GS:GS^2

GS but disputed

It is instead one measure from GS, with poor percentages. See figure 7.109b.

Figure 7.109b. Carter's *Eight Studies and a Fantasy for Woodwind Quartet* Mvt VII (Madden) (Beats and Measures)

Beats	73	124
Measures	18	31
	73 (.589)	51 (.411)
	18 (.581)	13 (.419)
	GS = 77	$GS^2 = 47$
	GS = 19	$GS^2 = 12$

Four beats (one measure) from GS with very poor percentages

Dallapiccola

Pozzi Escot provided a score and analysis of Luigi Dallapiccola's "Colore" from *Notebook for Annalibera.*[95] It is divided at GS^2 if the rest in measure 8 is included in the first section, as shown in figure 7.110.

Figure 7.110. Dallapiccola's *Notebook for Annalibera no. 9* "Colore" (Escot) (Eighth notes)

35	92
35 (.380)	57 (.620)
GS^2:GS	
GS^2	

Hoffman

Michael Hoffman's dissertation included a complete score and thorough analysis of his jazz composition *Suite: Three Jazz Spirals* (1995–6) constructed of phi durations. It consisted of three pieces, "Double Helix," "Smoke & Mirrors," and "Take the Time."

"Double Helix" used Fibonacci numbers in the pattern shown in figures 7.111.[96] It is not GS overall, but uses many Fibonacci numbers.

Figure 7.111. Hoffman's "Double Helix" from *Suite: Three Jazz Spirals* (Hoffman) (Measures)

21	26	60	94	115	120	154	175	188	196	205
21	5	34	34	21	5	34	21	13	8	5+2+2
94 (.459)				26 (.126)		85 (.415)				

Not GS, but full of Fibonacci numbers and one from GS without the coda

"Smoke & Mirrors" began with a sketch that applied GS to an arbitrarily chosen 120 measures.[97] See table 7.14.

Table 7.14. GS values for Hoffman's "Smoke & Mirrors from *Suite: Three Jazz Spirals*

120 = 1	17 = GS^4
74 = GS	11 = GS^5
46 = GS^2	7 = GS^6
28 = GS^3	4 = GS^7

This looks good, but close inspection shows that these numbers were applied to the beginnings of the sections, instead of to the durations.[98] Internally, the durations would be unchanged, but cumulatively they are all a measure off except the last. This is the same mistake cautioned against by several authors. Hoffman's analysis is shown in figure 7.112a, where nearly exact GS values are everywhere.[99] See our result in figure 7.112b.

Figure 7.112a. Hoffman's "Smoke & Mirrors" from *Suite: Three Jazz Spirals* (Hoffman) (Measures)

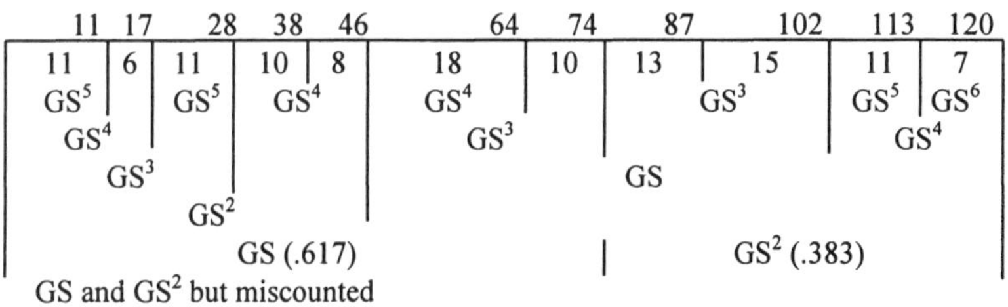

Figure 7.112b. Hoffman's "Smoke & Mirrors" from *Suite: Three Jazz Spirals* (Madden) (Measures)

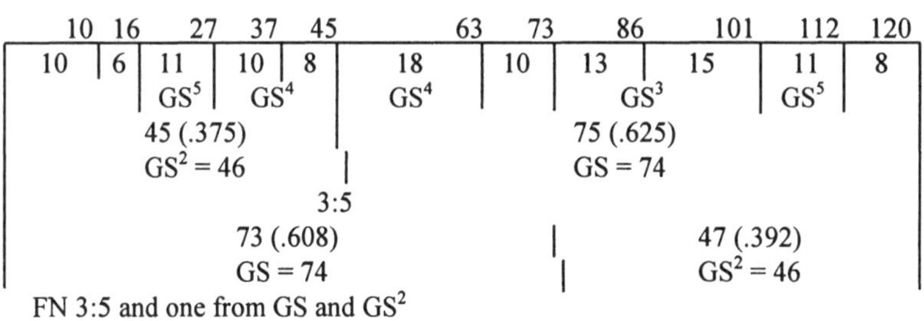

Roughly the same thing happened with "Take the Time." All the measure numbers are off by one except the last.[100] The corrected structure is shown in figure 7.113b. Only one subdivision is at GS.

Figure 7.113a. Hoffman's "Take the Time" from *Suite: Three Jazz Spirals* (Hoffman) (Measures)

19	51	65	83	101	117	129	145	165	177	195	213	229	241	257	269	289
19	32	14	18	18	16	12	16	20	12	18	18	16	12	16	12	20
					GS^6		GS^6					GS^6		GS^6		

19/51 (.373)
101/165 (.612) GS = 102
109/177 (.616) GS = 109
149/241 (.618) GS = 149
177/289 (.612) GS = 179
Highly GS but miscounted

Figure 7.113b. Hoffman's "Take the Time" from *Suite: Three Jazz Spirals* (Madden) (Measures)

18	50	64	82	100	116	128	144	164	176	194	212	228	240	256	268	289
18	32	14	18	18	16	12	16	20	12	18	18	16	12	16	12	21
					GS^6		GS^6					GS^6		GS^6		

18/50 (.360)
100/164 (.610) GS = 101
108/176 (.614) GS = 109
148/240 (.617) GS = 148
176/289 (.609) GS = 179
Only one value is at GS

Thus Hoffman used phi proportions to stimulate his thinking, but it is by no means accurate in every detail.

Johnston

Ben Johnston's *Quintet for Groups* (1966) was found by Barney Childs to contain phi proportions.[101] Childs stated, "The macro-rhythmic structure is based upon the Fibonacci series . . . and this determines all sectional, phrase, and phrase-group durations. At cue K the groups are moving in two simultaneous tempi . . . ; the new tempo (♩ = 135) begins a reverse Fibonacci series faster than the old tempo (♩ = 90) in a 3:2 ratio."

In *The Music of Ben Johnston*, Heidi von Gunden referred to Childs' article and added, "The sketch and the score also indicate that Johnston began by first plotting time. The score is carefully measured out with bar lines dividing measures according to his proportional scheme using the Fibonacci series and harmonic relations. The ratios are even labeled."[102] Although the score is not available, it seems certain from these authors' statements that Johnston knew that he was using phi.

Von Gunden studied Johnston's *String Quartet no. 6* and found that the piece "is a palindrome with the tempo of the retrograde two thirds faster than its original, [and that] the midpoint's climax occurs at about two thirds of the tonal [*sic*] duration—the golden mean."[103] She went on to say that, in a revision, Johnston added dynamics and articulations consistent with this structure. Although this score is also not available, it is possible to demonstrate from von Gunden's figure 8 and her discussion that she was correct, as shown in illustration 7.114. To clarify the math, if the first part has duration 1 and the second part (retrograde) has duration 2/3, the total duration is 5/3. Dividing 1 by 5/3 gives 3/5, or .600. Thus, the division is clearly Fibonaccian.

Figure 7.114. Johnston's *String Quartet no. 6* (Madden) (Proportions)

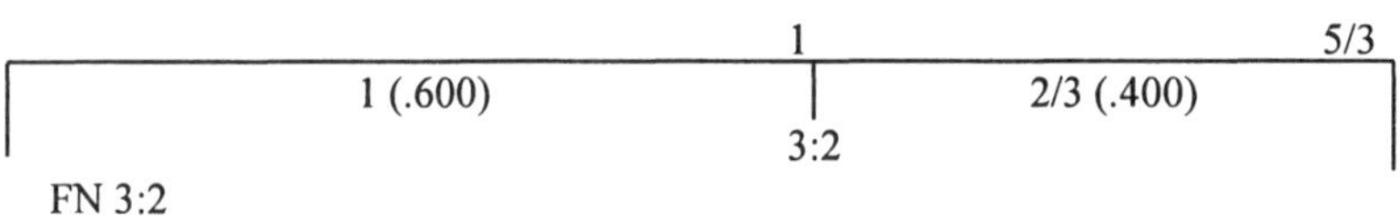

Kodály

Zoltán Kodály's *Prelude* (Mvt I) to the *Háry János Suite* (1926) was said by Lendvai to have a recapitulation consisting of 36½ measures ("with the fermata"), and that this was divided into 14 and 22½ measures. It does divide at GS^2 with or without the extra half measure, but it is only a partial movement. See figure 7.115.

Figure 7.115. Kodály's *Háry János Suite* Mvt I (Madden) (Measures)

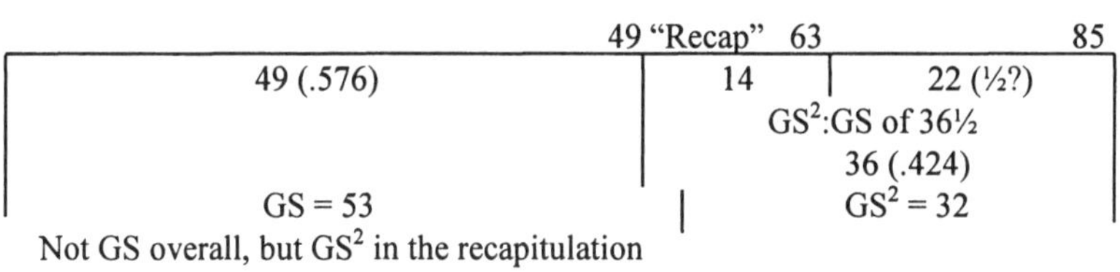

Adkins thought that the *Suite* has phi proportions throughout.[104] All are too far except the fourth movement. See figures 7.116–120.

Figure 7.116. Kodály's *Háry János Suite* Mvt II (Madden) (Measures)

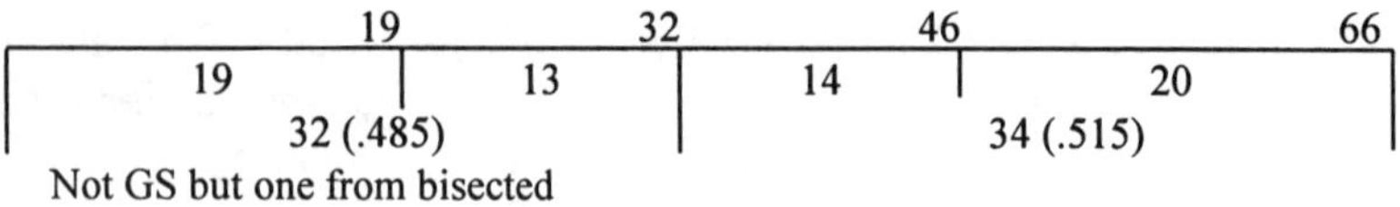

Figure 7.117. Kodály's *Háry János Suite* Mvt III (Madden) (Measures)

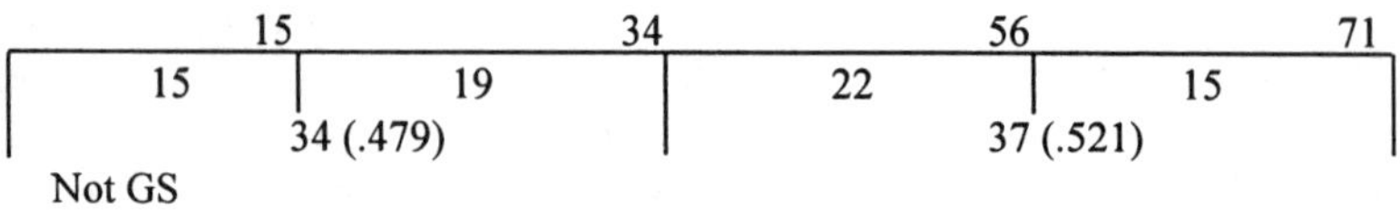

Figure 7.118. Kodály's *Háry János Suite* Mvt IV (Madden) (Measures)

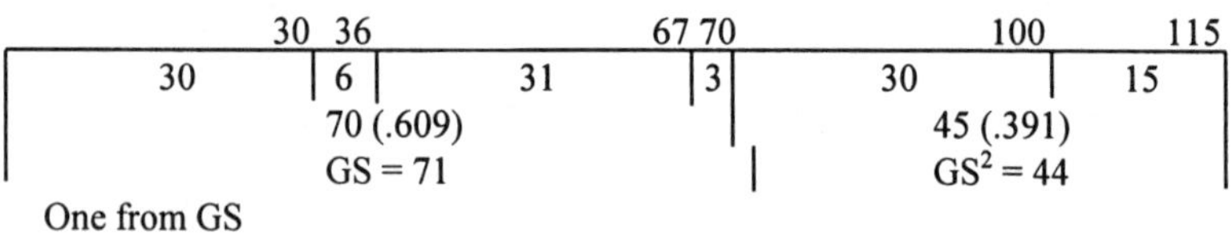

Figure 7.119. Kodály's *Háry János Suite* Mvt V (Madden) (Measures)

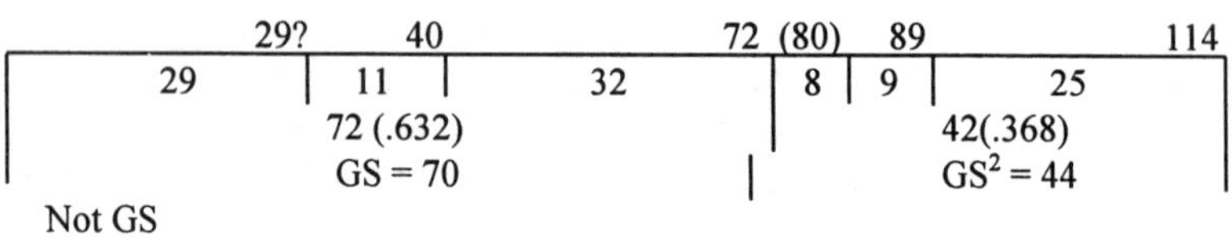

Figure 7.120. Kodály's *Háry János Suite* Mvt VI (Madden) (Measures)

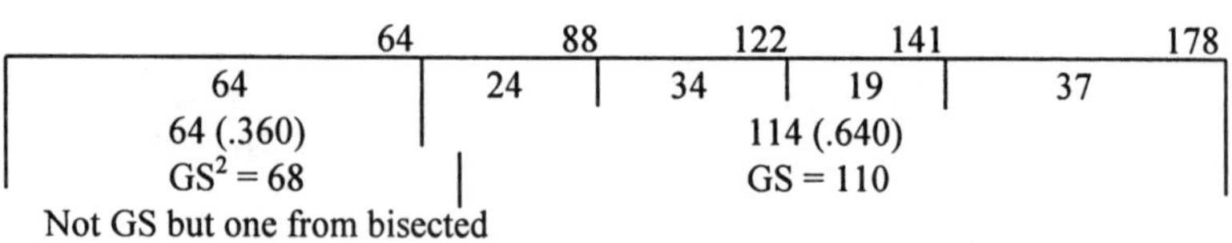

Lendvai also suggested that Kodály's *Psalmus Hungaricus* (1923)

"turns" at measure 245, but that is not structural.[105] There may be a structural point after 240, but this, too, is not phi. See figure 7.121.

Figure 7.121. Kodály's *Psalmus Hungaricus* (Madden) (Measures)

	114?	186	268	340	395
114	72		82	72	55
186 (.471)			209 (.529)		

Not GS

Kodály does not appear to have been a serious user of phi.

Ligeti

Escot suggested that Ligeti's *Continuum* (1968) is divided in phi proportions.[106] Although we disagree with her sub-sections, it does divide near GS. See figure 7.122. It should be noted that Ligeti stated that he did not use phi proportions with any precision after *Atmospheres* (1961).[107]

Figure 7.122. Ligeti's *Continuum* (Madden) ("Measures")

86	125	142.75	181.5	204
86	39	17.75	38.75	22.5
125 (.613)		79 (.387)		
GS = 126		GS^2 = 78		

One from GS

Tim Johnson studied Ligeti's *Apparitions* and *Atmospheres* and found phi.[108] He found that the first section of *Apparitions* (1957–9) ends in measure 71, with crochet (quarter note) number 144, and that the second section begins in measure 73. But it does not end until quarter note 154, which is just before the end of measure 73. Ligeti's assertion that he used phi thus cannot be accepted. See figure 7.123a and b. Also, the shorter parts do not have phi proportions as Ligeti stated.[109]

Figure 7.123a. Ligeti's *Apparitions* (Johnson) (Crochets)

144	234
144 (.615)	90 (.385)
GS = 145	GS^2 = 89

One from GS, but disputed

Figure 7.123b. Ligeti's *Apparitions* (Madden) (Quarter notes)

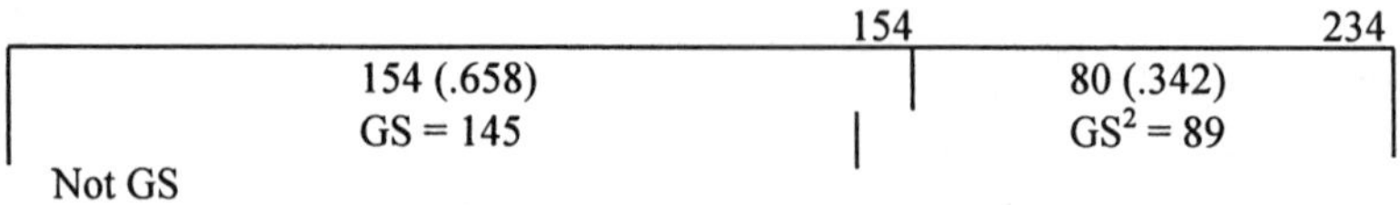

Johnson's analysis of *Atmospheres* (1961) lacks the example used in his talk and is sketchy. We found it to be one measure from GS^2 if tempo changes are ignored. This is shown in figure 7.124.

Figure 7.124. Ligeti's *Atmospheres* (Madden) (Measures)

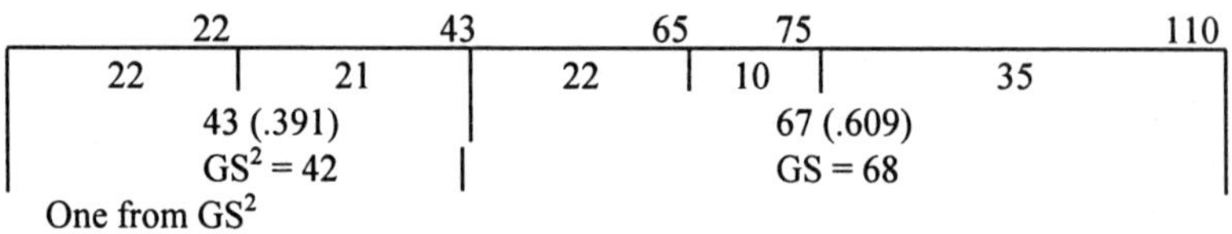

We conclude that, although Ligeti has spoken of using phi in his music, he did so only for a short time, and inaccurately.

Ravel

Maurice Ravel was claimed by Howat to have "made frequent use of GS."[110] The song *Les grands vents venus d'outremer*, for example, divides at one measure from both GS and GS^2, as shown in figure 7.125.[111]

Figure 7.125. Ravel's song *Les grands vents d'outremer* (Madden) (Measures)

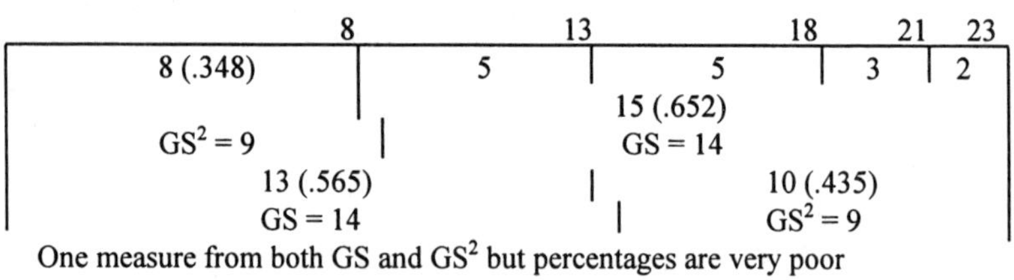

However, in the piano piece *A la manière de Borodin* the dynamic peak is too broad to show a clear relation to GS and the formal structure is not divided at or near GS, as shown in figure 7.126.

Figure 7.126. Ravel's *À la manière de . . . Borodine* (Madden)

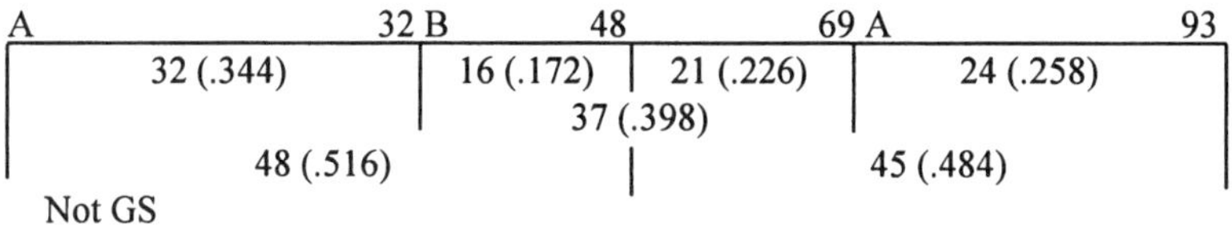

The second movement of the *Sonata for Violin and Cello* contains meter changes of 2/8 and 3/8, which must be considered inconsequential to a serious study of phi. However, the first movement is divided at GS^2, as shown in figure 7.127.[112]

Figure 7.127. Ravel's *Sonata for Violin and Cello* Mvt I (Madden)

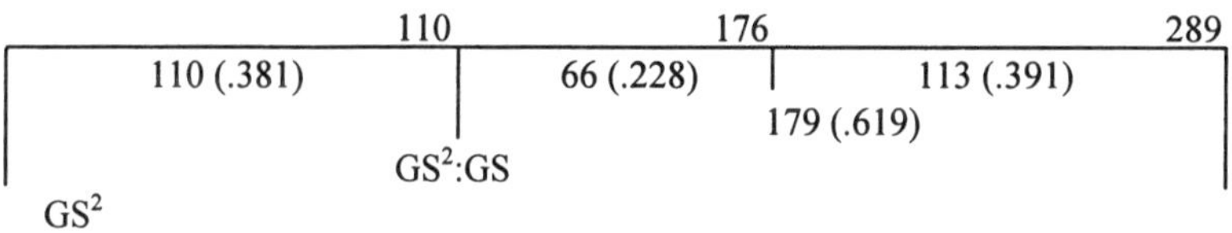

The *Sonata for Violin and Piano* is not phi, as shown in figure 7.128.[113]

Figure 7.128. Ravel's *Sonata for Violin and Piano* Mvt I (Madden)

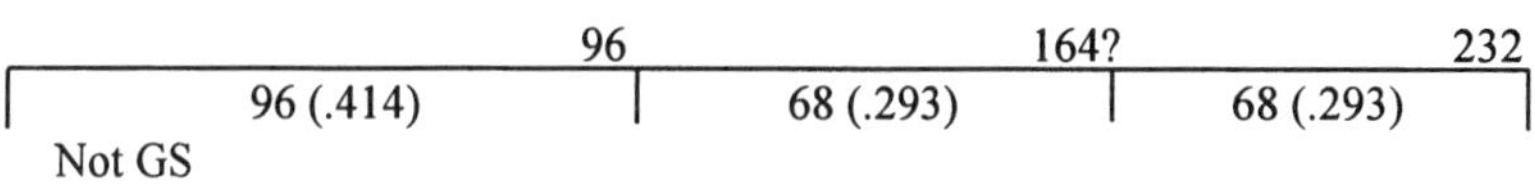

Although the "Fugue" (Mvt II) of *Le tombeau de Couperin* is divided at 21 and 34 bars, the piece is 61 bars long (not fib) and shows no other apparent usages of the golden proportion. Thus this is a superficial application of phi. See figure 7.129.[114]

Figure 7.129. Ravel's *Le tombeau de Couperin* "Fugue" (Madden)

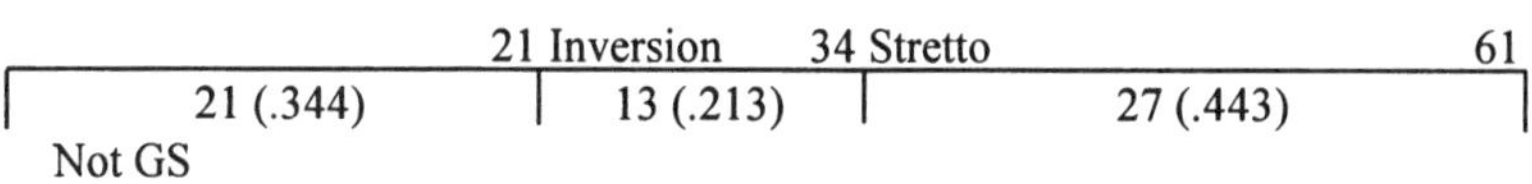

However, the first movement divides at one measure from both GS and

GS², as shown in figure 7.130a. It is also close to bisected with the repeat. See figure 7.130b.

Figure 7.130a. Ravel's *Le tombeau de Couperin* Mvt I (Madden) (W/out repeat)

4 33 37 57 94
4 | 29 | 4 | 20 | 37 (.394)
57 (.606)
GS = 58 | GS² = 36
37 (.394) | 57 (.606)
GS² = 36 | GS = 58
One measure from both GS and GS² and perfectly symmetrical

Figure 7.130b. Ravel's *Le tombeau de Couperin* Mvt I (Madden) (With repeat)

4 33 Rpt 62 66 86 123
4 | 33 | 29 | 4 | 20 | 37
62 (.504) | 61 (.496)
One half-measure from bisected

With such a mixed bag, we should look carefully at *Miroirs*.[115] The dynamic structure of "Oiseaux tristes" was mentioned in connection with Debussy and the formal structure is shown in figure 7.131. It is clearly divided at GS if Howat's count is correct, but it is more likely that section A returns after unit 71 instead of 75. It is matter of taste or expedience.

Figure 7.131a. Ravel's "Oiseaux tristes" from *Miroirs* (Structural) (Howat/Madden)

A 47 B 71? 75 A 93 Coda 121
47 (.388) | 28 (.232) | 18(.149) | 28 (.231)
75 (.620) | 46 (.380)
GS:GS²
GS but disputed

Figure 7.131b. Ravel's "Oiseaux tristes" from *Miroirs* (Structural) (Madden) (Units) (Corrected)

A 45 B 71 A 93 Coda 121
45 (.372) | 26 (.215) | 22(.182) | 28 (.231)
71 (.587) | 50 (.413)
Not GS

Howat mentioned that "Alborada del gracioso" has a phi structure, but we do not see the elaborate structure of Fibonacci numbers combining into Lucas numbers (5 + 3 + 3 = 11, 13 + 8 + 8 = 29, and 21 + 13 + 13 = 47) with the precision that he proposes.[116]

Thus, Ravel was a sometime—but knowing—user of phi proportions.

Risset

Jean-Claude Risset's *Computer Suite from Little Boy, Fall* is described by Robert Cogan as proportionately divided as 68/102 = .667.[117] Although he was correct that it is FN 2:3, he should have expressed it as 68/170 = .400, as shown in figure 7.132.

Figure 7.132. Risset's *Computer Suite from Little Boy,* "Fall" (Madden)

Schillinger

Joseph Schillinger wrote three huge volumes regarding mathematics and the arts. There has been continuing interest in them since their publication, despite much criticism. A review of *The Schillinger System of Musical Composition* that is of considerable interest here was written by acoustician John Backus.[118]

Backus complained that Schillinger's use of English was sometimes impenetrable (correct; the editors should have rewritten such passages); that the examples were often too profuse (the editors should have culled them); that Schillinger often confused mathematically trained readers by redefining mathematical symbols already in use instead of finding new ones (true, but a serious reader eventually becomes accustomed to the terms); and that there are several incorrect demonstrations of the sine and cosine functions (definitely a problem).[119]

Backus also pointed out that the diagram in figure 103 on page 330 of the *System* would not give the Fibonacci numbers as claimed. Figure 103 is an elaboration of figure 102, where r^1 is the diagonal of a $1:\sqrt{2}:\sqrt{3}$ triangle generated from a 1 by 1 square, as shown in figure 7.133a. Measurements have been added for clarification.

Figure 7.133a. Schillinger's figure 102 (with measurements added)

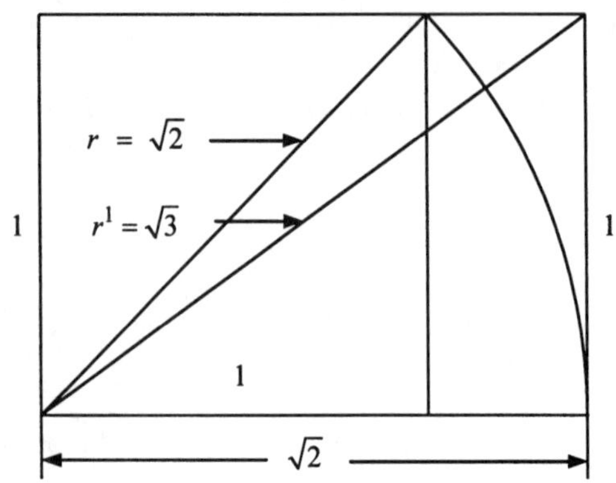

Schillinger claimed that diagonal r[1] would be the hypotenuse of a "pyramid triangle," that is, $1:\sqrt{\phi}:\phi$, but it clearly is not. To create a pyramid triangle, diagonal r would have to be that of a half square, not a square, making diagonal r[1] equal to phi, as shown in figure 7.133b.

Figure 7.133b. Schillinger's figure 102 as it should be

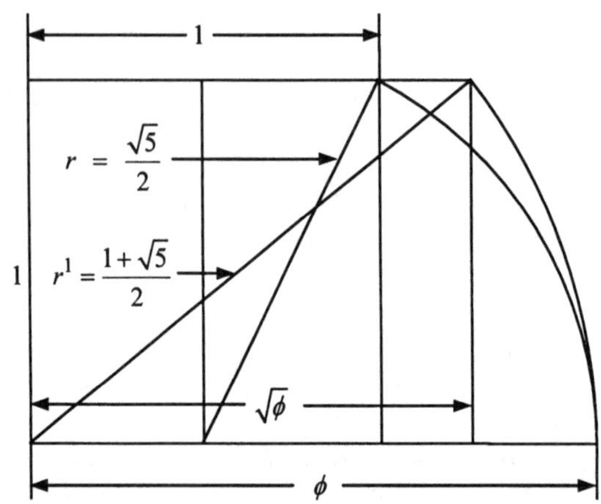

It is clear that Schillinger defined the pyramid triangle incorrectly. This is disappointing because at first it seemed like just a drafting error. But Backus was right: this is a serious error; combined with the obvious problems with the sine and cosine functions on the pages following 305, doubt is cast on the rest of the book. Unfortunately, *Fib and Phi* cannot give much attention to the rest of the mathematics because they have little to do with the golden proportion.[120]

Nonetheless, there are several good things to be said about this book. The system of graphic notation to combine attacks of differing patterns into a composite rhythm via "interference" of two or more rhythmic patterns to develop complex rhythms is good.[121]

Complicated rhythms can be built with this system. Some of the interference patterns based on Fibonacci numbers give the Latin rhythms known as rumba, tango, and so on. This seems to have been one of Schillinger's attractions for the jazz musicians of his day. In addition, Schillinger's discussion of jazz "swing," if correct, is insightful.[122]

Schillinger's *Study in Rhythm II* is analyzed by Xiaoman Zhang Wardell. However, it is just a rehash of Schillinger's own description of the piece, which uses a two-part contrapuntal pitch scheme set on a basic "attack-group" of 3 + 2 + 1 + 3 + 1 + 2 + 3 quarter notes permuted rotationally.[123] While it may be a good illustration of technique, it does not appear to be a very good piece.

Schillinger gave some thought to using phi (among other ratios) to calculate where to place the climaxes of a piece.[124] An example of this approach shows an accelerating sequence of climaxes after each of 13, 8, 5, 3, 2, and 1 units of time. This idea could have value for a composer willing to spend the necessary time and effort to understand it.

Sibelius

Jean Sibelius' *Finlandia* (1899–1900) was said by Adkins to have phi proportions, and counted the piece in beats of constant tempo.[125] See figure 7.134a.

Figure 7.134a. Sibelius's *Finlandia* (Adkins) (Beats) (Without repeat)

I 151 II 204 III 562

151 (.269) | 53 | 358 (.637)

204 (.363)

$GS^2 = 215$ | GS = 347

Not GS

This is too far from GS to suit us, but, as figure 7.134b shows, it is divided at one measure from GS^2 with the repeat and counted in measures.

Figure 7.134b. Sibelius's *Finlandia* (Madden) (Measures) (With repeat)

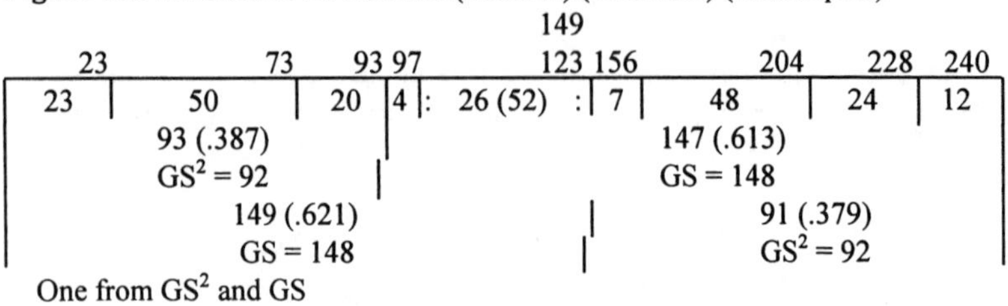

Jussi Jalas claimed that Sibelius's *Symphony V* (1915) contains phi in all three movements.[126] These claims are examined in figures 7.135 to 7.137. The units are timings in minutes of Jalas' performances. It seems that Jalas was right, but we should ask whether other conductors achieve similar timings.

Figure 7.135. Sibelius's *Symphony V* Mvt I (Jalas) (Minutes)

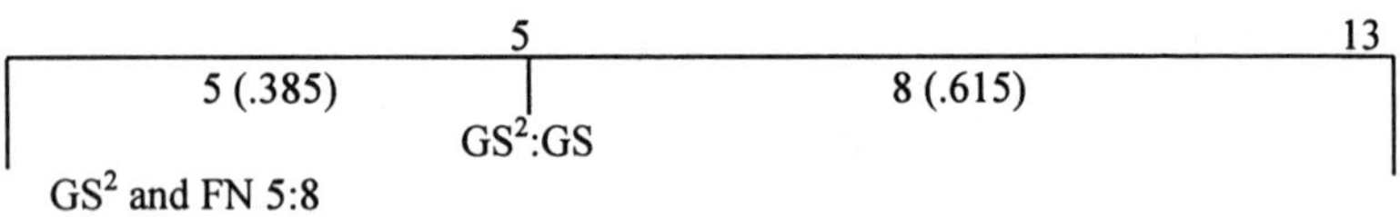

Figure 7.136. Sibelius's *Symphony V* Mvt II (Jalas) (Minutes)

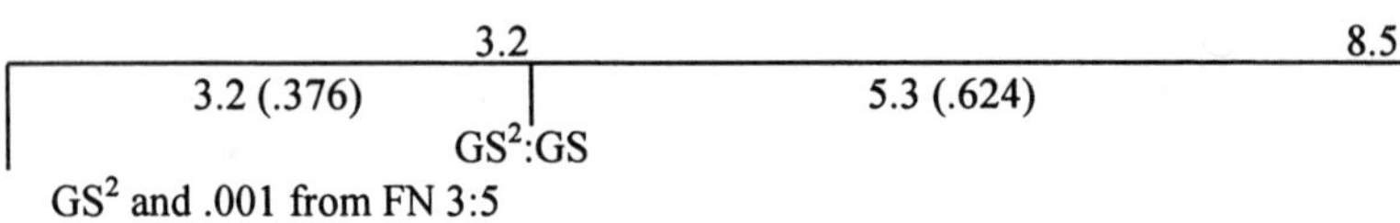

Figure 7.137. Sibelius's *Symphony V* Mvt III (Jalas) (Minutes)

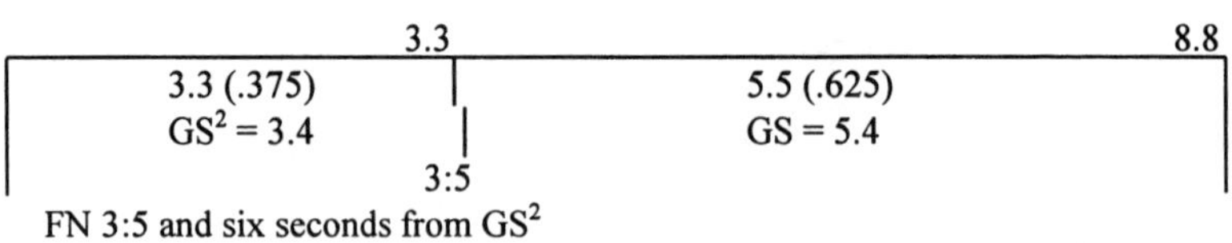

The timings of the three movements (13', 8.5', 8.8') do not give phi, as shown in figure 7.138.

Figure 7.138. Sibelius's *Symphony V* (Jalas) (Minutes)

13	21.5	30.3
13(.429)	8.5 (.281)	8.8 (.290)

Not GS. 13:8.5:8.8 is not FN

Sibelius is said to have spoken of phi proportions and obviously used them in these pieces. The date of *Finlandia* is quite early, so where did this knowledge come from?

Webern

Anton Webern's *Five Pieces for Orchestra op. 10 no. 4* was suggested by Pascoe to be divided at GS.[127] This must be accepted advisedly since there is a slight structural change after quarter note 12, but it is far from GS percentage-wise, as shown in figure 7.139.

Figure 7.139. Webern's *Five Pieces for Orchestra op. 10 no. 4* (Madden) (Quarter notes)

12	19
12 (.632)	7 (.368)

GS:GS^2

GS but percentage is poor and the piece may be too short to support it

Rogers' discussion of Webern's *Pieces for Orchestra op. 6 no. 1* covers only its wedge shape, and does not suggest that phi is present.[128] We agree, for the only "section break" that is not elided is after quarter note 12. The only other possibility we see is after 28 (Trp—elided) that would be one from GS^2 and very poor percentagewise. See figure 7.140.

Figure 7.140. Webern's *Pieces for Orchestra op. 6 no. 1* (Madden) (Quarter notes)

12 28	77
28 (.364) GS^2 = 29	49 (.636) GS = 48

One from GS^2 but percentage is poor

Other Claims

This section indicates some composers that either have been incorrectly claimed as users of phi proportions or for which information for verification was not available.

Since Aaron Copland's *Appalachian Spring* (1944) was judged by Adkins to be inconclusive, the piece is not examined here.[129]

George Gershwin's *Summertime* was thought by Wardell to use Fibonacci numbers throughout but her analysis was not persuasive and is not shown here.[130]

John Chowning is said to have used phi for both the pitch and formal structures of his electronic work *Stria*. However, Charles Dodge's illustration showed the formal structure dividing after 10 of 18 minutes.[131] This is not GS, as shown in figure 7.141.

Figure 7.141. Chowning's *Stria* (Dodge) (Minutes)

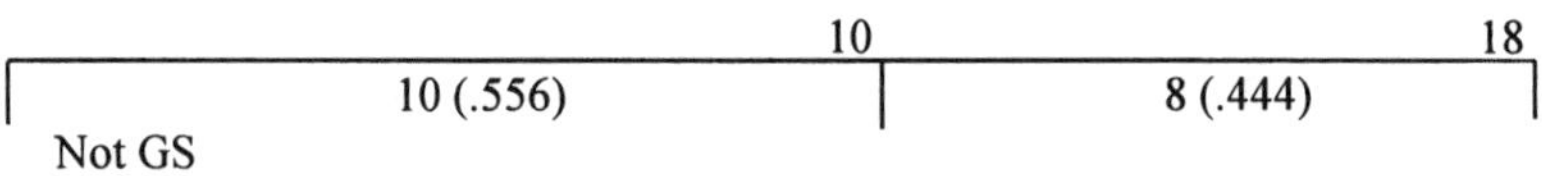

Lou Harrison's *Music Primer* was quoted by Hoffman, saying, "Melodic patterns might recur at Fibonacci related distances. Forms might be derived from Fibonacci related sections. Simultaneous Fibonacci related parts might form a polyphony and unequal ones could form entire movements by continuing the cross-metrics to their ends."[132] Hoffman did not cite music by Harrison that uses phi.

Mauricio Kagel's *Pas de cinq* was suggested by Schwartz as an example of phi proportion.[133] Although the score is constructed as a pentagram and is therefore physically in phi proportion, nothing about the music itself suggests that the durations have anything to do with phi. There are no fixed durations and approximate metronomic markings vary from 36 to 144 per minute, meaning that the paths would not be traversed in any pattern that would result in GS.

Luigi Nono's *Il Canto Sospeso* (1956) was said by Kramer and Berry to use a row of Fibonacci numbers which were permuted in serialist style and applied separately to the voices.[134] This may be true, but it is very subtle and does not provide any obvious phi structure.

Hugo Norden's main contribution to this book was his assertion in

1964 that the use of phi proportion in music was well known in his time. Since there was almost nothing in print, this seems doubtful unless there was an oral tradition. Kramer made non-specific comments that Norden had used phi in his compositions and had studied the use of phi by others, but scores are unavailable. Norden did, however, leave a glimpse of his method in his 1972 article "Proportions and the Composer."[135]

Krzysztof Penderecki's *Threnody for the Victims of Hiroshima* (1959–60) was said by Rogers and Kramer to be divided at GS, but it is not. Using Penderecki's timings and Morgan's structure (meas. 26 and 63–64), the analysis is as shown in figure 7.142.[136]

Figure 7.142. Penderecki's *Threnody for the Victims of Hiroshima* (Madden) (Seconds)

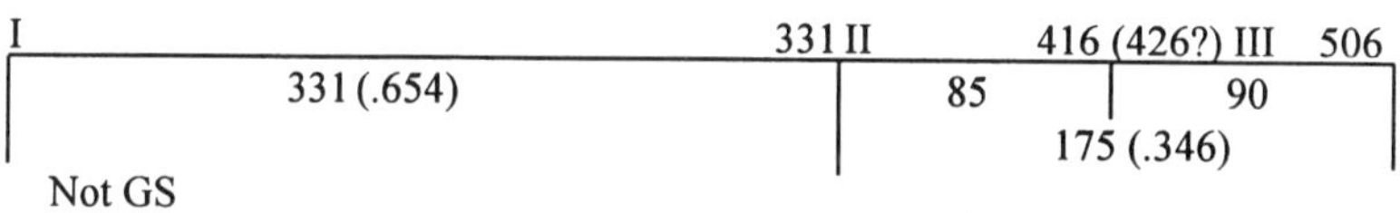

Sergei Rachmaninoff's *Symphony no. 2* (1901) Movement III was said by Rothwell to consist of 139 measures divided in GS at 86.[137] This would indeed be GS^2 if it were correct, as shown in figure 7.143a, but it should be 170 measures divided after 94, as shown in figure 7.143b.[138]

Figure 7.143a. Rachmaninoff's *Symphony no. 2* Mvt III (Rothwell) (Measures)

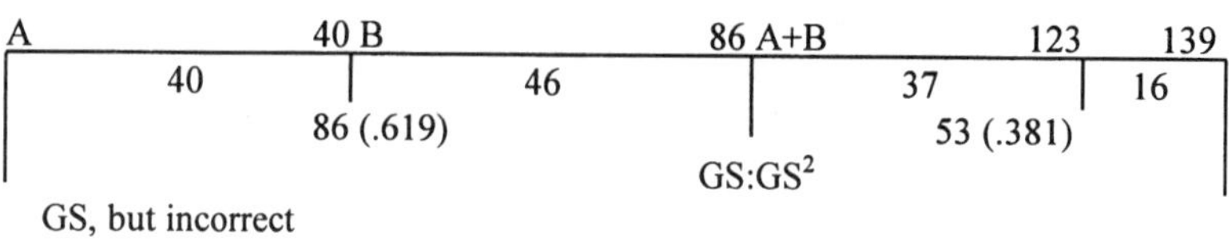

Figure 7.143b. Rachmaninoff's *Symphony no. 2* Mvt III (Madden) (Measures)

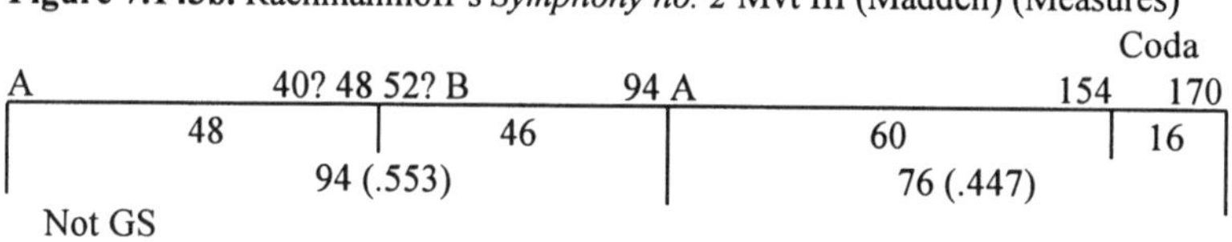

Iannis Xenakis was claimed by Jean-Pierre Dambricourt to have used phi proportions in his orchestral piece *Metastasis* (1955).[139] He took

Xenakis' frequent references to the architecture of Le Corbusier as evidence that phi proportions might be present in this piece, but gave no numbers to substantiate such a claim. In fact, there is no evidence for this in the music as shown by the measure count, shown in figure 7.144.

Figure 7.144. Xenakis's *Metastasis* (Madden) (Measures)

58	104	150	174	202	309	317	346
104	46	24	28	107	8	29	
150 (.434)				196 (.566)			

Not GS but one measure from bisected

Pierre Saby asked whether there were traces of Le Corbusier's Modulor system of building construction in Xenakis's *Nuits*.[140] Using his numbers, as shown in figure 7.145, we see no evidence of phi.

Figure 7.145. Xenakis's *Nuits* (Saby, Madden) (Measures)

I	69 II	131 III	168 IV	204 V	213 VI	263
69 (.262)	62	37	36	9	50	
168 (.639)			95 (.361)			

Not GS but one half-measure from bisected

Jonathan Kramer and others have left a few clues for future work, citing names but no pieces of some composers, and names and scarce pieces of others. Some of the composers on the following list, such as Gehlhaar, Huston, and Mott, were not listed in the eighth edition of *Baker's Biographical Dictionary of Musicians*. Kramer hinted that he might be tiring of the work in some dismissive remarks he made in *The Time of Music*, saying, "There is a mania to know whether the proportions were consciously planned. . . . The detective work is interesting, but the results are as inconclusive as they are unimportant."[141] To be fair, his concurrent large project about moment forms resulted in a very interesting book.

The names for which information was not available are listed below. Perhaps other researchers can follow these up.

Boris Blacher
    Ornamente für Klavier[142]
Richard Bunger
    Pianography: Fantasy on a Theme of Fibonacci[143]

Jean-Claude Eloy
Equivalences[144]
Rolf Gelhaar
Not specific[145]
Michael Gielen
Un vieux souvenir[146]
Scott Huston
Not specific[147]
Ernst Krenek
Fibonacci Mobile[148]
Robert Morris
Lorelei[149]
David Mott
Tube[150]
Per Nørgaard
Several works—not specific[151]
Third Symphony[152]
Elliott Sharp
Not Yet Time[153]
Fractal[154]
Larynx[155]

Conclusions

Many modern composers have used fib and phi. Some clearly knew they were doing so, but others only came close. Few authors have convincingly suggested why they did this and most composers have remained silent. We can only assume that the topic was being discussed and seemed like a good way to balance their works. A summary appears in table 7.15.

From this chart, it seems that there might be more profitable avenues of research than Debussy or Bartók. Barber, Carter, Dallapiccola, Hindemith, Prokofieff, Satie, Schoenberg, Sibelius, and Stravinsky spring to mind. Dallapiccola and Webern have not been thoroughly examined, but seem to offer promise. Varèse and Sibelius are the clear winners in the face-off conducted here, with 81.8% of their 11 pieces tested being within one measure of GS.

Table 7.15. Summary of 20th century composers' structural use of phi

Composer	Test-ed	% at GS	% at 1 meas.	% too far	No. FN	Comments
Barber	1		100.0			
Bartók	23	26.1	17.4	56.5	2	
Berg	2		50.0	50.0		
Cage	3			100.0	2	No GS
Carter	2	50.0	50.0			
Chowning	1			100.0		
Dallapiccola	1	100.0				
Debussy	13	35.7	28.6	35.7		
Hindemith	9	33.3	22.2	44.4	1	
Hoffman	3		33.3	66.7	2	
Johnston	2				2	
Kodály	7		14.3	85.7		
Ligeti	3		66.7	33.3		
Penderecki	1			100.0		
Prokofieff	8	42.9	21.4	35.7		
Rachmaninoff	1			100.0		
Ravel	7	14.3	28.6	57.1		
Risset	1			100.0	1	
Satie	11	54.5		45.5		
Schoenberg	11	27.3	27.3	45.4	1	
Sibelius	5	40.0	40.0	20.0	2	Timings
Stockhausen	3				3	No GS
Stravinsky	12	16.7	25.0	58.3	2	
Varèse	6	50.0	33.3	16.7		
Webern	2	50.0	50.0			Possible
Xenakis	2			100.0		

Chapter Notes

[1] [Kennedy p. 348]

[2] [Ar p. 93, 112] [Herz83 p. 110] The equation for the golden proportion was improperly formed by Argüelles as a/b = b/a + b on one page and correctly as a/b = b/(a + b) on the other. Watkins also misstated phi as b/a = a/a + b in Wat88 p. 96, whereas it should be b/a = a/(a + b).

[3] [Herz83 p. 111]

[4] [Pinc p. 58–9]

[5] [God95 p. 172–3] [Howat83 p. 163–73]

[6] [Marrs p. 315–25] [Bai83 p. 131] [Barr p. 99] [Howat83 p. 170] [Miz] [Olson p. 223–39] [Richardson] [Smith] [Wils] A strong case for the Priory of Sion was made by Baigent et al. Robert Anton Wilson, however, declared early on that it was a fraud, and David V. Barrett quoted Masonic historian John Hamill as saying there never was any mention of the Priory before the appearance of Baigent's books. Robert Richardson and Paul Smith have shown that the Priory was nothing more than an amusing distraction and, more recently, Carl E. Olson and Sandra Miesel have taken Dan Brown to task for publishing the Priory story in his novel *The Da Vinci Code* despite the fact that the French had known for some years that it was a hoax. Alas, our hopes for an easy explanation of Debussy have been dashed.

[7] [Par p. 210]

[8] [Howat83 p. 34, 173, 201–2]

[9] [Howat83 p. 140–1] [Howat88 p. 129] [DebP p. 165–75]

[10] [Howat83 p. 23–9, 194–200]

[11] [Howat83 p. 42, 203–208]

[12] [Howat83 p. 46–63, 209–221]

[13] [Par p. 213] [DebS p. 32–5]

[14] [Par p. 215] [DebS p. 74–7]

[15] [Par p. 216]

[16] [Par p. 219] [DebQ p. 1–15]

[17] [Par p. 242] [Bur p. 438–49]

[18] [Par p. 251] [DebN p. 102–42]

[19] [Par p. 236] [Howat83 p. 70–92] [DebM] We do not vouch for Howat's accuracy. He misstated measures 30 and 83 as 31 and 84.

[20] [Howat83 p. 110–27]

[21] [Howat83 p. 117] [DebM p. 65]

[22] [Howat83 p. 93–109] DebM]

---

[23] [Howat83 p. 97]

[24] Omitted is Howat's figure 8.2 showing a *dynamic* division *at* GS.

[25] [Howat83 p. 127–30]

[26] [Howat83 p. 105–9] Howat suggested from the evidence of the Sibley manuscript that the introduction did not originally use Fibonacci numbers and may have been altered by Debussy at a late date.

[27] [Len85] I have not been able to acquire Howat's article that led to this defense. It is "Bartók, Lendvai, and the Principles of Proportional Analysis" in *Music Analysis,* 1983.

[28] [Len71 p. 27–28] [Len83 p. 51] [Howat77] [Howat88 p. 131] [Sol] [Lowm] [BartMus]

[29] [Howat77] [Howat88 p. 132]

[30] [Len71 p. 28–9] [BartMus] Lendvai's identification of 285 as the second theme is doubtful, as it is different from the theme after 76.

[31] [Len71 p.18–22] [BartT]

[32] [Lowm p. 527] [BartT]

[33] [Len83 p. 209] [Gri p. 118] [BartP] Italics are Lendvai's.

[34] [Len83 p. 683] [BartP]

[35] [Gri p. 118]

[36] [Len83 p. 681] [BartQ]

[37] [Loc p. 322–3] [BartQ]

[38] [Pas p. 128] [Per p. 207] [Bat p. 147] [BartQ]

[39] [Rothwell p. 159] [BartQ]

[40] [Len71 p. 27] [Len83 p. 253] [Kra73] [Kra88 p. 306] [BartA] Kramer misstated the sequence as 5, 8, 5, 5, 3, 13, 8.

[41] [Len71 p. 18] [Len83 p. 681] [BartMik]

[42] [Len71 p. 18] [Len83 p. 681] [BartMik]

[43] [Len83 p. 436–7] [BartV]

[44] [Len71 p. 18] [Len83 p. 681] [BartD]

[45] [Pas p. 124] [BartR]

[46] [Len71 p. 18] [Len83 p. 681] [BartC]

[47] [Howat83 p. 187]

[48] [Neum p. 66n] [HindL]

[49] [Rothwell p. 139] [HindL]

[50] [Rothwell p. 140–44] [HindP]

[51] [Pas p. 185–6] [HindS]

[52] [Pas p. 181–5]

[53] [Adkins p. 41–2] [ProkS]

[54] [Adkins p. 42–3] [ProkL]

[55] There is no phi activity in Mvt IV.

[56] [Rothwell p. 152] [ProkP]

---

[57] [Rothwell p. 154] [ProkP] Rothwell called it "*Two Pieces,*" but in our edition there are four.

[58] [Khait p. 129] [Gillmor p. xx, 59] [Pinc p. 59, 142] [God95 p. 170–3]

[59] [Adams p. 242–52] [Sat]

[60] [Gillmor p. 87–8] [Sat]

[61] [Schoe83 p. 127] There is also a reference to phi on page 132, but this relates to key relationships, and is not a correct use of the golden proportion.

[62] [War p. 20–1] [Bur p. 465–9] [Mor p. 64–70]

[63] [Pas p. 194–5] [Mor p. 30–44]

[64] [Pas p. 196–7] [Mor p. 45–59]

[65] [Pas p. 197–200] [SchoeQ]

[66] [Webster p. 245]

[67] [SchoeSQ]

[68] [Kra73 p. 121–5] [Kra88 p. 313–5] [StoK]

[69] I have attempted to reconcile this without success.

[70] [Kra73 p. 126] [Kra88 p. 315] I was unable to get Stockhausen's analysis.

[71] [Kra73 p. 125–6] [Kra88 p. 315] [StoA]

[72] [Cope p. 142–3] [StoS]

[73] [Kra88 p. 289–303] [StravS] [StravSTP] [StravA]

[74] The graph is unbalanced due to the large amount of information that has to go into the right-hand side.

[75] [Rothwell p. 148] [StravO]

[76] [Pas p. 205] [StravC]

[77] Rehearsal numbers 16 and 56.

[78] [Wat94 p. 263] [StravT]

[79] [Wat94 p. 263] [StravI]

[80] [Mad p. 67] [Mor p. 187–215]

[81] [VarO]

[82] [VarI]

[83] [VarA]

[84] [VarE]

[85] [VarD]

[86] [Rothwell p. 156–7] [Barber]

[87] [Pas p. 149] [Mor p. 235–48]

[88] [Mor p. 222–34] Also known as *Five Orchestral Songs, op. 4.*

[89] [Brin p. 48] [Rog p. 49]

[90] [CageP]

[91] [CageD]

[92] [CageF]

[93] [Hof p. 66] [Cart]

[94] [Rog p. 97] [Bur p. 549–50]

[95] [Esc p. 109]
[96] [Hof p. 93]
[97] [Hof p. 89]
[98] [Hof p. 162]
[99] [Hof p. 186]
[100] [Hof p. 192, 222]
[101] [Childs p. 117]
[102] [Gun p. 100, 103]
[103] [Gun p. 169]
[104] [Len83 p. 36] [Adkins p. 39–40] [KodH]
[105] [Len83 p. 35] [KodP]
[106] [Esc p. 113–6] [LigC]
[107] [LigL p. 44]
[108] [JohnsonT p. 2] [LigAp] [LigAt]
[109] [LigL p. 43] [Wal p. 2] It is disconcerting to go up against a composer's own statements, but I am open to discussion. If someone can explain this, please let me know. Wallin stated that Ligeti's constructions have little to do with fractal theory.
[110] [Howat83 p. 191]
[111] [RavS]
[112] [RavVC]
[113] [RavVP]
[114] [RavT]
[115] [Howat83 p. 189–90] [RavM]
[116] [Howat83 p. 191] [RavM]
[117] [Cog p. 108–11]
[118] [Back]
[119] Comments are mine.
[120] For more information on the Schillinger system and some possible insights to the mathematics, see Lou Pine's massive bibliography on Schillinger expected to go on line at www.peabody.jhu.edu/. As this is not on line at the time of writing, I offer John Backus's review noted in the bibliography herein and Charles Samuel Smith's 1951 Master's thesis "An Analysis of Selected Mathematical Aspects of Schillinger's Approach to Music," available from University Microfilms at www.umi.com/hp/Products/Dissertations.html.
[121] [SchilS p. 24–6]
[122] [SchilS p. 85–8] There seems to be no reason to doubt it, but I am not qualified to comment.
[123] [War p. 59] [SchilS p. 1388–99]
[124] [SchilS p. 1362]
[125] [Adkins p. 101]

[126] [Jal p. 119]
[127] [Pas p. 101–5] [Nort Exp. Ed. p. 782]
[128] [Rog p. 96–7] [Webern6]
[129] [Adkins p. 36–7]
[130] [War p. 35]
[131] [Dod p. 137] I was not able to obtain a complete analysis of this piece and Chowning did not respond to a request for information.
[132] [Hof p. 69] I was not able to obtain this book. The reference would be Lou Harrison. *Music Primer*. New York: C. F. Peters, 1971, p. 42.
[133] [Schwa p. 157] [Kagel]
[134] [Kra73] [Kra88 p. 315] [Berr p. 409] [Nono]
[135] [Kra88 p. 317] [Nord72]
[136] [Rog p. 98] [Kra88 p. 317] [Mor p. 393–412]
[137] [Rothwell p. 135–7]
[138] [Rach]
[139] [Damb p. 33] [XeM]
[140] [Saby p. 139]
[141] [Kra88 p. 318]
[142] [Berr p. 385]
[143] [Kra73] [Kra88 p. 316]
[144] [Kra88 p. 317]
[145] [Kra88 p. 317]
[146] [Kra88 p. 317]
[147] [Kra88 p. 317]
[148] [Mor p. 263]
[149] [Kra73] [Kra88 p. 316]
[150] [Kra88 p. 316]
[151] [Kra88 p. 316–7]
[152] [Schwa p. 483] [Rog p.49]
[153] [Hof p. 68–9]
[154] [Hof p. 68–9]
[155] [Hof p. 68–9]

Chapter 8

Conclusions

Introduction

In the Preface, we said that there were certain questions that needed to be researched. Although we responded to most of them in the text, certainly they do not all have clear answers. We will attempt to clarify some of them here, but any that remain must be left to those who follow.

Comments

What and when did musicians know about fib and phi?

By the time of Machaut the Fibonacci numbers must have been well known. At any rate, they are found frequently in the rondeaux. And wherever fib lies, phi is nearby. We are not certain that this could not happen as a result of alternating patterns of A and B sections without any conscious thought, but the great concentration of phi in this music is certainly provocative.

It is virtually certain that J. S. Bach knew about phi and possible that Vivaldi did also. Many twentieth-century composers knew about it and used it, such as Varèse, Sibelius, and Stockhausen. Other composers are less certain, but show a possible knowledge, such as Mozart, Haydn,

Chopin, Brahms, Mendelssohn, Barber, Hindemith, Dallapiccola, Stravinsky, Schoenberg, and others. Some who certainly did not use phi are Penderecki, Wagner, Xenakis, and numerous others.

Was phi known and used in ancient times?

What we know about the Egyptians is that they could not have had the same understanding about phi that we do because their mathematics was not sophisticated enough. Nonetheless, they had some rule of thumb that produced the pyramid triangle in some of the pyramids. This is one of the more perplexing things about the Egyptians. No one has been able to determine conclusively how this came about.

Greek writings do not clearly show knowledge of phi until Euclid in the third century BC, after the golden age from which pottery has been found that may have contained golden proportions. This is still hotly debated and the case is by no means settled.

How was the knowledge transmitted through time?

Most authors are quite sure that phi was not used with any accuracy in medieval buildings. Euclid's extreme and mean ratio was rediscovered in the Renaissance and widely discussed. Fibonacci numbers also were known, either from Nicomachus or Fibonacci himself. Frankly, it seems unlikely that the stone masons or any secret organizations had anything to do with the transmission of phi.

Was it strictly mathematical, or did it carry esoteric meaning?

Although the Greeks are known to have revered geometry in general and numbers in particular, this reverence seems to have been different than it is today. Each number had its own meaning: 1 was masculine, 2 was feminine, 3 was the union of 1 and 2, and so on, and 10 was the sum of 1, 2, 3, and 4.

To the Pythagoreans, ratios of small numbers were especially interesting, such as the acoustic ratios 2:1, 3:2, 4:3, and so on that were used in tuning systems then and are still used now. Elaborate schemes were devised using them, and they were used in most medieval buildings.

The esoteric meaning of Fibonacci ratios such as 5:3, 8:5, 13:8, etc., seems to be much more recent, in response to Pacioli's *De Divina Proportione* and Kepler and Bernoulli's work relating them to Euclid's extreme and mean ratio. The reverence for these numbers now seems to be restricted to religious sects such as the Rosicrucians and Pythagoreans.

Is there support for using phi in science, or is it only used in cults?

Phi is found in the mathematics of fractal geometry and chaos theory, specifically in the complex plane portion of the Mandelbrot set where it is the slowest parameter to resolve to the chaotic boundary. Its function at that particular point is not entirely understood, but it relates directly to turbulent dynamics, which of course is a scientific application in the real world. In addition, fib and phi are found in the structures of plants, as described in chapter 1 herein, and may have stimulated the imaginations of artists and musicians.

Is phi used carefully and consciously, or strictly by intuition?

Fib and phi have been used carefully and consciously by several composers. See the chart at the end of this chapter.

Intuitive use is hard to demonstrate. How can we tell if an approximate use of phi is intentional or accidental? As a magical principal occurring intuitively, it eludes clear definition. In the work of those composers whose structures vary a great deal—such as Haydn—such use would seem to be accidental. But Herz-Fischler was of the opinion that phi does not occur by accident. Thus, this question does not yet have a clear answer.

How reliable were Jay Hambidge's claims that the Greeks used phi?

Hambidge's theories were widely regarded as valuable, but the material has not endured. Scanning his books (when they can found!) reveals that the theory was very complex and that practically anything could have some aspect of phi.

But artists do not generally work this strictly and would be badly hampered by Hambidge's principles. Furthermore, many illustrations of phi are badly distorted and the principles abused. As used by photographers, phi is only approximate to a loose 1/3–2/3 relationship, calculated on the fly, too fast to have any real accuracy. Don't most painters work more or less the same way?

Some analyses of paintings and of Greek vases contain elaborate arrays of circles and lines that are nearly impossible to interpret. This approach applied by Powell to Dufay's music borders on the absurd and adds little meaning to the analyses.

Did Pacioli, Fechner, Zeising, Hambidge, and other scholars influence the musicians?

Well, somebody did, because most of the musicians studied here obviously knew about phi, whether or not they used it regularly. So far, none of the correspondents have cited any sources for their knowledge. Why did Debussy and Satie—both Rosicrucians—use phi sometimes, where Bartók—not a Rosicrucian—also used it sometimes? Was Bartók influenced, as Lendvai claims, by the botanists Church, Thompson, et al?

Where did Varèse get the idea, and Stockhausen? The latter's use was limited to Fibonacci numbers and shows no evidence of GS proportions.

Is intuitive use of phi simply a question of "a little more than half"?

This term used by Rothwell in a paper of the same name seems to describe the main principle used by many of the composers studied here, along with notions of symmetry. Haydn's proportions seem to be spread over one to four or more measures from GS or GS^2 in approximately equal measure. Thus, if some pieces fall *at* GS, they would seem to do so by accident or intuition. How do we demonstrate this? Alternatively, although most of Beethoven's works are divided far from phi, how do we demonstrate that he avoided it? We cannot dogmatically declare that it is one way or the other.

Can intuitive use of phi be clearly demonstrated?

A study of intuitive use of phi would have to take a different approach than has been used here. It would have to involve heavy use of statistical averages, with massive databases. Even if we could show an average near phi, what would that tell us about the composers' thought processes? And would it show that there is something magical about this number in music? This seems doubtful. Certainly, an average would peak somewhere between .25 and .75, for very few pieces have major structural points outside those limits. Values .33, .50, and .67 would be good candidates, as well as .382 and .618. How would we decide the superiority of one over the other? There is probably no spike at any of these numbers in a statistical average. This is a project for analysts more advanced than we.

Is the presence of exact Fibonacci numbers evidence of knowing use?

Yes, and we have used it as such in the chart that follows, along with values occurring within one point of GS. The latter gives .16% accuracy, which was the stated goal in chapter 1. Many of the composers studied in this book had at least one piece with this accuracy, as shown in the chart.

Do the golden proportions found in the structures of some plants and animals mean that phi has some mystical significance?

Probably not. It is just a natural way for plants and animals to get the best use out of the materials at hand. It is a natural process that applies consistently to any particular species if damage does not occur in the growth process. In music, of course, it is most likely an artificial process, to the extent that composers even think about it.

Final Words

We conclude with a list of our judgments regarding all the composers studied in this book. Some of these entries are arbitrary (namely, *Liber Usualis,* which does not fit comfortably in any category) and none has been scientifically proven. Readers may disagree with some of these statements, but where a value is within one point of GS, or FN is exact, there is little to dissuade us. It should be remembered that GS or FN was not used in every piece by any of these composers.

We have used four categories to distinguish between the composers.

Must Have Known: Had more than one piece falling within .001 of GS or GS^2 or used exact Fibonacci numbers

Might Have Known: Had at least one piece within one measure of GS or GS^2 or .001 of FN

Probably Did Not Know: Had at least one piece within one measure of GS or GS^2 but with poor accuracy

Did Not Use Phi: We found no examples of GS or FN by this composer.

Table 8.1. Table of phi usage by all the composers in this volume

Must Have Known	Might Have Known	Probably Did Not Know	Did Not Use Phi
Medieval			
Machaut		Liber Usualis	
Renaissance			
Binchois	Dufay	Des Prés	Compère
De Lantins	Franck	Marenzio	Maffoni
Dunstable	Glogauer Liederbuch	Palestrina	

Must Have Known	Might Have Known	Probably Did Not Know	Did Not Use Phi
Gibbons	Hassler		
	Hofhaimer		
	Janequin		
	Lasso		
	Obrecht		
	Ockeghem		
	Paumann		
	Schlick		
	Sermisy		
Baroque			
J. S. Bach	W. F. Bach	Scarlatti	
	Chambonnières		
	Handel		
	Landi		
	Monteverdi		
	Vivaldi		
Classic			
Haydn	Beethoven		
Mozart			
Romantic			
Brahms	Berlioz	Rimsky-Korsakov	Saint-Saëns
Chopin	Schubert		Wagner
Mendelssohn			
20th Century			
Bartók	Barber	Webern	Chowning
Debussy	Berg		Penderecki
Hoffman	Cage		Rachmaninoff
Johnston	Carter		Xenakis
Prokofieff	Dallapiccola		
Satie	Hindemith		
Schoenberg	Kodály		
Stockhausen	Ligeti		
Stravinsky	Ravel		
Varèse	Risset		
	Sibelius		

Selected Bibliography

Listed are only those books, articles and scores actually consulted in preparation of this book. Most of these abbreviations are found in the chapter notes.

[Adams] Adams, Courtney. “Erik Satie and Golden Section Analysis.” *Music and Letters* 77, no. 2 (May, 1996): 242–52.

[Adkins] Adkins, Donald Glenn. “Golden Mean Durational Analysis as a Guide to Orchestral Repeats and Tempos.” D.M.A. diss., University of Missouri–Kansas City, 1987.

[Ar] Argüelles, José A. *Charles Henry and the Formation of a Psycho-physical Aesthetic*. Chicago: The University of Chicago Press, 1972.

[Ax] Axelrod, Alan. *The International Encyclopedia of Secret Societies & Fraternal Orders*. New York: Facts on File, 1997.

[BachA] Bach, Johann Sebastian. *The Art of the Fugue & A Musical Offering*. Originally published, Leipzig: Breitkopf & Härtel, 1885. Reprint, New York: Dover, 1992.

[BachC] ———. *Chaconne for Violin* from *Partita no. 2, BWV 1004. Neue Ausgabe Sämtlicher Werke*. New York: Bärenreiter Kassel, 1958.

[BachE] ———. *Englische Suiten*. Budapest: Könemann Music, 1995.

[BachF] ———. “Saint Anne’s Fugue,” *BWV 552. Dritter Teil der Klavier-Übung*. New York: Bärenreiter Kassel, 1969.

[BachI] ———. *Two and Three Part Inventions*, ed. by Hans Bischoff. New York: Edwin F. Kalmus, 1943.

[BachO] ———. *Toccatas, Fantasias, Passacaglia and Other Works for Organ*. New York: Dover Publications, Inc., 1987.

[BachV] ———. *Violin Concerto BWV 1042. Neue Ausgabe Sämtlicher Werke*. New York: Bärenreiter Kassel, 1986.

[BachW] ———. *The Well-Tempered Clavier*. Originally published, Leipzig: Bach-Gesellschaft, 1866. Reprinted with corrections and explanation of ornaments, New York: Dover, 1983.

[Bachmann] Bachmann, Tibor, and Peter J. Bachmann. "An Analysis of Béla Bartók's Music Through Fibonaccian Numbers and The Golden Mean." *The Musical Quarterly* 65, no. 1 (January 1979): 72–82.

[Back] Backus, John. “Pseudo-Science in Music.” *Journal of Music Theory* 4 (1960): 221–32.

[Bai83] Baigent, Michael, Richard Leigh, and Henry Lincoln. *Holy Blood, Holy Grail*. New York: Dell Publishing, 1982, 1983.

[Bai86] ———. *The Messianic Legacy*. New York: Dell Publishing, 1986.

Bibliography

[Bai89] Baigent, Michael and Richard Leigh. *The Temple and the Lodge*. New York: Arcade Publishing, Inc., 1989.

[Bak94] Baker, Theodore. *The Concise Edition of Baker's Biographical Dictionary of Musicians*. Rev. Nicolas Slonimsky. 8th ed. New York: Schirmer Books, 1994.

[Bak95] ———. *Pocket Manual of Musical Terms*. Ed. Theodore Baker. 4th ed. rev. Nicolas Slonimsky. 5th ed. rev. Laura Kuhn. New York: Schirmer Books, 1995.

[Barber] Barber, Samuel. *Complete Piano Music*. N.p.: G. Schirmer, Inc., n.d.

[Barbera] Barbera, Charles Andre Barbera. "The Persistence of Pythagorean Mathematics in Ancient Musical Thought." Ph.D. diss., University of North Carolina at Chapel Hill, 1980.

[Bard] Bardet, Dominique Pascale. "Logos as Number and Proportion in Amiens Cathedral and in the Kyrie of the Messe de Nostre Dame by Guillaume de Machaut." Ph.D. diss., Ohio University, 1999.

[Barr] Barrett, David V. *Secret Societies: From the Ancient and Arcane to the Modern and Clandestine*. London: Blandford, 1999.

[BartA] Bartók, Béla. *Allegro Barbaro*. Originally published, Vienna: Universal Edition A.G., 1918. Reprint, n.p.: Boosey & Hawkes, Inc., n.d.

[BartC] ———. *Contrasts*. New York: Boosey & Hawkes, 1942.

[BartD] ———. *Divertimento*. New York: Boosey & Hawkes, 1940.

[BartMik] ———. *Mikrokosmos*. New York: Boosey & Hawkes, 1995.

[BartMus] ———. *Music for String Instruments, Percussion and Celesta*. London: Universal Edition, 1937.

[BartP] ———. *Sonata for Piano*. Originally published, Boosey & Hawkes, 1966. Revised by Peter Bartok. N.p.: Boosey & Hawkes, 1992.

[BartQ] ———. *The String Quartets of Béla Bartók*. Boosey & Hawkes, 1945.

[BartR] ———. *First Rhapsody*. New York: Boosey & Hawkes, 1952.

[BartT] ———. *Sonata for Two Pianos and Percussion*. New York: Boosey & Hawkes, 1942.

[BartV] ———. *Violin Concerto no. 2*. New York: Boosey & Hawkes, 1946.

[Bat] Bates, Karen Anne. "The Fifth String Quartet of Béla Bartók: An Analysis Based on the Theories of Ernő Lendvai." Ph.D. diss., University of Arizona, 1986.

[BeethSon] Beethoven, Ludwig van. *Complete Piano Sonatas*. 2 vols. Originally published, Leipzig: Universal Edition A. G., 1923. Reprint, New York: Dover Publications, Inc., 1975.

[BeethStr] ———. *Complete String Quartets and Grosse Fuge*. Originally published, Leipzig: Breitkopf & Härtel, n.d. Reprint, New York: Dover Publications, Inc., 1970.

[BeethSym] ———. *Symphonies*. 3 vols. Originally published, Braunschweig: Henry Litolff's Verlag, n.d. Reprint, Mineola, NY: Dover Publications, Inc., 1989.

Bibliography

[Benj] Benjamin, Thomas. *Counterpoint in the Style of J. S. Bach.* New York: Schirmer Books, 1986.

[BergL] Berg, Alban. *Fünf Orchesterlieder, op. 4.* Also known as *Altenberg Lieder.* Wien: Universal Edition, 1966.

[BergS] ———. *Lyric Suite.* Vienna: Universal Edition, 1927.

[Berl] Berlioz, Hector. *New Edition of the Complete Works.* London: Bärenreiter, 1972 and 2001.

[Bern] Bernard, Jonathan W. "Space and Symmetry in Bartók." *Journal of Music Theory* 30, no. 2 (fall 1986): 185–201.

[Berr] Berry, Wallace. *Structural Functions in Music.* New York: Dover Publications, Inc., 1987.

[Boyer] Boyer, Carl B. *A History of Mathematics.* 2nd ed. New York: John Wiley & Sons, Inc., 1991.

[Brah] Brahms, Johannes. *Complete Symphonies in Full Score.* Originally published, Leipzig: Breitkopf & Härtel, n.d. Reprint, New York: Dover Publications, Inc., 1974.

[Brin] Brindle, Reginald Smith. *The New Music: The Avant-garde Since 1945.* New York: Oxford University Press, 1987.

[Bru] Brunés, Tons. *The Secrets of Ancient Geometry and its Use.* Copenhagen: Rhodos, 1967.

[Bur] Burkhart, Charles. *Anthology for Musical Analysis.* 4th ed. Fort Worth: Harcourt Brace Jovanovich College Publishers, 1986.

[CageD] Cage, John and Lou Harrison. *Double Music.* New York: C.F. Peters, 1961.

[CageF] Cage, John. *First Construction (in Metal).* New York: Henmar Press, 1939.

[CageP] ———. *Piano Works, 1935–1948.* New York: Henmar Press, 1998.

[Camf] Camfield, William A. "Juan Gris and the Golden Section." *The Art Bulletin* 47, no. 1 (March, 1965): 128–134.

[Camp] Camp, Elinor Jane Perry. "Temporal Proportion: A Study of Sonata Forms in the Piano Sonatas of Mozart." Ph.D. diss., Florida State University, 1968.

[Capt] Capt, E. Raymond. *A Study in Pyramidology.* Muskogee, OK: Artisan Sales, 1996.

[Cart] Carter, Elliott. *String Quartet no. 2.* 2nd ed. New York: Associated Music Publishers, Inc., 1962.

[Case] Case, Paul Foster. *The True and Invisible Rosicrucian Order.* York Beach, ME: Samuel Weiser, Inc., 1989.

[Childs] Childs, Barney. "Ben Johnston: *Quintet for Groups.*" *Perspectives of New Music* 1968: 110–21.

[ChopB] Chopin, Frédéric. *Complete Ballades, Impromptus, & Sonatas.* Originally published, New York: G. Schirmer, 1885–6. Reprint, Mineola, NY: Dover Publications, Inc., 1998.

[ChopN] ———. *Nocturnes and Polonaises*. Originally published, New York: G. Schirmer, 1894. Reprint, Mineola, NY: Dover Publ., Inc., 1998.

[ChopP] ———. *Preludes & Etudes*. Originally published, New York: G. Schirmer, 1895. Reprint, Mineola, NY: Dover Publications, Inc., 1998.

[Clen] Clendinning, Jane Piper. "The Pattern-Meccanico Compositions of György Ligeti." *Perspectives of New Music* 31 (1993): 192–234.

[Cog] Cogan, Robert. *New Images of Musical Sound.* Cambridge, MA: Harvard University Press, 1984.

[CogEsc] Cogan, Robert and Pozzi Escot. *Sonic Design: The Nature of Sound and Music*. Cambridge, MA: Publication Contact Intl, 1976, 1984.

[Col] Colman, Samuel. Ed. by C. Arthur Coan. *Nature's Harmonic Unity: A Treatise on its Relation to Proportional Form*. New York: Putnam's, 1912.

[Cond] Condat, Jean-Bernard, ed. *Golden Section and Music*. New York: Verlag Peter Lang, 1988.

[Cone] Cone, Edward T. *Musical Form and Musical Performance*. New York: W. W. Norton, 1968.

[Cook] Cook, Theodore Andrea. *The Curves of Life*. Originally published, London: Constable and Company, 1914. Reprint, New York: Dover Publications, Inc., 1979.

[Cope] Cope, David. *New Directions in Music*. 6th ed. Prospect Heights, IL: Waveland Press, Inc., 1993.

[Cowan] Cowan, Henry J. *Science and Building: Structural and Environmental Design in the Nineteenth and Twentieth Centuries*. New York: Wiley-Interscience, 1978.

[Cowen] Cowen, Painton. *Rose Windows.* London: Thames and Hudson, 1979.

[Damb] Dambricourt, Jean-Pierre. "Le nombre d'or et la fin du pythagorisme musical." In *Golden Section and Music*, ed. by Condat: 27–38.

[Dar] Daraul, Arkon. *A History of Secret Societies*. Secaucus, NJ: Citadel Press, 1997.

[DebM] Debussy, Claude. *La Mer*. Reprint from "early French edition." Mineola, NY: Dover Publications, Inc., 1997.

[DebN] ———. *Nocturnes.* In *Three Great Orchestral Works.* New York: Dover Publications, Inc., 1983.

[DebP] ———. *Piano music.* New York: Dover Publications, Inc., 1972–3.

[DebQ] ———. *String Quartets by Debussy and Ravel.* New York: Dover Publications, Inc., 1987.

[DebS] ———. *Songs of Claude Debussy,* vol. 1. Milwaukee: Hal Leonard Corporation, 1993.

[Dem] Demaree, Robert William, Jr. "The Structural Proportions of the Haydn Quartets." Ph.D. diss., Indiana University, 1973.

[Doc] Doczi, György. *The Power of Limits: Proportional Harmonies in Nature, Art, and Architecture*. Boston: Shambhala, 1994.

[Dod] Dodge, Charles and Thomas A. Jerse. *Computer Music, Synthesis, Composition, and Performance.* 2d ed. New York: Schirmer Books, 1997.

[Dor] Dorfman, Allen Arthur. "A Theory of Form and Proportion in Music." Ph.D. diss., University of California, Los Angeles, 1986.

[Dufay] Dufay, Guglielmi. *Opera Omnia.* Rome: American Institute of Musicology in Rome, 1948.

[Dunlap] Dunlap, Richard A. *The Golden Ratio and Fibonacci Numbers.* Singapore: World Scientific, 1997.

[Dunlop] Dunlop, Ian. *The Cathedrals Crusade: The Rise of the Gothic Style in France.* New York: Taplinger Publishing Company, 1982.

[Dunn] Dunn, Christopher. *The Giza Power Plant: Technologies of Ancient Egypt.* Santa Fe: Bear & Company, Inc., 1998.

[Dunst] Dunstable, John. *Complete Works: Music Britannica.* London: Stainer and Bell Ltd for the Royal Musical Association and the American Musicological Society, 1953.

[Esc] Escot, Pozzi. *The Poetics of Simple Mathematics in Music.* Cambridge, MA: Publication Contact International, 1999.

[Fie] Field, Michael, and Martin Golubitsky. *Symmetry in Chaos: A Search for Pattern in Mathematics, Art and Nature.* Oxford: Oxford University Press, 1992.

[Frey] Frey, Ronald Jan. "Attractiveness of the Golden Section in the Aesthetics of Sound Perception." M.A. thesis, University of Toronto, 1994.

[Gad97] Gadalla, Moustafa. *Egyptian Cosmology: the ABSOLUTE Harmony.* Erie, PA: Bastet Publishing, 1997.

[Gad99] ———. *Historical Deception: The Untold Story of Ancient Egypt.* Greensboro, NC: Tehuti Research Foundation, 1999.

[Garl87] Garland, Trudi Hammel. *Fascinating Fibonaccis: Mystery and Magic in Numbers.* Palo Alto, CA: Dale Seymour Publications, 1987.

[Garl95] ———. *Math and Music: Harmonious Connections.* N.p.: Dale Seymour Publications, 1995.

[Garl97] ———. *Fibonacci Fun: Fascinating Activities with Intriguing Numbers.* White Plains, NY: Dale Seymour Publications, 1997.

[Gau] Gauldin, Robert. *Eighteenth-Century Counterpoint.* Prospect Heights, IL: Waveland Press, Inc., 1988.

[Gaz] Gazalé, Midhat J. *Gnomon: From Pharaohs to Fractals.* Princeton, NJ: Princeton University Press, 1999.

[Ghy] Ghyka, Matila. *The Geometry of Art and Life.* Originally published, New York: Sheed and Ward, 1946. Reprint, New York: Dover Publications, Inc., 1977.

[Gillies] Gillies, Malcolm, ed. *The Bartók Companion.* Portland, OR: Amadeus Press, 1994.

[Gillings] Gillings, Richard J. *Mathematics in the Time of the Pharaohs.* Originally published, Cambridge, MA: The MIT Press, 1972. Reprint, Mineola, NY: Dover Publications, Inc., 1982.

[Gillmor] Gillmor, Alan M. *Erik Satie.* New York: W. W. Norton & Co., 1988.

[God93] Godwin, Joscelyn, ed. *The Harmony of the Spheres.* Rochester, VT: Inner Traditions International, 1993.

[God95] ———. *Music and the Occult: French Musical Philosophies 1750–1950.* Rochester, NY: University of Rochester Press, 1995.

[Gou] Gould, Glen. *The Art of the Fugue.* CD. New York: Sony Music Entertainment, 2001.

[Gre] Green, Douglass M. *Form in Tonal Music: An Introduction to Analysis.* 2d ed. Fort Worth, TX: Harcourt Brace Jovanovich College Publishers, 1979.

[Gri] Griffiths, Paul. *Bartók.* London: J.M. Dent & Sons Ltd, 1984.

[Gun] Gunden, Heidi von. *The Music of Ben Johnston.* Metuchen, NJ: The Scarecrow Press, Inc., 1986.

[HAM1] Davison, Archibald T. and Willi Apel. *Historical Anthology of Music: Oriental, Medieval and Renaissance Music.* Rev. ed. Cambridge, MA: Harvard University Press, 1966.

[HAM2] ———. *Historical Anthology of Music: Baroque, Rococo, and Pre-Classical Music.* Cambridge, MA: Harvard University Press, 1950.

[Hamb26] Hambidge, Jay. *The Elements of Dynamic Symmetry.* Originally published, n.p.: Brentano's, 1926; Reprint, n.p. Yale University Press, 1948. Reprint, New York: Dover Publications, Inc., 1967.

[Hamb32] ———. *Practical Applications of Dynamic Symmetry.* New Haven: Yale University Press, 1932.

[Hand] Händel, Georg-Friedrich. *Hallische Händel-Ausgabe.* Leipzig: Deutscher Verlag für Music, 1959.

[HaydSon] Haydn, Joseph. *Complete Piano Sonatas.* 2 vols. Originally published, Leipzig: Breitkopf & Härtel, 1907. Reprint, New York: Dover, 1984.

[HaydStr] ———. *String Quartets.* 5 vols. Originally published, London: Ernst Eulenberg Ltd., n.d. Reprint, New York and Mineola: Dover Publications, Inc., 1979, 1980, 1982, 1985, 2000.

[HaydSym] ———. *Symphonies 88–104.* 3 vols. Originally published, Boston: Haydn Society, 1951 and other unnamed editions. Reprint, New York and Mineola: Dover Publications Inc., 1983, 1997, 1999.

[Hayl] Haylock, Derek W. "The Golden Section and Beethoven's Fifth." *Mathematics Teaching* (Sept 1978). Reprinted in *Golden Section and Music,* ed. by Condat: 121–124.

[Herz83] Herz-Fischler, Roger. "An Examination of Claims Concerning Seurat and 'The Golden Number.'" *Gazette des Beaux-Arts* 101 (March 1983): 109–12.

[Herz98] ———. *A Mathematical History of the Golden Number*. Originally published as *A Mathematical History of Division in Extreme and Mean Ratio*, Ontario, Canada: Wilfrid Laurier University Press, 1987. Reprint, Mineola, NY: Dover Publications, Inc., 1998.

[Herz00] ———. *The Shape of the Great Pyramid*. Waterloo, Ontario: Wilfrid Laurier University Press, 2000.

[Hicks] Hicks, Michael. "Interval and Form in Ligeti's *Continuum* and *Coulée*." *Perspectives in New Music* 31 (1993): 172–82.

[HindL] Hindemith, Paul. *Ludus Tonalis*. New York: Associated Music Publishers, Inc., 1943.

[HindP] ———. *Klaviermusik II*. Sämtliche Werke. Mainz: B. Schott's Söhne, 1981.

[HindS] ———. *Sonata for Clarinet and Piano*. New York: Schott, 1939.

[Hof] Hoffman, Michael. "Fibonacci Proportions in Composition: An Analysis of an Original Work for Jazz Orchestra." Ph.D. diss., New York University, 1999.

[Howard] Howard, Michael. *The Occult Conspiracy: Secret Societies—Their Influence and Power in World History*. Rochester, VT: Destiny Books, 1989.

[Howat77] Howat, Roy. "Debussy, Ravel and Bartók: Towards Some New Concepts of Form." *Music & Letters* 58, no. 3 (July 1977): 285–93.

[Howat83] ———. *Debussy in Proportion: A Musical Analysis*. Cambridge: Cambridge University Press, 1983.

[Howat85] ———. "Letter to the Editor: Remarks on Lendvai's 'The limits of Musical Analysis' in NHQ XXVI." The New Hungarian Quarterly 26, (Winter 1985): 263.

[Howat88] ———. "Debussy, Bartók et les formes de la nature." In *Golden Section and Music,* ed. by Condat: 125–138.

[Hum] Humphreys, David. *The Esoteric Structure of Bach's Clavierübung III*. Cardiff [Wales]: University College Cardiff Press, 1983.

[Hun] Huntley, H. E. *The Divine Proportion: A Study in Mathematical Beauty*. New York: Dover Publications, Inc., 1970.

[Jal] Jalas, Jussi. "Sibelius's Symphonies and the Golden Section." In *Golden Section and Music,* ed. by Condat: 117–120.

[Janequin] Janequin, Clement. *Chansons Polyphoniques*. Editions de L'oiseau-lyre. Monaco: Les Remparts, n.d.

[JohnsonR] Johnson, R. Stanley. *Cubism & La Section D'Or: Reflections on the Development of the Cubist Epoch: 1907–1922*. Chicago: Klees/Gustorf Publishers, 1991.

[JohnsonT] Johnson, Tim. "Lendvai Comes to Cologne: György Ligeti in the 1950s and '60s." European Music Analysis Conference, Bristol University, 2002. www.btinternet.com/~tim.johnson77/rambler/LendvaiInCologne.html.

[Kagel] Kagel, Mauricio. *Pas de cinc*. London: Universal Edition, 1967.

[Kennedy] Kennedy, Michael. *The Oxford Dictionary of Music*. New York: Oxford University Press, 1985.

[Kennan] Kennan, Kent. *Counterpoint Based on Eighteenth Century Practice*. Englewood Cliffs, NJ: Prentice-Hall, Inc., 1959.

[Kep] Kepes, Gyorgy, ed. *Module, Proportion, Symmetry, Rhythm*. New York: George Braziller, 1966.

[Khait] Khaitzine, Richard. *Fulcanelli et le cabaret du Chat Noir: Histoire artistique et secret de Montmartre.* Villeselve: Editions Ramuel, 1997.

[Kirk] Kirk, Kenneth Patrick. "The Golden Ratio in Chopin's Preludes, Opus 28." Ph.D. diss., University of Cincinnati, 1987.

[Kli53] Kline, Morris. *Mathematics in Western Culture*. New York: Oxford University Press, 1953.

[Kli67] ———. *Mathematics for the Nonmathematician.* Originally published as *Mathematics for Liberal Arts*, Reading, MA: Addison-Wesley Publishing Company, Inc., 1967. Reprint, Mineola, NY: Dover Publications, Inc., 1985.

[Kli72] ———. *Mathematical Thought From Ancient to Modern Times*. 3 Vols. New York: Oxford University Press, 1972.

[Kno] Knorr, Wilbur Richard. *The Ancient Tradition of Geometric Problems*. Originally published, Boston: Birkhäuser, 1986. Reprint, Mineola, NY: Dover Publications, Inc., 1993.

[KodH] Kodály, Zoltán. *Háry János Suite*. New York: Boosey & Hawkes, 1939.

[KodP] ———. *Psalmus Hungaricus*. Rev. ed. London: Universal Edition, 1997.

[Kok] Kokkonen, Joonas. "Les Sonates pour Piano de Mozart et la Section d'Or." In *Golden Section and Music,* ed. by Condat: 111–116.

[Kra73] Kramer, Jonathan. "The Fibonacci Series in Twentieth-Century Music. *Journal of Music Theory* 17, no. 1 (spring 1973): 110–48.

[Kra88] ———. *The Time of Music: New Meanings, New Temporalities, New Listening Strategies*. New York: Schirmer Books, 1988.

[Lar] Larson, Paul. "The Golden Section in the Earliest Notated Western Music." *Fibonacci Quarterly* 16, no. 6 (December 1978): 513–15. Reprinted in *Golden Section and Music,* ed. by Condat: 49–51.

[Law] Lawlor, Robert. *Sacred Geometry: Philosophy and Practice.* London: Thames and Hudson, 1982.

[Leh] Lehner, Mark. *The Complete Pyramids: Solving the Ancient Mysteries*. London: Thames and Hudson Ltd., 1997.

[Len66] Lendvai, Ernő. "Duality and Synthesis in the Music of Béla Bartók." In *Module, Proportion, Symmetry, Rhythm*, ed. by Gyorgy Kepes: 174–193.

[Len71] ———. *Béla Bartók: An Analysis of his Music*. Originally published, London: Kahn & Averill, 1971. Rev. 1979. First publ. in U.S., White Plains, NY: Pro/Am Music Resources Inc., 1991

[Len81] ———. "The Quadrophonic [*sic*] Stage of Bartók's *Music for Strings, Percussion and Celesta*." *The New Hungarian Quarterly* 22, no. 84 (1981): 70–85.

[Len83] ———. *The Workshop of Bartók and Kodály*. Budapest: Editio Musica, 1983.

[Len85] ———. "The Limits of Musical Analysis." The New Hungarian Quarterly 26, no. 97 (Spring 1985): 201–7.

[Les] Lester, Joel. *Analytic Approaches to Twentieth-Century Music*. New York: W. W. Norton & Company, 1989.

[Liber] *Liber Usualis*. Great Falls, MT: St. Bonaventure Publications, 1997.

[LigAp] Ligeti, György. *Apparitions*. Wien: Universal Edition, 1971.

[LigAt] ———. *Atmospheres*. Wien: Universal Edition, 1980.

[LigC] ———. *Continuum*. Mainz: B. Schott's Söhne, 1970.

[LigL] ———. *Ligeti in Conversation*. London: Eulenburg Books, 1983.

[Liv] Livio, Mario. *The Golden Ratio: The Story of Phi, the World's Most Astonishing Number*. New York: Broadway Books, 2002.

[Loc] Locke, Derek. "Numerical Aspects of Bartók's String Quartets." *The Musical Times* (June 1987): 322–25.

[Lowi] Lowinsky, Edward E. *Secret Chromatic Art in the Netherlands Motet*. Transl. from the German by Carl Buchman. Originally published, New York: Columbia Univ. Press, 1946; Reissued, Russell & Russell, 1967.

[Lowm] Lowman, Edward A. "Some Striking Proportions in the Music of Béla Bartók." *The Fibonacci Quarterly* 9, no. 5 (December 1971): 527–37.

[Mach] Machaut, Guillaume de. *Oevres completes*. Monaco: Éditions de l'Oiseau-Lyre, 1977.

[Mad] Madden, Charles. *Fractals in Music: Introductory Mathematics for Musical Analysis*. Salt Lake City: High Art Press, 1999.

[Mark] Markowsky, George. "Misconceptions About the Golden Ratio." *College Math Journal* 23, no. 1 (Jan 1992): 2–19.

[Marrs] Marrs, Jim. *Rule by Secrecy: The Hidden History that Connects the Trilateral Commission, the Freemasons, and the Great Pyramids*. New York: HarperCollins Publishers, Inc., 2000.

[MenSong] Mendelssohn, Felix. *Songs Without Words*. N.p.: Kalmus, n.d.

[MenSym] ———. *Symphonies [Nos. 3, 4, 5]*. 3 vols. Leipzig: Breitkopf & Härtel, n.d. Reprint, Mineola, NY: Dover Publications, Inc., 1994, 1997, 1999.

[Mich69] Michell, John. *The New View Over Atlantis*. Rev. ed. Originally published, n.p.: Sago Press, 1967; Garstone Press, 1972; Abacus, 1973. Revised and reprinted, London: Thames & Hudson Ltd., 1983.

[Mich77] ———. *A Little History of Astro-Archaeology*. Originally published as *Secrets of the Stones*, n.p., 1977, 1989. Reprint, New York: Thames & Hudson, 2001.

[Mich01] ———. *The Dimensions of Paradise: The Proportions and Symbolic Numbers of Ancient Cosmology*. Kempton, IL: Adventures Unlimited Press, 2001.

[Miz] Mizrach, Steven, Dr. "Priory of Sion: The Facts, the Theories, the Mystery." www.fiu.edu/~mizrachs, n.d.

[Mor] Morgan, Robert P, ed. *Anthology of Twentieth-Century Music.* New York: W. W. Norton, 1992.

[Moss] Moss, Lawrence. "A Compositional Analysis." In *Readings in Schenker Analysis and Other Approaches*, ed. by Maury Yeston. New Haven: Yale University Press, 1977.

[MozSon] Mozart, Wolfgang Amadeus. *Complete Sonatas and Fantasies.* Originally published in separate unnamed editions. Reprint, Mineola: Dover Publications, Inc., 1996.

[MozStr] ———. *Complete String Quartets*. Originally published, Leipzig: Breitkopf & Härtel, 1881–2. Reprint, New York: Dover Publications, Inc., 1970.

[MozSym] ———. *Symphonies*. 3 vols. Originally published, Leipzig: Breitkopf & Härtel, 1880–2. Reprint, New York and Mineola: Dover Publications, Inc., 1974, 1991, 2000.

[Mur] Murray, Stephen. *Notre-Dame: Cathedral of Amiens*. New York: Cambridge University Press, 1996.

[Nataf] Nataf, André. *The Occult.* New York: Chambers, 1991. First published as *Les maîtres de l'occultisme*. Paris: Bordas, 1988.

[Neug] Neugebauer, O. *The Exact Sciences in Antiquity*. Originally published, n.p.: Brown University Press, 1957. Reprint, New York: Dover Publications, Inc., 1969.

[Neum] Neumeyer, David. *The Music of Paul Hindemith*. New Haven: Yale University Press, 1986.

[New] Newman, Rochelle, and Martha Boles. *Universal Patterns: The Golden Relationship: Art, Math & Nature.* Rev. ed. Bradford, MA: Pythagorean Press, 1992.

[Nono] Nono, Luigi. *Il Canto Sospeso*. Mainz: Ars Viva Verlag, 1957.

[Nord64] Norden, Hugo. "Proportions in Music." *Fibonacci Quarterly* 2, no. 3 (October 1964): 219–22.

[Nord72] ———. "Proportions and the Composer." *Fibonacci Quarterly* 10, no. 3 (April 1972): 319–23.

[Nort] Norton Scores, The. 2 vols., Standard Edition, Revised; Expanded Edition. Ed. Roger Kamien. New York: W. W. Norton & Company, Inc., 1968, 1970, 1972.

[O'Don] O'Donnell, John. "And Yet They are not Three Fugues; but One Fugue." *Diapason* 68, no. 1 (December 1976): 11–13.

[Olsen] Olsen, Scott Anthony. "The Pythagorean Plato and the Golden Section: A study in Abductive Inference." Ph.D. diss., Univ. of Florida, 1983.

[Olson] Olson, Carl E. and Sandra Miesel. *The Da Vinci Hoax: Exposing the Errors in the Da Vinci Code.* San Francisco: Ignatius Press, 2004.

[Orl] Orledge, Robert. *Satie the Composer*. New York: Cambridge University Press, 1990.

[Pal] Palestrina, Giovanni Pierluigi da. *Mottetti*. Vol. 4. *Le opere complete.* Roma: Fratelli Scalera, 1939–.

[Par] Parks, Richard S. *The Music of Claude Debussy*. New Haven: Yale University Press, 1989.

[Pas] Pascoe, Clive Brownley. "Golden Proportion in Musical design." D.M.E. diss., University of Cincinnati, 1973.

[Pei92] Peitgen, Heinz-Otto, Hartmut Jürgens, and Dietmar Saupe. *Chaos: New Frontiers of Science.* New York: Springer-Verlag, 1992.

[Pen] Pennick, Nigel. *Sacred Geometry: Symbolism and Purpose in Religious Structures*. Chieveley, Berks: Capall Bann Publishing, 1994.

[Per] Perle, George. "The String Quartets of Béla Bartók." In *A Musical Offering: Essays in Honor of Martin Bernstein,* ed. by Edward H. Clinkscale and Claire Brook. New York: Pendragon Press, 1977, p. 193–210.

[Picknett] Picknett, Lynn and Clive Prince. *The Templar Revelation: Secret Guardians of the True Identity of Christ*. New York: Touchstone, 1998.

[Pinc] Pincus-Witten, Robert. *Occult Symbolism in France: Joséphin Peladan* [*sic*] *and the Salons de la Rose-Croix*. New York: Garland Publishing, Inc., 1976.

[Pow] Powell, Newman. "Fibonacci and the Gold Mean: Rabbits, Rumbas, and Rondeaux." *Journal of Music Theory* 23, no. 2 (fall 1979): 227–73.

[ProkL] Prokofieff, Sergei. *Lieutenant Kijé*. Boca Raton, FL: E.F. Kalmus, 198?

[ProkP] ———. *Selected Works for the Piano.* New York: G. Schirmer, n.d.

[ProkS] ———. *Symphony No. 1 (Classical).* Boca Raton FL: E.F. Kalmus, 199-?.

[Pru] Prusinkiewicz, Przemyslaw and Aristid Lindenmayer. *The Algorithmic Beauty of Plants*. New York: Springer-Verlag, 1990.

[Rach] Rachmaninoff, Sergei. *Symphony No. 2*. Melville, NY: Belwin Mills, n.d.

[Råd] Råde, Lennart, and Bertil Westergren. *Beta: Mathematics Handbook*. 2d ed. Boca Raton: CRC Press, 1990.

[Rat] Ratner, Leonard G. *Classic Music: Expression, Form, and Style*. New York: Schirmer Books, 1980.

[RavM] Ravel, Maurice. *Miroirs*. N.p.: G. Schirmer, Inc., n.d.

[RavS] ———. *Douze chants*. Paris: Durand, 1909.

[RavT] ———. *Le Tombeau de Couperin*. Paris: Durand, 1918.

[RavVC] ———. *Sonate pour Violon et Violoncelle*. Paris: Durand, 1922.

[RavVP] ———. *Sonate pour Violon et Piano*. Paris: Durand, 1927.

[Rhind] Rhind Mathematical Papyrus. Vol. 1. Oberlin, OH: Mathematical Association of America, 1927.

[Rich] Richardson, Robert. *The Priory of Sion Hoax. Gnosis* 51 (1999): 49–55. Www.alpheus.org/html/articles/esoteric_history/richardson1.html.

[Rim] Rimsky-Korsakov. *Capriccio Espagnol*. Mineola: Dover, 1998.

[Rog] Rogers, Michael Richard. "The Golden Section in Musical Time: Speculations on Temporal Proportion." Ph.D. diss., University of Iowa, 1977.

[Ros88] Rosen, Charles. *Sonata Forms*. Rev. ed. New York: W. W. Norton, 1988.

[Ros95] ———. *The Romantic Generation*. Cambridge, MA: Harvard University Press, 1995.

[Ros97] ———. *The Classical Style: Haydn, Mozart, Beethoven*. Rev. ed. New York: W. W. Norton & Company, 1997.

[Rothwell] Rothwell, James Austin. "The Phi Factor: Mathematical Proportions in Musical Form." D.M.A. diss., Univ. of Missouri–Kansas City, 1977.

[Run] Runion, Garth E. *The Golden Section*. Palo Alto, CA: Dale Seymour Publications, 1990.

[Saby] Saby, Pierre. "Organisation formelle dans *Nuits* de Iannis Xenakis: Traces de Modulor?" In *Golden Section and Music,* ed. by Condat: 139–45.

[Saint] Saint-Saëns, Camille. *Symphony No. 3* transcribed for piano. Boston: O. Ditson, 1930.

[San] Sandresky, Margaret V. "The Golden Section in Three Byzantine Motets of Dufay." *Journal of Music Theory* 25, no. 2 (fall 1981): 291–306. Reprinted in *Golden Section and Music*, ed. by Condat: 53–67.

[Sat] Satie, Erik. *All Piano Works Published by Salabert*. Paris: Editions Salabert S.A., 1989.

[Scar] *The Graded Scarlatti*. New York: G. Ricordi & Co., ed. by Marthe Morhange Motchane, n.d.

[SchilM] Schillinger, Joseph. *The Mathematical Basis of the Arts.* Originally published, New York: Philosophical Society, 1948. Reprint, New York: Da Capo Press, 1976.

[SchilS] ———. *The Schillinger System of Musical Composition.* 2 vols. Originally published, New York: Carl Fischer, Inc., 1941, 1942, 1946. Reprint, New York: Da Capo Press, 1978.

[Schn] Schneider, Michael S. *A Beginner's Guide to Constructing the Universe: The Mathematical Archetypes of Nature, Art, and Science.* Originally published, HarperCollins Publishers, 1994. Reprint, New York: HarperPerennial, 1995.

[Schoe83] Schoenberg, Arnold. *Theory of Harmony.* Transl. Roy E. Carter. Berkeley: University of California Press, 1983.

[Schoe95] ———. *The Musical Idea.* New York: Columbia Univ. Press, 1995.

[SchoeQ] ———. *Quintet, Op. 26.* London: Universal Edition, n.d.

[SchoeSQ] ———. *Sämtliche Werke.* Wien: Universal Edition, 1999.

[Schub] Schubert, Franz. *Four Symphonies [Nos. 4, 5, 8, 9].* Originally published, Leipzig: Breitkopf & Härtel, 1884–5. Reprint, New York: Dover Publications, Inc., 1978.

[Schum] Schumann, Robert. *Complete Symphonies.* Originally published, Leipzig: Breitkopf & Härtel, 1881–9. Reprint, New York: Dover Publications, Inc., 1980.

[Schwa] Schwartz, Elliot and Daniel Godfrey. *Music Since 1945: Issues, Materials, and Literature.* New York: Schirmer Books, 1993.

[Sermisy] Sermisy, Claudin de. *Opera omnia.* Vol. III, No. 12. N.p.: American Institute of Musicology, 1974.

[Sib] Sibelius, Jean. *Finlandia* arr. for piano. New York: G. Schirmer, 1905.

[Sid] Sidorowicz, Alexander E. "The Proportional and Spatial Analysis of the First Movement Sonata-Allegro Form of Mozart, Haydn, and Beethoven." Ph.D. diss., Kent State University, 1981.

[Sim] Simson, Otto von. *The Gothic Cathedral: Origins of Gothic Architecture and the Medieval Concept of Order.* Originally published, New York: Bollingen Foundation Inc., 1956. 2nd ed., Bollingen Foundation Inc., 1962. 3rd ed., Princeton, NJ: Princeton University Press, 1988.

[Smith] Smith, Paul. Several articles about the Priory of Sion. Http://priory-of-sion.com.

[Sol] Solomon, Larry. "Symmetry as a Compositional Determinant." Excerpt from *Larry Solomon's Dissertation.* Http://wc.pima.edu/~lsolomon/diss7.htm. 1973, rev. 2002.

[Som] Somfai, László. *Béla Bartók: Composition, Concepts, and Autograph Sources.* Berkeley and Los Angeles: Univ. of California Press, 1996.

[Star] Starbird, Margaret. *The Woman with the Alabaster Jar: Mary Magdalen and the Holy Grail.* Rochester, VT: Bear & Company, 1993.

[Stecc] Stecchini, Livio Catullo. "Notes on the Relation of Ancient Measures to the Great Pyramid." In *Secrets of the Great Pyramid* by Peter Tompkins: 287–382.

[StoA] Stockhausen, Karlheinz. *Adieu*. Wien: Universal Edition, 1988.

[StoK] ———. *Klavierstück IX*. London: Universal Edition Ltd., 1967.

[StoS] ———. *Stop*. Vienna: Universal Edition, 1972.

[StravA] Stravinsky, Igor. *Agon*. New York: Boosey & Hawkes, 1957.

[StravC] ———. *Concerto in D*. Rev. ed. New York: Boosey & Hawkes, 1961.

[StravI] ———. *In Memoriam Dylan Thomas*. N.p.: Boosey & Hawkes, n.d.

[StravO] ———. *Octet for Wind Instruments*. Rev. ed. New York: Boosey & Hawkes, 1952.

[StravSTP] ———. *Sonata for Two Pianos*. Mainz: Schott, 1945.

[StravT] ———. *Three Pieces for String Quartet*. New York: Boosey & Hawkes, n.d.

[StravS] ———. *Symphonies of Wind Instruments*. New York: Boosey & Hawkes, 2001.

[Tam] Tame, David. *Beethoven and the Spiritual Path*. Wheaton, IL: The Theosophical Publishing House, 1994.

[Tat] Tatlow, Ruth. *Bach and the Riddle of the Number Alphabet*. Cambridge: Cambridge University Press, 1991.

[Tchai1] Tchaikovsky, Peter Ilyitch. *Symphonies No. 1, 2, and 3*. Originally published, Moscow: Editions de Musique de l'URSS, 1946. Reprint, New York: Dover Publications, Inc., 1992.

[Tchai4] ———. *Fourth, Fifth and Sixth Symphonies*. Originally published, Leipzig: Breitkopf & Härtel, n.d. Reprint, New York: Dover Publications, Inc., 1979.

[Teague] Teague, Walter Dorwin. *Design This Day: The Technique of Order in the Machine Age*. New York: Harcourt, Brace and Company, 1949.

[Thom] Thompson, D'Arcy Wentworth. *On Growth and Form*. Originally published as *On Growth and Form: A New Edition*, Cambridge, England: Cambridge University Press, 1942. Reprint, Mineola, NY: Dover Publications, 1992.

[Tom] Tompkins, Peter. *Secrets of the Great Pyramid*. New York: Galahad Books, 1971.

[Tov35] Tovey, Donald Francis. *Essays in Musical analysis*. Vol. 1. London: Oxford University Press, 1935

[Tov45] ———. *Beethoven*. New York: Oxford University Press, 1945.

[Vaj] Vajda, Steven. *Fibonacci & Lucas Numbers, and the Golden Section: Theory and Applications*. Chichester: Ellis Horwood, 1989.

[VarA] Varèse, Edgard. *Arcana*. New York: Franco Colombo, Inc., 1927.

[VarD] ———. *Deserts*. New York: G. Ricordi & Co., 1954.

[VarE] ———. *Ecuatorial*. New York: G. Ricordi & Co., 1934.

[VarI] ———. *Intégrals*. New York: Colfranc Music Publishing Corp., 1925.
[VarO] ———. *Offrandes*. New York: Franco Colombo, Inc., 1921.

[Wag] Wagner, Richard. *Die Meistersinger von Nürnberg, Vorspiel*. Scarsdale, NY: E. F. Kalmus, n.d.
[Wal] Wallin, Rolf. "Fractal Music—Red Herring or Promised Land?" Lecture given at the Nordic Symposium for Computer Assisted Composition. Http://www.notam.uio.no/~rolfwa/Fractalarticle.html. 1989.
[War] Wardell, Xiaoman Zhang. "An Examination of Selected Contemporary Works Composed by Means of Numbers." D.M.A. diss., Claremont Graduate School, 1996.
[Wat88] Watkins, Glenn. *Soundings: Music in the Twentieth Century*. New York: Schirmer Books, 1988.
[Wat94] ———. *Pyramids at the Louvre: Music, Culture, and Collage from Stravinsky to the Postmodernists*. Cambridge, MA: The Belknap Press, 1994.
[Webern5] Webern, Anton. *5 Stücke für Orchester*. Vienna: Universal Edition, 1951.
[Webern6] ———. *6 Stücke für Orchester*. Vienna: Universal Edition, 1961.
[Webster] Webster, J. H. Douglas. "Golden Mean Form in Music." *Music and Letters* 31 (July 1950): 238–48.
[Wilk] Wilkins, Nigel. "Rondeau." Vol. 21. *The New Grove Dictionary of Music and Musicians,* ed. Stanley Sadie. 2nd ed. London: Macmillan Publishers Limited, 2001.
[Wils] Wilson, Robert Anton. "The Priory of Sion." *Gnosis* 6 (winter 1988): 30–9.

[Yes] Yeston, Maury, ed. *Readings in Schenker Analysis and Other Applications*. New Haven: Yale University Press, 1977.

[XeF] Xenakis, Iannis. *Formalized Music: Thought and Mathematics in Music*. Rev. ed. Stuyvesant, NY: Pendragon Press, 1992.
[XeM] ———. *Metastasis*. New York: Boosey & Hawkes, n.d.

Index

Bolded numbers are summaries